Fearless Cross-Platform Development with Delphi

Expand your Delphi skills to build a new generation of Windows, web, mobile, and IoT applications

David Cornelius

BIRMINGHAM—MUMBAI

Fearless Cross-Platform Development with Delphi

Group Product Manager: Richa Tripathi
Publishing Product Manager: Alok Dhuri
Senior Editor: Rohit Singh
Content Development Editor: Kinnari Chohan
Technical Editor: Maran Fernandes
Copy Editor: Safis Editing
Project Coordinator: Deeksha Thakkar
Proofreader: Safis Editing
Indexer: Pratik Shirodkar
Production Designer: Joshua Misquitta

First published: September 2021

Production reference: 2151121

Published by Packt Publishing Ltd.
Livery Place
35 Livery Street
Birmingham
B3 2PB, UK.

ISBN 978-1-80020-382-2

www.packt.com

To my wife, Terresa, for her incredible patience and support during the writing of this book.

– David

Contributors

About the author

David Cornelius is a software engineer in the Pacific Northwest of the US with over 30 years of experience writing applications for education, research, finance, inventory management, and retail. He has been the coordinator for the Oregon Delphi User Group for two decades and keeps active in various online forums. David is the founder and principal developer at Cornelius Concepts, LLC. and an Embarcadero MVP.

In his spare time, he likes to ride his motorcycle, escape the city with his RV, play the tuba or bass guitar, and play strategy board games with friends.

I would like to thank, first of all, my technical reviewers, Jonathan Eaton and Marco Breveglieri, and the editors at Packt Publishing who helped me see things from the perspective of the reader and caught project errors and omissions I submitted to GitHub (which I tried to pass off as challenges to the readers). They certainly saved me much embarrassment and helped ensure the book you have makes sense—and the sample programs actually compile!

I would also like to thank attendees of the Oregon Delphi User Group, a small but loyal band of awesome Delphi programmer friends, who have been cheerleaders for my writing efforts.

Finally, I very much appreciate the love and support that friends and family have given me even when they didn't fully understand what I was writing about. They helped more than they know.

About the reviewer

Jonathan Eaton is a Texan who loves living in Oregon. He has been an avid professional user of Delphi ever since he used it for a Y2K project way back in 1999. Since then, he has used Delphi to design, develop, and update software applications for a variety of industries, including chemical mixing, mineral asset accounting, insurance form processing, commercial real estate appraisal, and the development of software components of medical devices. In 2013, Jonathan became an ASQ Certified Software Quality Engineer and has been a strong proponent for the application of the principles of quality management to software development ever since. He's also written and self-published three novels and several collections of short stories.

Marco Breveglieri is a software and web developer. He started programming when he was 14 and got his first home personal computer, a beloved Commodore 16. Going forward—and getting serious—years later he attended secondary school focusing on computer science, continuing that learning path and taking his first steps from BASIC to Pascal and C++. Today, Marco continues to work for his own IT company using Delphi to create any kind of application. He also uses Visual Studio to build websites, using the Microsoft web stack based on the .NET Framework. He often takes part in technical conferences and holds training courses about programming with all these tools, especially Delphi, C#, and the web standards HTML5, JavaScript, and CSS3.

Table of Contents

3
A Modern-Day Language

Section 2: Cross-Platform Power

4
Multiple Platforms, One Code Base

5

Libraries, Packages, and Components

6

All About LiveBindings

7

FireMonkey Styles

8

Exploring the World of 3D

Section 3: Mobile Power

9
Mobile Data Storage

10
Cameras, the GPS, and More

11
Extending Delphi with Bluetooth, IoT, and Raspberry Pi

Section 4: Server Power

12
Console-Based Server Apps and Services

13
Web Modules for IIS and Apache

14
Using the RAD Server

15

Deploying an Application Suite

Assessments

Other Books You May Enjoy

Index

Preface

A favorite author once penned, "Resistance to change leads to catastrophic change." I fully believe that and, in this field, learning and growing and yes, changing, is good—anything else is career death.

Delphi has been around for over 25 years, but it is anything but old. The language, libraries, and toolset are not stagnant and applications built with this suite of compilers rival those of competing products. Continual updates, frequent webinars, lively forum debates, a recent explosion of books, and a myriad of blog sites are evidence of a rich and active developer community.

This book takes you on a journey. This journey will build upon and extend the hours you've invested in creating beautiful desktop applications and teach you how to craft new, globally usable mobile apps and backend servers that power today's interconnected platforms. Technology has exploded with ways to communicate and share data between apps, web services, and devices of all kinds. People demand choices—choice of operating system, choice of style, choice of screen size, and choice of location.

Companies everywhere have risen to the challenge of providing options to satisfy these customer demands with unique tools that market to a new generation of developers. Are you feeling left behind, wondering how to move forward?

Reading this book will expand your Delphi programming skill set in a step-by-step manner, explaining what you need to know to handle the variety of challenges addressing new platforms will bring. And we'll have fun along the way. We'll build database applications that have sample data built in to make prototyping easier than ever. We'll build a 3D game you can play on your smartphone. We'll create a full-featured mobile app you can extend for your own creative uses. We'll also build powerful backend REST servers with virtually no code.

Take this journey with me to multiple platforms—fearlessly!

Who this book is for

This book is written mostly for the Delphi programmer who is confident on the Windows platform with the VCL but unsure what mobile development will mean for them. If you're relatively new to Delphi programming, this book will be a quick jump start into the rich and wonderful integrated development environment we all love—you'll figure it out quickly!

The experienced cross-platform developer will likely seek me out on a forum and point out obvious things I missed. I'm sorry in advance—this was a journey for me as well!

What this book covers

Chapter 1, Recent IDE Enhancements, will help you get up to speed quickly, starting with an understanding of the many parts of the IDE, and then explaining "What's new?" in each of the last five major versions.

Chapter 2, Delphi Project Management, simplifies the bewildering choice of where to start! With so many platforms and project templates, deciding how to create a new app can itself be overwhelming. We'll explore build configurations, shortcuts to managing options and project groups, and show you how to use the command-line compiler for automation.

Chapter 3, A Modern-Day Language, clears up any misunderstanding that Delphi is not capable of building cross-platform apps. We'll showcase enhancements made through the years to the Pascal-based syntax that facilitated the flexible and powerful language it is today.

Chapter 4, Multiple Platforms, One Code Base, is where we'll really start diving into cross-platform topics. The first step in that direction is learning about FireMonkey—we will take you carefully through several differences compared with the VCL in some simple apps so you can get comfortable quickly. Then we introduce various form factors of mobile devices and show how to manage them in Delphi. Finally, we cover several conditional compilation constants you'll need to know in order to separate code for specific platforms.

Chapter 5, Libraries, Packages, and Components, is an important chapter as writing a simple DLL isn't so simple when you have to deploy to platforms other than Windows—we cover several gotchas that you need to know. After we build both a dynamic library and a package (and why you might need one over the other), we turn a package into a cross-platform component.

Chapter 6, All About LiveBindings, explains how this expression-based data connector technology is more than just a replacement for data-aware controls in Delphi. The LiveBindings Designer is explained with tips on organizing with layers. We demonstrate how the LiveBindings Wizard can create components for you already hooked up to your data. Finally, we use custom formatting and parsing—and end with code that creates custom LiveBinding methods installed in Delphi.

Chapter 7, FireMonkey Styles, uncovers the nuances of styling a FireMonkey application and how to use and customize FireMonkey styles effectively for a distinctive appearance. We will build a simple app with a variety of controls and four different styles that you can run on each of your devices to see the differences.

Chapter 8, Exploring the World of 3D, is a fun chapter that demonstrates how you can use your favorite programming tool to utilize popular GPU engine libraries with ease. We build a simple app with a variety of 3D shapes, textures, lighting, and cameras for a broad overview of the capabilities. We end the chapter by using some of these techniques and more to build an escape game you can play on your phone!

Chapter 9, Mobile Data Storage, shows how well FireDAC works with different database products on multiple platforms, comparing one application that uses InterBase ToGo and a similar one using SQLite. We'll answer questions on licensing, explore free management tools, and offer tips for working with touchscreen devices.

Chapter 10, Cameras, the GPS, and More, demonstrates the power of the FireMonkey library to encapsulate various platform capabilities in simple components that you can readily use in your apps. We'll also build a database-enabled mobile app that utilizes techniques learned in previous chapters but that will grow in functionality for the rest of this book.

Chapter 11, Extending Delphi with Bluetooth, IoT, and Raspberry Pi!, explores various types of Bluetooth technologies, including BLE, which is what beacons are based on. We dive further and show how BLE is the basis for all IoT components and explain how to use ThingConnect components you can get from GetIt. We conclude this chapter on small devices by demonstrating how to deploy a Delphi app to a Raspberry Pi.

Chapter 12, Console-Based Server Apps and Services, switches gears from small, mobile devices to server technologies. We create both Windows services and Linux daemons and also show how to take an open source project and modify it for our own needs when we need to implement custom logging.

Chapter 13, Web Modules for IIS and Apache, continues the server discussion by concentrating on web server modules for the two most popular web servers – IIS on Windows, and Apache for both Windows and Linux. We bring over the data module tested in our console-based server and use WebBroker to build a simple but nice-looking web page that displays data in a grid with very little code.

Chapter 14, Using the RAD Server, is a big chapter covering a big product available for Delphi Enterprise and up that serves as another type of platform, one that provides a REST server in a box where you only add your custom business methods. We teach what RAD Server brings to the table, why it can pay for itself, and how to write modules for it. Then we go one step further and modify our sample mobile app to use it.

Chapter 15, Deploying an Application Suite, culminates the application development process by covering important aspects of deploying to production with discussions about externally defined configuration files, various security concerns, application icons and identity, testing, the installation of server backends, and mobile app submission.

To get the most out of this book

You will need Embarcadero Delphi 10.4 Sydney Professional or Enterprise on Windows 64-bit with access to one or more mobile platforms on which you can deploy and test the sample applications:

Software/hardware covered in the book	Operating system requirements
Delphi 10.4 Sydney	Windows 10 64-bit
Linux applications	Ubuntu 14.04/16.08/18.04 or RedHat Enterprise 7
Mac development	macOS 10.13 High Sierra – 11 Big Sur
iOS development	iOS 11, 12, or 13
Android development	Android 6 Marshmallow – Android 10.0

You will need the Enterprise edition of Delphi to build Linux apps or RAD Server modules. Also, while the Community Edition may be used for many of the examples, you may be restricted from downloading from the GetIt portal.

If you are using the digital version of this book, we advise you to type the code yourself or access the code from the book's GitHub repository (a link is available in the next section). Doing so will help you avoid any potential errors related to the copying and pasting of code.

Download the example code files

You can download the example code files for this book from GitHub at
`https://github.com/PacktPublishing/Fearless-Cross-Platform-Development-with-Delphi`. If there's an update to the code, it will be updated
in the GitHub repository.

We also have other code bundles from our rich catalog of books and videos available at
`https://github.com/PacktPublishing/`. Check them out!

Download the color images

We also provide a PDF file that has color images of the screenshots and diagrams used
in this book. You can download it here: `https://static.packt-cdn.com/downloads/9781800203822_ColorImages.pdf`.

Conventions used

There are a number of text conventions used throughout this book.

`Code in text`: Indicates code words in text, database table names, folder names,
filenames, file extensions, pathnames, dummy URLs, user input, and Twitter handles.
Here is an example: "Back in the form unit, `uCardPanel.pas`, it starts with the `unit`
keyword and has both an `interface` and `implementation` section, each with a
`uses` clause."

A block of code is set as follows:

```
procedure TfrmPeopleList.lbPeopleClick(Sender: TObject);
var
  APerson: TPerson;
begin
  if lbPeople.ItemIndex > -1 then begin
    APerson := lbPeople.Items.Objects[lbPeople.ItemIndex]
                as TPerson;
    lblPersonName.Caption := APerson.FirstName + ' ' +
                     APerson.LastName;
    lblPersonDOB.Caption  := FormatDateTime('yyyy-mm-dd',
                          APerson.DateOfBirth);
  end;
end;
```

When we wish to draw your attention to a particular part of a code block, the relevant lines or items are set in bold:

```
constructor TfrmPeopleList.TPerson.Create(
                        NewFN, NewLN string; NewDOB: TDate);
begin
  FFirstName  := NewFN;
  FLastName   := NewLN;
  FDateOfBirth := NewDOB;
end;
```

Bold: Indicates a new term, an important word, or words that you see on screen. For instance, words in menus or dialog boxes appear in **bold**. Here is an example:

"Under the **Develop** section of Delphi's default **Welcome** page, click the **Open a sample project...** link and drill down through the **Object Pascal**, **VCL**, and **CardPanel** folders, and then open the **CardPanel** project."

> Tips or important notes
> Appear like this.

Get in touch

Feedback from our readers is always welcome.

General feedback: If you have questions about any aspect of this book, email us at customercare@packtpub.com and mention the book title in the subject of your message.

Errata: Although we have taken every care to ensure the accuracy of our content, mistakes do happen. If you have found a mistake in this book, we would be grateful if you would report this to us. Please visit www.packtpub.com/support/errata and fill in the form.

Piracy: If you come across any illegal copies of our works in any form on the internet, we would be grateful if you would provide us with the location address or website name. Please contact us at copyright@packt.com with a link to the material.

If you are interested in becoming an author: If there is a topic that you have expertise in and you are interested in either writing or contributing to a book, please visit authors.packtpub.com.

Share Your Thoughts

Once you've read *Fearless Cross-Platform Development with Delphi*, we'd love to hear your thoughts! Scan the QR code below to go straight to the Amazon review page for this book and share your feedback.

https://packt.link/r/1-800-20382-9

Your review is important to us and the tech community and will help us make sure we're delivering excellent quality content.

Section 1: Programming Power

This section introduces Delphi as a powerful programming tool for fully embracing cross-platform development, explains why it's so much more than just a legacy Windows compiler, and helps get you up to speed with the latest IDE and language enhancements that accelerate development and decrease the time to market. This foundational knowledge sets the stage for the rest of the book, helping you to navigate the IDE, write efficient code, drop to the command line if you need, and become familiar with new programming constructs.

This section comprises the following chapters:

- *Chapter 1, Recent IDE Enhancements*
- *Chapter 2, Delphi Project Management*
- *Chapter 3, A Modern-Day Language*

1
Recent IDE Enhancements

Delphi is not just a language, nor just an **Integrated Development Environment** (IDE), nor just a toolset or programming environment, but all of that and more. Starting in the early 1980s with Turbo Pascal shipping on a single floppy disk, it has grown to be a powerful suite of libraries, tools, connected services, and integrated components that support virtually every computing platform. And as with most technological tools, Delphi is constantly growing, improving, and adding features.

This first chapter will give you a quick recap of all that the IDE encompasses and what new features you may have missed in the last few versions. Some of these are just for convenience, such as quickly locating a unit in Windows Explorer, or jumping to a method with the navigation toolbar. There are visual improvements with high DPI support, a new dark mode, and structural highlighting options. You'll be more efficient with faster loading times, greatly improved Code Insight features, and editor tricks such as bookmark stacks. Finally, you will understand your code better with add-ons, such as Project Statistics and Toxicity Metrics.

You will read about all of these and more in the following sections of this chapter:

- Understanding the Delphi IDE
- Delphi 10 Seattle
- Delphi 10.1 Berlin
- Delphi 10.2 Tokyo
- Delphi 10.3 Rio
- Delphi 10.4 Sydney

Understanding the Delphi IDE

The idea of combining the editing, compiling, and managing of project files within one integrated programming application started way back in the 1970s, but didn't really catch on in the PC arena for quite some time. Borland Pascal, the pre-cursor to Delphi, pioneered many facets of an IDE in the creation of **Disk Operating System** (**DOS**) applications and took many of its features and user interface constructs with it to the Windows desktop.

At its core, an IDE needs to assist the developer in managing a software project's many moving parts: editing source code, crafting user interfaces, managing project parameters, compiling the application and resources, testing and debugging, and preparing a deployable application. These vary from language to language, so an IDE needs to understand the tools that it supports very well—and the developers who use it.

Delphi's IDE, like many other modern tools, has a rich code editor with syntax checking and color schemes, resizable window panes, structure and object views, data and source repository connections, and the ability to save various configurations for different needs. Here are the various windows and views you'll find in Delphi:

- **Code Editor**: The main focal point for writing code.
- **Project Manager**: Combine files necessary for compiling a project.
- **Object Inspector**: Access properties and events of components.
- **Structure Pane**: View and organize components hierarchically.
- **Tool Palette**: Component list.
- **Message Window**: Compilation output, search results, and more.
- **To-Do List**: Parsed TODO comments from the code are neatly displayed here.
- **Templates**: Macros for expanding frequently typed patterns of code.

- **LiveBindings Designer**: Visually build data connections between components.

- **Class Explorer**: Tree view of the classes in the current unit.

- **Data Explorer**: View data and manage connections from dbExpress or FireDAC sources without leaving the IDE.

- **Model View**: Manage object models of your data.

- **Other Debugger views**: Call Stack, watches, local variables, breakpoints, Threads, events, Modules, and CPU windows.

The following screenshot shows these windows and views in the Delphi IDE:

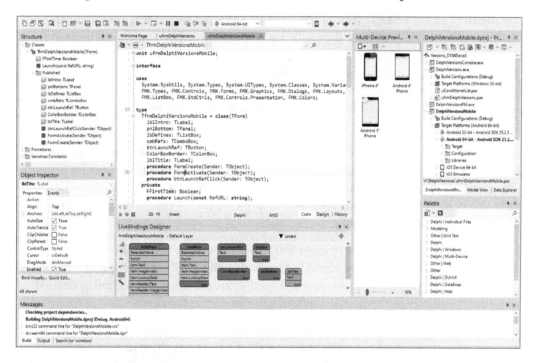

Figure 1.1 – The Delphi 10.4 Sydney IDE showing several available window panes

Let's look at some of these in a little more detail.

Like most popular code editors that developers use these days, there is color syntax highlighting, hotkeys for moving code around, bookmarks, class completion, and template expansion. There is a lot of keyboard backward-compatibility in Delphi (even dating back to the old WordStar days) but also a lot of customization to set up the key combinations that work best for you. Classes and methods can be folded (or "collapsed"), and you can create your own foldable regions. Error insight can alert you to syntactical errors in your code as you type.

The **Project Manager** can contain a single project or a group of projects, and there are buttons to manage aspects of them with ease—as a group or individually. You can determine where compiled objects will be placed, establish host applications to debug modules, and manage multiple platform configurations here. More on this will be covered in *Chapter 2, Delphi Project Management*.

The **Object Inspector** allows you to initialize properties and hook up events of components at design time to save you the trouble of having to write all that out in code. If you build custom components, you can publish your own properties and events, which will also show up here. Plus, you can register your own property editors to provide even greater functionality.

The **Structure** pane is a handy view of elements that changes depending on the context. When you're in the code editor, it shows the list of classes, methods, and used units in the current unit. When you're in the form designer, it shows a tree view of the components on the form that can be rearranged by dragging them with the mouse to a different place on the hierarchy (with some obvious parent-container restrictions; for example, a panel can contain a label, whereas a label cannot contain a panel). Double-clicking one of these elements takes you to that element.

When you're designing a form or data module, you can place items on it from the **Tool Palette**. This is also context-sensitive, thereby only listing the elements that can be placed on the current form. For example, only FireMonkey components will be available for a FireMonkey form; but database connection components can be placed on a VCL form, a FireMonkey form, or a data module.

When building a project or running other processes, the output will be shown in the **Message** window. This is often collapsed but expands temporarily at appropriate times. Like most other windows, this pane can be pinned to keep it open and resized as desired. Search results and console application output will show up as tabs here.

The Delphi code editor watches for specific comments that start with TODO: and builds a list from these found in all the units of your project and shows them in a **To-Do List** window with a clickable checkbox. This allows you to see all the places in your code where you've marked items you need to finish. By double-clicking on the to-do list item, it will take you to that line in the code. When you are finished with the task, click the checkbox in the to-do list and it will mark it as complete by changing the comment from **TODO** to **DONE**. (You can also uncheck an item marked **DONE** and it will turn the comment back to **TODO**.)

Live templates (shown in the window pane that is simply titled **Templates**) can speed up coding considerably. They are code macros that can insert text and allow you to fill in the blanks to complete larger constructs with very few keystrokes. They can be set to automatically expand after a certain sequence of letters are typed or manually activated by hitting *Ctrl + J*. You can create your own or edit the built-in ones.

Some of the first code templates in Delphi were implemented by the popular free IDE plugin, GExperts (sort of a mix of Code Librarian and Code Proofreader) but other products soon introduced their own version. Live templates were added in Delphi 2006.

We'll talk about the **LiveBindings Designer** in greater detail in *Chapter 6, All About LiveBindings*, but as a brief overview, this is a visual way to link data between various entities. For example, a form that displays an edit box to type in a name has to load the value from somewhere and also save it when the user is done. Typing code to load and save such elements can become quite tedious. This lets you do it visually.

The **Class Explorer** (not shown in the previous screenshot) is similar to the **Structure** pane when viewing code in that they both show a tree view of classes. But while the **Structure** pane shows more than just classes, it has few options for actually dealing with the classes. That's where the **Class Explorer** shines—it allows you to view *and* manage methods, fields, and properties of classes with the mouse. Think of it as a simplified version of ModelMaker Code Explorer if you've ever used that product.

The IDE allows you to save the arrangement of these panes as desktop profiles. You can have several desktop profiles and use different ones for different projects. You can select one to be the default for standard editing and another as the default for debugging—the IDE will automatically switch to the debug desktop when you start an application and switch back to the default one when it's finished.

Here's an example of what the IDE could look like while debugging:

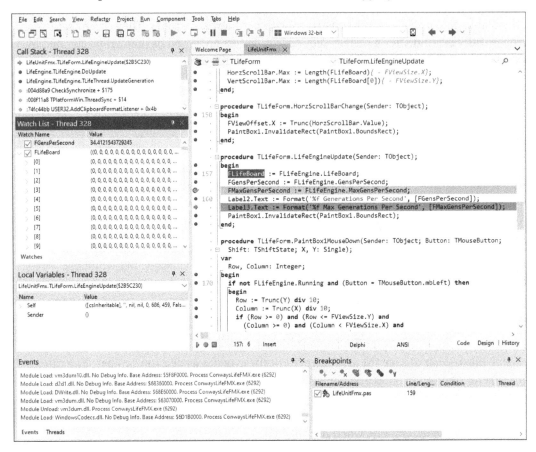

Figure 1.2 – Debugging an application in Delphi 10.4

As you can see, the Delphi IDE is center-stage in the development process of creating and managing Delphi projects. Since no two developers work the same way, its flexibility, customization, and the wide variety of built-in tools help make this environment a powerful tool.

Now that you've had a glimpse of what the IDE is, let's step back a few versions, highlighting some of the changes it has undergone.

Delphi 10 Seattle

The Delphi XE series broke ground in many ways, not least of which was the FireMonkey GUI to support mobile devices. Updates were fast and furious, and some complained it was hard to keep up with them. XE8 was the last of this line and introduced a new **Welcome** screen that has carried through to the current version. Delphi 10 Seattle started the "10" series in August 2015 and was named, at least in part, to coincide and align itself with Windows 10.

There weren't any ground-breaking IDE features introduced in the first of the Delphi 10 series, but several important improvements were made in project loading speed, support for large project groups, and high DPI support, especially notable when working with forms at different DPIs. Several menu items were moved for better organization and simplification, and an **Editor** submenu was added to contain the editor's context menu items. The nicest new feature of the IDE was a convenient **Show in Explorer** menu item, which will open File Explorer for a file or folder in the Project Manager, as shown:

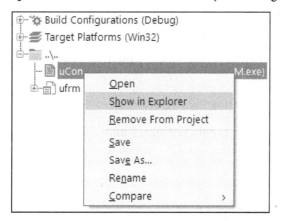

Figure 1.3 – Delphi 10 Seattle "Show in Explorer" from Project Manager

If you ever purchased the Castalia for Delphi suite of IDE tools from TwoDesk Software, you may recall several nifty features that they added to Delphi. Embarcadero had just acquired Castalia and quickly incorporated many of those features into the editor. These include the following:

- **MultiPaste**: Pops up a window allowing you to add text before and after each line before pasting (really handy for copying SQL or HTML into your code).

- **Project Statistics**: Informs you of the time spent in various parts of the IDE, such as designing and editing.

- **Navigation Toolbar**: A nice way to view and jump to files in the project and sections within the file.

- **Sync Prototypes**: Make a change to the parameters in the declaration of a class method and it applies those same changes to the implementation.

- **Structural Highlighting**: Draws lines in the editor to visually depict blocks of code.

- **Smart Keys**: Adds editor shortcuts, including a fast way to surround code with braces and parentheses.

All of these features have added settings in the options pages.

Other additions included are the following:

- Automatic recovery of files if the IDE crashes while editing

- Being able to change the font size quickly with *Ctrl + Num+* and *Ctrl + Num-*

- The ability to enable **High-DPI Awareness** in your VCL application

- Several improvements to the **Object Inspector**, structure view, and the **Select Directory** dialog

- A new option to hide non-visual components

- Improved memory management allowing the IDE to use up to 4 GB of RAM (up from 2 GB previously)

- Better support for importing old projects even as far back as Delphi 1

Finally, there were several mobile enhancements, as well including a new **Android Service** project type, background execution on iOS, and an option to allow iOS 9 applications to access non-SSL URLs.

Delphi 10.1 Berlin

The IDE for Delphi 10.1 Berlin released in April 2016 had quite a few enhancements. In all previous versions of Delphi, units with forms shared the same IDE window—you had to switch between code and form design mode. Finally, in Delphi 10.1 Berlin, using an **Embedded Designer** is now an option in the **Form Designer** section of the **Tools | Options** dialog. Switching this option off allows you to move a floating window for both the VCL and FireMonkey form designers separately from the code editing window.

FireUI Live Preview is an amazing new tool that gives FireMonkey application designers the ability to preview and debug their user interfaces in real time on remote devices. The IDE acts as a "client" that communicates with a **Platform Assistant Server** you need to install on the target device, which connects over a local network or a connected USB cable. The server capability is built into the Windows IDE, and PAServer for Windows, Linux, and OS X is included with the Delphi installation. Details on setting this up will be explained in *Chapter 4, Multiple Platforms, One Code Base.*

GetIt Package Manager was introduced a few versions back and allows quick access to open source, trial, and commercial components, program templates, and styles. In Delphi 10.1 Berlin, the **Project Options** dialog window gained a new section called **GetIt Dependencies**. Here, you can check off one or more GetIt packages that are used in your project. Then, when you open a project where one or more of those marked GetIt packages are not installed, you'll get a message that the project has dependencies that are not installed.

> **NOTE**
>
> If any of these dependencies are not installed when you try to build your application and you get an MSBuild error stating the GetItCmd task failed unexpectedly, you need to add an environment variable (**Tools | Options | Environment Variables**): **New Variable: Name** = "BDSHost" and **Value** = "true".

If you ever purchased the Castalia for Delphi suite of IDE tools from TwoDesk Software, you may recall several nifty features they added to Delphi. Embarcadero acquired Castalia in 2015 and has incorporated many of those features into Delphi 10 Seattle, as previously noted. One of my favorites is **bookmark stacks**. This allows you to drop a temporary bookmark in your code, go to another section of code, or even another file, then recall the temporary bookmark to jump right back to where you were coding, cleaning up the temporary bookmark. However, you may want to customize the default keys for this as it uses a difficult key sequence to drop (*Ctrl + K* and *Ctrl + G*) and pick up (*Ctrl + Q* and *Ctrl + G*) the bookmarks.

> **NOTE**
>
> In the *Delphi 10.3 Rio* section, you'll learn about the **Navigator** plugin that has been added that provides nice keyboard shortcuts for this feature.

If you're most efficient when keeping your hands on the keyboard, you may really like **Selection Expansion**. This allows you to click *Ctrl + W* in the code, and with repeated clicks of *Ctrl + W*, expand the selected text to increasingly larger code blocks, starting with the current identifier and moving up to statement, line, block, method, and so on.

There's a new menu item under **Projects** called **Method Toxicity Metrics** that gives you statistics about the procedures and functions in the active project and that may indicate good candidates for refactoring. You can export these metrics and customize thresholds under **Tools | Options | Toxicity Metrics**.

Other improvements include more support for Android services, Android smart watches, iOS ad hoc applications, CPU view support for iOS and Android, a new **File Associations** page on the **Options** window, and the ability to show or hide the **Navigation Toolbar**. Finally, the IDE is now DPI-aware.

Delphi 10.2 Tokyo

The frequency of updates finally started slowing back to a reasonable pace with the release of Delphi 10.2 Tokyo in March 2017, and a long-anticipated feature request finally made it into the second release 9 months later: a **dark theme**! While you could change the colors of many of the IDE components in previous versions (and, in fact, there were "dark themes" already available in the popular Delphi Theme Editor), they mostly affected the code editor—the core of the IDE was still based on the typical Windows standard look with a white background. With the dark theme enabled, menus, dialogs, edit boxes, and more are themed around a dark set of colors, which many people find reduces eye strain when working at a computer for long hours or late at night when there's less light. If you want to switch themes for different hours of the day, it's a quick mouse click from the desktop toolbar:

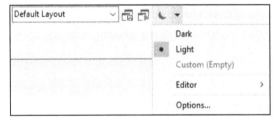

Figure 1.4 – The Delphi 10.2 Tokyo desktop toolbar showing quick toggle between the Dark and Light themes

The third release of Delphi 10.2 improved both the light and dark themes with cleaner lines and aligned controls. Several items in the **Options** box were moved for better categorization and the whole interface became more readable. The currently focused area was made more prominently displayed and some window panes were renamed:

Figure 1.5 – Delphi 10.2 Tokyo using the Dark theme

The preceding screenshot shows how the IDE looks in dark mode.

Delphi 10.3 Rio

In November 2018, the first release of Delphi 10.3 Rio was made available. Besides improvements in themes, cleaner lines, aligned controls, better tab-readability, and clearer window focus changes, there were no major changes to the IDE in any of the three releases of Delphi 10.3 Rio. That's not to say this was not a significant version, as there were several language and library features added, which we will cover later.

One nice improvement can be seen in the **New Items** list (from the menu, go to **File | New | Other**). It changed to present a scrollable list of items with full descriptions and larger icons and titles giving more information about what you're about to create. Some items are now in multiple categories, such as the FireMonkey Metro UI application, which is in both the **Multi-Device** and the **Windows** categories. These are great aids in helping you find the item you want.

GetIt Package Manager was improved in a similar fashion to show items in a scrollable list. Other aspects of its interface were improved as well:

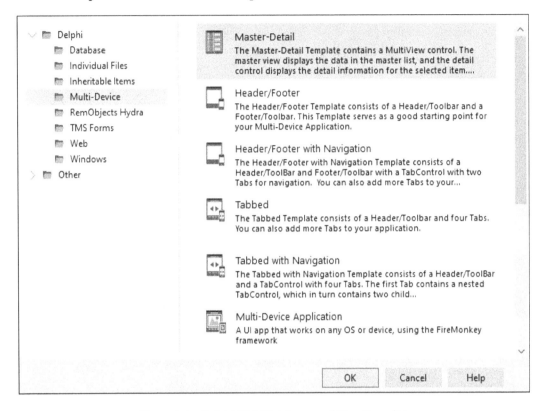

Figure 1.6 – Delphi 10.3 Rio's improved New Items list

The categories in both the general and project options have been reworked, with some new ones introduced and others moved around. For general IDE options, this has shortened the list, but more importantly, the full list of categories is no longer fully expanded, making it much more manageable. One big change to the recategorization is that the **Delphi** section (with library paths and **Type Library** options) has been pulled out of **Environment** and is now under its own **Language** category (which actually makes much more sense).

Many Delphi users have included Andreas Hausladen's IDE Fix Pack as part of their Delphi install routines over the years. Some of those fixes are now part of Delphi 10.3 Rio.

Finally, some speed enhancements have been made, including loading forms with LiveBindings.

There were a couple of nice IDE plugins added to GetIt Package Manager that should be mentioned: **Bookmarks** and **Navigator**. Both of these were acquired by Embarcadero from Parnassus.

Bookmarks extends Delphi's bookmark stacks feature introduced in 10.1 Berlin with unlimited temporary bookmarks and much more convenient hotkeys (*F2* to set, and *Esc* to pop back and remove), along with a dockable window to list them within context.

Navigator provides a hotkey (*Ctrl + G*) that pops up a list of quick places you might need to jump in your code, such as the interface **Uses** section or the declaration for a field. With a few keys of incremental search or a couple of down arrows, you can hit *Enter* to go there immediately, then after you're done and want to get back to where you were, simply hit *Esc*! It's a beautiful marriage with the **Bookmarks** feature. But wait, there's more! Additionally (and optionally), **Navigator** also provides a resizable mini-map of your code on the right-hand side of the code editor that not only gives you a glimpse of where you are in the current unit but also allows you to click and drag to view a different portion of the unit without changing your current cursor position.

> NOTE
>
> One important note before we move on to the latest version of Delphi is that the **Integrated Translation** tools will no longer be improved and developers are warned to migrate to a different set of translation tools if they are being used.

Delphi 10.4 Sydney

After 2 years of updates to the previous version of Delphi, 10.4 Sydney was finally released in May 2020. One important change for this version is that it now *requires a 64-bit version of Windows 10*. Running the IDE is no longer supported on any prior version of Windows (what this means is that while it might be technically possible to install Delphi 10.4 on an earlier 64-bit version of Windows, it is not recommended and you may encounter problems in some aspects of working with it—you have been warned!). For most developers, this won't require a new computer but it is a big step for the IDE—even though at its core, it is still a 32-bit application!

You can still target 32-bit operating systems of Windows, but 32-bit macOS and iOS devices have been removed (although when using the iOS simulator, it's still 32-bit). Android 64-bit and 32-bit devices are supported for the Android API versions 6 through 10.

Delphi 10.4 is now only supported on 64-bit Windows 10:

Developer	32-bit	64-bit
Windows 10	No	Yes
Windows 8.x	No	No
Windows 7	No	No

Here's a list of target devices supported by Delphi 10.4:

Targets	32-bit	64-bit
Windows 10	Yes	Yes
Windows 8	Yes	Yes
Windows 7	Yes	Yes
Windows Server 2016	Yes	Yes
Windows Server 2012 R2	Yes	Yes
macOS 10.13 High Sierra	No	Yes
macOS 10.14 Mojave	No	Yes
macOS 10.15 Catalina	No	Yes
iOS 13, iOS 12, iOS 11	No	Yes
Android 6-10	Yes	Yes
Ubuntu Linux 16.04 LTS	No	Yes
Ubuntu Linux 18.04 LTS	No	Yes
RedHat Enterprise Linux 7	No	Yes

One significantly improved feature is **Code Insight**. This is a set of useful typing helpers in the code editor and includes Code Completion, Code Parameter Hints, Code Hints, Block Completion, Help Insight, Class Completion, Error Insight, and Code Browsing. Often, parts of it are disabled because it can slow down coding, especially on large projects or slow machines as it precompiles and looks up identifiers in related units to help you type faster and catch errors on the fly.

In Delphi 10.4, these tools have been off-loaded from bogging down the editor to a separate server process that communicates with the IDE asynchronously using a **Language Server Protocol** (**LSP**). The **Code Insight** settings are now defined per language in the **Options** windows. What's more is that **Code Insight** now also works while debugging!

Reporting bugs must be done manually now. If you get an IDE exception, you can save the error report to a file, then go to `https://quality.embarcadero.com` and submit a problem discussion, along with the saved error report.

If you have ever gotten tired waiting for the **Options** screen to open, you will be happy to know that it opens faster now. There's a constructor for each page, which is now delayed until the page is shown.

Scrolling in GetIt Package Manager has been greatly improved with a single tweak. The list position now stays on the item just installed, so you can immediately go to the next one in the list instead of scrolling back down from the top as in previous versions. This was a great annoyance when installing a long list of styles.

There are three new convenience features in Delphi 10.4's code editor status bar:

- **Syntax Highlighter**: If you work in multiple languages, the syntax checker for the current language is displayed and you can click it and switch to a different language. This supports more than just Delphi and C++. The IDE might become your default editor to work on SQL, PHP, INI, CSS, HTML, and other supporting files, centralizing your editing tasks:

Figure 1.7 – Override the default syntax checker selected for the current file

- **File Encoding**: Your current file's encoding is now displayed and can be changed. This could be quite handy when working with XML files—another aid to keep all your files in the IDE:

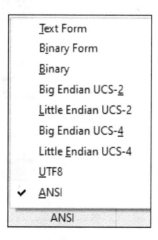

Figure 1.8 – Switch file encoding quickly

- **Font size**: There's now a trackbar in the status bar with increase/decrease buttons to allow you to adjust the font size. Clicking *Ctrl + Num+* and *Ctrl + Num-* still do the same thing:

Figure 1.9 – Visually change the font size in the editor

Another notable change is that the clipboard history (added in Delphi 10 Seattle as part of the Castalia integration) has been removed, partially for security reasons and partially because if you copy very large amounts of data to the clipboard, it could potentially crash the IDE.

The **Welcome** page, which was drastically changed in Delphi XE8 and was cluttered with advertising, events, and more, now shows three columns of links:

- Creating new projects, opening existing ones, and links to **Platforms and Extensions Manager** and **GetIt Package Manager**

- Favorite and recent project listings

- **YouTube Video Channel**

Here is what it looks like:

Figure 1.10 – The default Welcome screen in Delphi 10.4

If this is still too much, you can modify the web page and take out what you feel is unnecessary. The **Welcome** page is simply a local web page in your Delphi installation's `Welcomepage` folder. Choose the subfolder for your language (for example, `en` for English) and make sure you run your editor as **Administrator** if Delphi is installed under the standard `Program Files (x86)` protected folder structure. Then, after making a backup of the files there, simply modify the HTML, JavaScript, and CSS to your heart's content!

There are several other minor improvements in speed and control-painting and placement, in addition to several bugs fixed. The IDE now uses native VCL controls, which resolves a few display issues.

Summary

In this chapter, we reviewed everything Delphi's IDE has to offer, then, after touching on the XE series, highlighted the improvements in Delphi versions 10 Seattle, 10.1 Berlin, 10.2 Tokyo, and 10.3 Rio, and finally covered the major enhancements to the newest version of Delphi, 10.4 Sydney.

The Delphi IDE is a powerful developer tool and grows in capability with every version. As technology continues to evolve, the feature set changes accordingly to embrace the software build processes for today's complex needs and the plethora of devices to support.

With a thorough knowledge of the IDE comes increased productivity through efficient use of the keyboard, proper window placement, customized colors and fonts, handy tools, and useful add-ons. Those who take the time to develop proficiency in this toolset will be valuable assets to their employers and clients as they build great products.

Now we're ready to use this tool as we create projects of all kinds. Keep reading to learn about the possibilities!

Questions

1. What is the difference between the **Structure** pane and the **Class Explorer**?
2.   How can the IDE help locate a project file in Windows Explorer?
3. Where are auto-recovery files saved?
4. How do you hide non-visual components?
5. When should you add GetIt dependencies?
6. Can you quickly switch between light and dark modes?
7. What is LSP?

Further reading

- What's new in Delphi 10 Seattle: `http://docwiki.embarcadero.com/RADStudio/Seattle/en/What's_New#IDE`

- What's new in Delphi 10.1 Berlin – IDE: `http://docwiki.embarcadero.com/RADStudio/Berlin/en/What%27s_New#IDE`

- What's new in Delphi 10.2 Tokyo – IDE: `http://docwiki.embarcadero.com/RADStudio/Tokyo/en/What%27s_New#IDE`

- What's new in Delphi 10.3 Rio – IDE: `http://docwiki.embarcadero.com/RADStudio/Rio/en/What%27s_New#IDE`

- New Delphi features: `http://docwiki.embarcadero.com/RADStudio/Sydney/en/What's_New#IDE`

- FireUI live preview `http://docwiki.embarcadero.com/RADStudio/Rio/en/FireUI_Live_Preview`

- Delphi 10.4 Sydney – key IDE enhancements: `http://docwiki.embarcadero.com/RADStudio/Sydney/en/What%27s_New#Key_IDE_Enhancements`

- Code Insight and LSP: `http://docwiki.embarcadero.com/RADStudio/Sydney/en/Code_Insight_Reference`

- Embarcadero acquires Castalia: `https://www.embarcadero.com/press-releases/embarcadero-acquires-castalia-and-usertility-from-twodesk-software`

- DPI-Awareness: `https://docs.microsoft.com/en-us/windows/win32/win7appqual/user-interface---high-dpi-awareness`

- Dark theme arrives in Delphi 10.2.2: `https://community.idera.com/developer-tools/b/blog/posts/new-in-10-2-2-dark-ide-theme`

- Delphi Theme Editor: `https://github.com/RRUZ/delphi-ide-theme-editor/`

- IDE Fix Pack: `https://www.idefixpack.de/`

- **Bookmarks** and **Navigator**: `https://parnassus.co/bookmarks-and-navigator-acquired-by-embarcadero/`

2
Delphi Project Management

There are many types of projects that Delphi can build: standalone desktop applications for Windows and Mac, mobile apps for Android and iOS smartphones and tablets, Linux daemons, Android packages and services, and even plugin packages to add functionality within Delphi itself; all types are possible. If all you've ever built are Windows **Visual Component Library** (**VCL**) applications, it may be daunting to consider other platforms. But Delphi 10.4 is a very capable toolset, having simplified the steps you need to take, which allows you to concentrate on what you do best—building great software! Most of the heavy lifting is done for you when supporting other platforms, so it's time to extend your skills to these devices.

To help you down the right path, you need to know about the project types, target platforms, and starting templates available and how to best utilize them. After reading this chapter, you will have no difficulty knowing where to start. Additionally, you'll learn when to use debug and release configurations to separate testing and deployment scenarios, how to integrate version control right within the IDE, and how to set up automation both inside the IDE with build events and outside with command-line tools. These topics will be covered in the following sections:

- Exploring project types and target platforms
- Using build configurations effectively
- Working with related projects

- Managing source modifications
- Using the command-line tools for build automation

Technical requirements

This chapter simply requires using Delphi on a Windows 10 machine to explore different project types. There will be screenshots of short example applications running on different platforms but they will be brief and only exist to showcase what lies ahead. The code for a couple of small Windows projects, including a batch script for compiling several projects is on GitHub at: `https://github.com/PacktPublishing/Fearless-Cross-Platform-Development-with-Delphi/tree/master/Chapter02`

Exploring project types and target platforms

As we look at each project type in this section, there will be different target platforms available. For example, a Windows VCL application can only be built on Windows platforms (either 32-bit or 64-bit), but a multi-device application can have several different target platforms. This is shown in the **Projects** window as the second sub-item under the project name and can be expanded to show the currently supported platforms added for that project. To add support for other platforms, right-click on **Target Platforms** and select from the list available. For example, a new multi-device application comes supporting six targets by default, but you can also add **iOS Simulator**:

Figure 2.1 – Adding and selecting a target platform

As we talk about the different project types and what platforms are available for each, this is how you add support for platforms that are not initially added to your project. You can only compile for one target at a time; the currently selected target is shown in parentheses next to the **Target Platforms** header.

Creating a console application

If you don't need a rich interface with complex graphical controls, you may need nothing more than a console application. This is the simplest project type and is often used for small utility programs, automated unit testing, and prototyping Windows services or IIS modules. It can write output to the console using Writeln and pause for input using Readln but does not include any graphical libraries. Console applications can be built on Windows, Linux (Enterprise or Architect edition), or Mac OS X—and even though you can add platforms for Android and iOS, there seems to be no documentation or support on how to run console apps on mobile platforms.

To create a console application, select **File | New | Other** from the menu and select **Console Application**. A single source file is created with the `program` keyword and a compiler directive of `{$APPTYPE CONSOLE}`. All the sources can be contained in this one file and the program flow is simple and contained within the `begin-end` block created for you. The compiled application on Windows will be `.EXE`.

Running this will bring up a console window, typically with a black background and white text, and any output you've added to the application is displayed in a scrolling window within this box. You can add code to read text entered from the keyboard using Readln—which, if you're running it from the IDE, you will need to do in order to keep it from finishing and closing so quickly that you miss seeing any output.

Here's what a simple console application looks like when running:

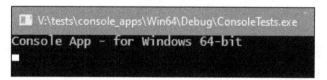

Figure 2.2 – Windows 64-bit console application

We will use console apps to demonstrate language concepts and to test Windows services in later chapters.

Building a Windows VCL application

A Windows VCL application has been the most common project type since the first version of Delphi. In fact, the main reason Delphi was created was to make it easier to develop Windows applications by replacing the **Object Windows Library** (**OWL**) framework used in Borland Pascal as wrappers around the Windows API. The VCL hides a lot of complexity and yet provides an extensible hierarchy of object-oriented components to ease the development of any type of Windows application.

To start one of these project types, select **File | New | Other** and select **Windows VCL Application**. A project source file is created with the program keyword and a "form" unit is added, which comprises two physical files: a .DFM, or Delphi, form, and .PAS, the source for this form with the unit keyword at the top of the file. The extension of the compiled application will be .EXE, for "executable."

Without doing anything else, simply building and running this project immediately launches a Windows application with one form that can be moved around the screen and closed like any other Windows application.

With a quick addition of a label and a button, this is what that might look like:

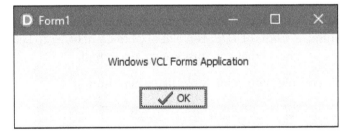

Figure 2.3 – Windows 32-bit VCL forms application with a label and a button added

Let's now have a rundown on starting a multi-device application.

Starting a multi-device application

If you want to write an application but you don't want it limited to the Windows platform, you should create a multi-device application. Similar to Windows VCL applications, it creates standalone executables with one or more form units but instead of using the VCL, it uses the FireMonkey framework, a cross-platform set of controls that allow one code base to build applications that look and feel like they were built separately for Windows, Android tablets and phones, Macs, and iOS devices.

From the **File | New | Other** menu, select **Multi-Device Application**. Before a project is created, you're presented with a choice of several template projects that will get you started quickly:

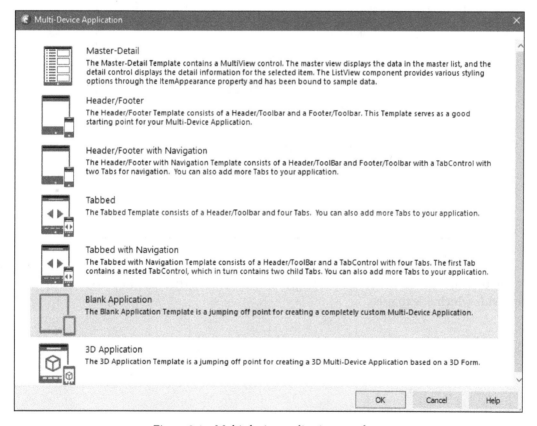

Figure 2.4 – Multi-device application templates

Each of these, after selecting a directory for the new application, creates a project source file with the `program` keyword, and a form unit that comprises two physical files: a `.FMX`, or FireMonkey, form, and `.PAS`, the source for this form with the `unit` keyword at the top of the file. There are also seven target platforms added for you, which you can see if you expand the list by clicking the little > symbol to the left of **Target Platforms** in the Project Manager:

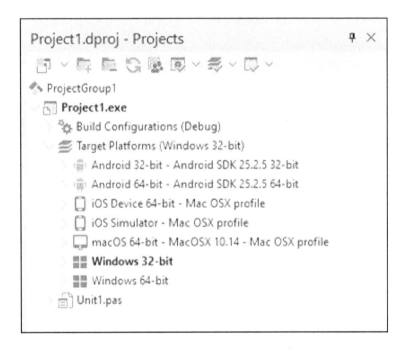

Figure 2.5 – Target platforms added to a new multi-device application

Let's go through the various templates to give you an idea of what they look like to help you decide which one to use.

The first one, **Master-Detail**, shows an example of laying out a list of records on the left side of the screen with thumbnails and names of people in a ListView control, and as you click each person, a bigger image with their name, title, and details show up in labels and a memo on the right. All controls are arranged using a layout control. It supports scrolling the list of records up and down with a finger drag as you would expect on a mobile device. The records and details are linked using LiveBindings. Without any modifications, here's what it looks like on a Kindle Fire (we'll show how to do this in *Chapter 4*, *Multiple Platforms, One Code Base*):

Figure 2.6 – "Master-Detail" multi-device template app, running on an Android tablet

The **Header/Footer** and **Header/Footer with Navigation** templates are very similar. Both application templates show a header at the top and a footer at the bottom that you can customize. The area between in the latter template contains a TabControl for your content with an ActionList for moving forward or backward in the list of tabs. Here's how it looks, unmodified, on an Android phone:

Figure 2.7 – "Header/Footer with Navigation" multi-device template app on an Android phone

The **Tabbed** and **Tabbed with Navigation** templates are also similar to each other except that the first has its title in the header with four tabs below, while the second has the four tabs along the top with the content of each tab containing its own title; plus, the first tab has a set of sub-tabs with navigation buttons. Here's what the **Tabbed** template looks like on Windows and the **Tabbed with Navigation** template looks like on a Mac:

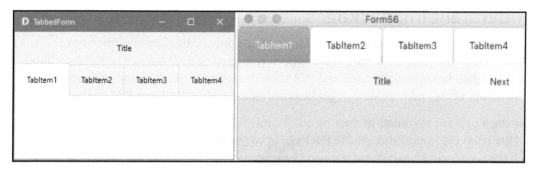

Figure 2.8 – "Tabbed" and "Tabbed with Navigation" multi-device template apps on Windows and Mac

If you don't need headers or footers, a master/detail list of records, or tabbed navigation, or if you just want to build your mobile app from scratch, select the **Blank Application** template—just as it says, there are no controls on the form.

A **3D Application** template also gives you a blank slate with no controls but has a different set of units included that set up the 3D framework. Additionally, the form inherits from TForm3D instead of TForm like the other multi-device application template forms do. We will cover 3D apps in detail in *Chapter 8, Exploring the World of 3D*.

Working with dynamic libraries

Delphi has long enjoyed being one of the few development tools that can ship a complete application in a single executable without the need for installing supporting libraries and resource files at the destination. But this is not always an advantage, especially if you need to ship separate executables that use the same code. That code could be put into a **Dynamically Loaded Library** (**DLL**) and shared among them. This is, in fact, how most operating systems and many of today's large applications are put together—lots of interconnected modules that work together. When there's an update for that code, you don't have to replace the whole system but just that one DLL (this is sometimes all a "patch" in a software update does).

Of course, Delphi can support these as well. Simply select **File | New | Other** from the menu, and choose **Dynamic Library** from the list. A single source file is created with the `library` keyword. All the sources can be contained in this one file and a begin-end block is created for you—although it is likely you won't use it. Compiling a dynamic library on Windows generates a DLL.

A dynamic library cannot run by itself but is called from another library or an application. As such, a library is, for the most part, simply a published list of procedures or functions (libraries can also contain published resources). We will cover creating and using libraries in *Chapter 5, Libraries, Packages, and Components*.

Understanding packages

A Delphi package is a special kind of library that can only be used by applications or libraries written in Delphi or within the Delphi IDE itself. The advantage is that it contains metadata that makes sharing code easier. A more important feature is that it allows you to write components and plugins that get installed in the Delphi IDE—or RAD Server.

Starting a package is similar to starting a Dynamic library. When you select **File** | **New** | **Other** from the menu, and choose **Package** from the list, a single source file is created with the `package` keyword and a `requires` section that lists other packages that must be present to support the functionality that this package provides. Compiling a package generates a file ending with `.BPL` (**Borland Package Library**). (Borland was the original publisher of the Turbo Pascal, Borland Pascal, and now Delphi line of products.)

There are two types of packages: `run-time` and `design-time`. Runtime packages are a great way to modularize your application as they are similar to dynamic libraries in that they can provide a shared set of libraries, functions, and resources to be used by other applications or libraries. They can also be loaded dynamically as libraries can. However, as mentioned previously, they are not as versatile as they can only be called from code written with the same version of Delphi. Dynamic libraries, on the other hand, can be shared between applications and modules written in other languages—or other versions of Delphi.

Design-time packages allow you to augment functionality within the Delphi IDE itself, typically to support components with custom property editors or provide special functionality that hooks into the Delphi IDE, such as providing features in the editor or adding menu items for launching new functionality. We will discuss these capabilities in *Chapter 5, Libraries, Packages, and Components.*

Dealing with Android services

As you get more acquainted with writing applications for the mobile platform, you'll learn that there are many more differences than just the user interface. On modern desktop PCs, you can assume that there is a lot more memory and disk space available than what you have on a hand-held device. You must take these application design considerations into account. For Android applications, it can be quite advantageous to offload some of the work your application must perform to a background process known as an Android service. Yes, Delphi has a project type to support these, and you do that by selecting **File** | **New** | **Other** from the menu, and choose **Android Service** from the list.

Before any project files are created, you are presented with a choice of four Android service types:

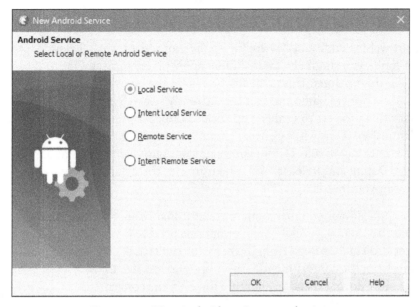

Figure 2.9 – New Android service type selection

Once one is selected, a project source file is created with the `program` keyword, and a `form` unit is added that comprises two physical files: a `.DFM`, or Delphi, form, and `.PAS`, the source for this form with the `unit` keyword at the top of the file. This looks surprisingly similar to a Windows VCL application, but notice that the extension of the compiled output is `.SO` instead of `.EXE`. Additionally, the `uses` clauses of both the project and unit are obviously geared to the Android platform.

Customizing your favorite project types

This section has instructed you to use the **File | New | Other** menu selection for each project type because that menu item is always there—the other menu items you see after first installing Delphi can be added, removed, or rearranged. Just select **File | New | Customize** and modify the list to your liking. For example, if you never work on Windows VCL applications, you can remove that item from the list and add other project or file types that you use more often. You can also add separators to help organize the items into groups that make sense to you. Any items you add here that can be added as part of the current project will show up in the context menu when you right-click the project and select **Add New**. It should be noted also that if you *remove* items from this customizable list, they will also be removed from the **Add New** context menu.

Using build configurations effectively

Before we start writing code and developing applications, there are a few more project management topics you should be aware of that apply to any project. One of those is the use of **Build Configurations**. This is the first sub-item that shows in the **Project Manager** window under the project name and lists the currently selected build configuration in parentheses for that project. You can set up various project options and save them as a build configuration, then switch between them by expanding the **Build Configurations** list and double-clicking on one. Delphi comes preconfigured with two standard configurations: **Debug** and **Release**. You can customize the default configurations or add other ones to suit your needs.

When testing applications, you may want to step through your code, set breakpoints, and watch variables. To do this, several compiler options need to be enabled in the project options (select **Project | Options** from the menu or right-click on the project name and select **Options** from the context menu). These options add size to the compiled file and, in some cases, slow down execution slightly. However, the convenience of inspecting your code at runtime can save hours.

When you're satisfied with your work and are ready to deploy your application, you can switch the selected build configuration to **Release**, which turns off those debug compiler options and enables optimization, then rebuild your project in order to give your customer the best experience when they run your code.

Some organizations may only allow one build configuration to be used in both development and deployment out of concerns that differences may make it difficult to exactly replicate a reported problem. However, you might still want to use build configurations and set up all the compiler and linking options the same to produce identically compiled code in order to use other facets of the build configurations that don't affect the compiled files: **build events**.

Build events

Build events are the closest thing to automation within the IDE. We'll talk about scripting your whole build process from the command line a little later, but with build events, you can launch external tools to sign your code, generate installers, make copies of your packages to another folder, and anything else a Windows batch file can do. Plus, Delphi makes macros available that are replaced with various parts of your project, such as the output directory, the active build configuration, the extension of the file being compiled, and more.

NOTE

The statements in build events are not called the same way that batch files are but are concatenated into one long statement with ampersands (&) to ensure they are executed in order. Therefore, `statement1` and `statement2` will be executed as if you had entered `statement1 & statement2`. If you embed `if` statements in your build events, be sure to use parentheses to enclose statements that should not affect the execution of further statements.

To configure the build events for your project, open the **Project | Options** menu (or right-click on your project name and select **Options**), then expand **Building** and click on **Build Events**.

You'll notice that build events are specific to the target platform. That makes sense as the tools you use or the actions you perform may very well be different for different platforms. But you can also set up tools that are common to sets of platforms, or even all platforms, by choosing one of the base configurations (for example, **All**, **Debug**, or **Release**), then when you select a specific platform (for example, Windows 32-bit platform), it will inherit the list of commands from its base configuration and allow you to add additional ones for the specific platform. For example, you might want to initialize test data for all configurations but generate installers only for **Release** builds.

There are three categories of events:

- **Pre-build events**: Commands to be run before a build starts. If there's an error in this process, you can optionally halt the entire project build process by setting the **Cancel on error** checkbox.

- **Pre-link events**: These commands will be run after the units are compiled but before the linking phase starts (more applicable to C++ than Delphi). Again, you can halt the build process with the **Cancel on error** checkbox.

- **Post-build events**: This set of commands will be run after the project has been successfully built. This type of build event has an extra option where you can always launch the commands or only if the target is out of date.

For each of these types, click on the ellipsis button to enter one or more commands. For example, the following post-build event will sign the code with an SSL certificate (using the Windows SDK's SignTool), then generate an installation executable (using InnoSetup, in this example), and finally copy the result to a shared folder on the network (assuming paths are set up so that the SignTool and setup builder applications can be found):

```
signtool sign /f "MyCert.pfx" /p "MyPW" "$(OUTPUTPATH)"
iscc.exe $(PROJECTNAME).iss
copy "$(OUTPUTDIR)\..\output\setup.exe" \\server\deploys
```

Remember, this is similar to writing a batch file but has the advantage of using project-specific macros provided by the Delphi environment so that if you rename a project, you don't have to change the build event script.

Now, let's expand our focus to work with several projects at once.

Working with related projects

You may find yourself often switching between multiple projects, either as part of an application suite or simply having a list of assigned projects to maintain. Instead of closing one and opening a different one, you can put multiple projects into a project group for quicker accessibility. From the **Project** menu (or from the context pop-up menu after right-clicking on the project group name), select **Add Existing Project…** (or **Add New Project…**) and follow the steps for opening or creating a project. Once you have one or more projects loaded, you can right-click at the top of the projects list in the **Project Manager** window and select **Save Project Group As…** to save the list of projects. Now, you can open and close a group of projects as easily as you can a single project.

You may already be well-versed in managing groups of projects, but a couple of toolbar buttons added back in Delphi XE2 in the **Project Manager** simplify actions you can take on multiple projects at once.

For instance, let's say you maintain a large product that has a main executable project, several libraries, and maybe a non-visual component set. This application suite runs in both Windows 32-bit and 64-bit and on Mac 64-bit, and a problem has been reported for the Windows 32-bit version—and you're not sure which module is causing the problem. You need to switch all projects to the Win32 platform and **Debug** mode, then recompile them all and start testing.

After loading the project group, instead of tediously going to every project and setting the platform and build configuration, simply go to the project window's toolbar, drop down the **Set Active Configuration** menu button, and select **Debug**. Then, drop down the **Set Active Platform** menu button and select **Win32**, and all the projects in the project group are ready to debug on Win32.

Here's a screenshot of the toolbar with the **Configurations** button dropped down and the **Platforms** button to its immediate right:

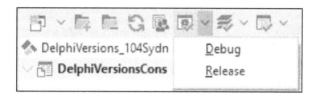

Figure 2.10 – The project group toolbar

After finding and resolving the problem, you can again use the **Set Active Configuration** toolbar button to set all projects back to **Release** mode, and right-click on the project group, and select **Build All** to recompile all the projects.

If you don't remember the buttons' functions by icon, you can right-click on the toolbar and activate **Text Labels**, but if the project window is too narrow, not all the buttons will show—and this toolbar is not customizable.

Once we've got a handle on managing multiple projects, we need to keep track of all the changes we're making to them.

Managing source modifications

Team collaboration invariably involves some sort of source code repository – no longer is a shared network folder good enough. Version histories with trackable changes are part of the daily life of the professional programmer. Everyone has their favorite tools for viewing changes and checking in code, but you might want to try out the built-in repository support in the Delphi IDE as another way of getting updates or committing your work. We'll start by looking at a part of the IDE you may not have used much, the code editor's **History** tab, and then show how integrating source repositories can make that view much more useful.

Looking through your code history

The discussion on how version control works in Delphi starts with a look at the **History** tab of the code editor window:

Figure 2.11 – The code editor's History tab

When you're typing code, you're usually in the **Code** tab. When working on a data module or form, you can switch between **Code** and **Design** with the *F12* key (by default). Toggling between code and design can also be done by switching tabs. The rightmost tab of these code editor tabs is the **History** tab and shows recent changes made to the file you're working on. Once you switch to the **History** tab, there are three page tabs:

- **Contents**: Displays selectable revisions in the top pane and the contents of the corresponding revision in the bottom pane. The revision fields shown include revision name or description, an optional label, the date it was made, and the author. By default, the author is simply the username of the current Windows user.

- **Information**: In the top pane, the only difference from the **Contents** view is the addition of an optional comment. The bottom pane shows additional comments if they exist.

- **Differences**: The top pane presents two copies of the revisions for the current file: **From**, and **To**. Once you select a **From** revision and a **To** revision, the bottom pane shows the differences between them, indicated with color-coded lines and *plus* and *minus* signs in the left column.

In the following screenshot, the **Differences** pane of the **History** tab is shown with the original version of a file selected in the **From** revision list and the most recent changes in the **To** revision list. The bottom pane shows the changes that were made to the file, including the addition of an image and the removal of the **private** and **public** sections:

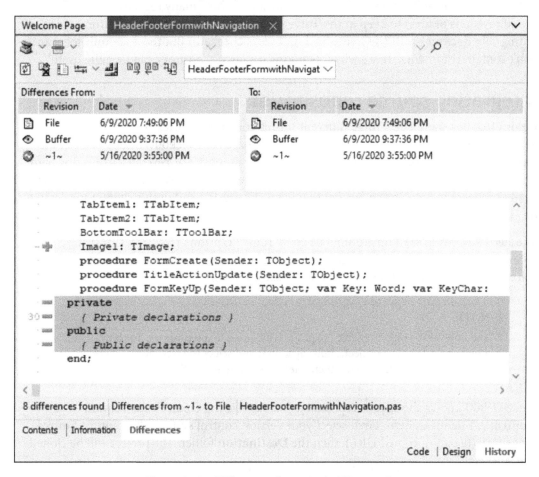

Figure 2.12 – Differences shown in the History tab

There are several toolbar buttons above the **Differences** pane that help you manage the revisions, including reverting changes to a file and traversing through the differences. There are also several icons to help you identify the type of revision for each line, such as a local backup file or a file stored in a source repository. This functionality is built into Delphi before you add version control. You can specify how many local backups (locally stored revisions of files) to keep in the **Tools | Options | User Interface | Editor** page by setting **File Backup Limit** (if **Create backup files** is checked). But these are only local and don't contain comments. That's why integrating a source repository can be quite useful.

Integrating source repositories

Delphi 10.4 has support for three different source code repositories: **SVN**, **Git**, and **Mercurial**. They all support the basic operations: **clone** (or checkout) a copy of a remote repository to a local folder, **add** files and projects to a repository, **refresh** a local copy with remote changes, and **commit** or check in modifications—but they differ slightly, so you'll need to follow the links in the *Further reading* section at the end of this chapter for details on your particular repository type.

The steps for setting each up are similar: go to **Tools | Options | Version Control**, select your repository type, then fill in the requested information, and optionally, customize the colors that will affect the **Log** view of the repository history.

> **NOTE**
> The **Log** view is only available for projects assigned to a repository. To show a repository log for a project, right-click on it and select the repository you're using from the context menu, then select **Show Log** in the fly-out menu.

Once Delphi knows about your repository, you can use the **Open From Version Control... File** menu item. First, select your version control system, then enter a **Source** repository (usually a copied URL), then the **Destination** folder. The project will be cloned and opened.

Now, when you right-click on a project or folder, there's a context menu for your repository type that lets you commit or revert changes, and other management functions appropriate for your repository.

Back to the **History** tab, a project under version control will show repository comments mixed in with the local backups. Here's an example of some changes made to a file in a Git repository:

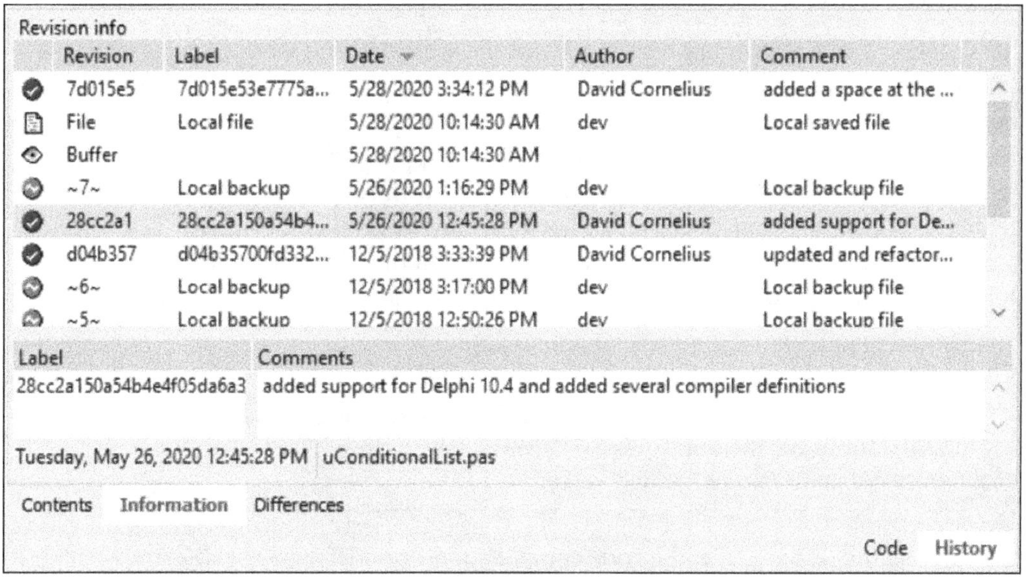

Revision info					
Revision	Label	Date ▾	Author	Comment	
7d015e5	7d015e53e7775a…	5/28/2020 3:34:12 PM	David Cornelius	added a space at the …	⌃
File	Local file	5/28/2020 10:14:30 AM	dev	Local saved file	
Buffer		5/28/2020 10:14:30 AM			
~7~	Local backup	5/26/2020 1:16:29 PM	dev	Local backup file	
28cc2a1	28cc2a150a54b4…	5/26/2020 12:45:28 PM	David Cornelius	added support for De…	
d04b357	d04b35700fd332…	12/5/2018 3:33:39 PM	David Cornelius	updated and refactor…	
~6~	Local backup	12/5/2018 3:17:00 PM	dev	Local backup file	
~5~	Local backup	12/5/2018 12:50:26 PM	dev	Local backup file	⌄

| Label | Comments |
| 28cc2a150a54b4e4f05da6a3 | added support for Delphi 10.4 and added several compiler definitions |

Tuesday, May 26, 2020 12:45:28 PM   uConditionalList.pas

Contents   Information   Differences

Code   History

Figure 2.13 – Information page of the History tab showing both local backups and repository commits

All these changes were made by the same person, but the local backups took the username from the Windows user, whereas the repository commits were based on the repository user settings. The bottom pane shows the hash of the Git commit in the **Label** field and the full comment submitted.

Now, let's switch gears a little and move outside of the IDE.

Using the command-line tools for build automation

We used build events to add automation to the IDE, but we still have to start the IDE, load a project, and manually start the build. What if you want a dozen libraries and projects automatically recompiled every night? Delphi has this covered, too.

Using the IDE hides a lot of details. It's nice that most of the time, all you need to know is that there are multiple target platforms and building one produces the desired file in the right places. But a look at the **Build** page of the **Messages** pane reveals very long commands being called.

What's more is that a temporary resource script file is created and a resource compiler is called to put any icons, images, version information, and so forth into binary format. When using just the IDE, these resources are defined in various project options pages and all the details are handled for you. But when building applications and libraries yourself using the command-line tools, you have to write your own resource script file to pull these together, then call the resource compiler. Looking at what is called during the IDE's build process in the **Messages** pane, you can use the same call the IDE uses, **CGRC.EXE**, or call **BRCC32.EXE** directly, the main resource compiler used for Delphi applications.

> **NOTE**
> There may be occasions to use the Microsoft SDK Resource Compiler (**RC. EXE**) instead, but there are several differences to be aware of. Read more in the link listed in the *Further reading* section.

After building the resource file, the code compiler appropriate for your platform is called:

Platform	Compiler
32-bit Windows	DCC32.EXE
64-bit Windows	DCC64.EXE
Linux	DCCLINUX64.EXE
64-bit iOS	DCCIOSARM64.EXE
iOS Simulator	DCCIOS32.EXE
Android	DCCAARM.EXE

The command-line options passed to these toolchains stem from which platform is being built and how your environment is configured. If you run one of those programs from the command line without any command arguments, it'll show you all the options available to specify such things as search paths, debug options, namespaces, record alignment, optimization types, compatibility settings, and many other things you may never need to worry about. Delphi does a good job of setting up a lot of defaults for different targets, but also provides a nice, organized presentation of these in your project's options.

Before we look at building one of the projects from the IDE, let's take the simplest possible project and compile it from the command line just to try it out. This would be a console app, of course, and Delphi adds a couple of things in there that we can take out (otherwise, we'll have to specify library paths, which we don't want to do for this quick test). Here's the code, which you can also access at https://github.com/PacktPublishing/ Fearless-Cross-Platform-Development-with-Delphi/blob/master/ Chapter02/00_SmallestConsoleApp/TinyConsoleApp.dpr :

```
program TinyConsoleApp;
{$APPTYPE CONSOLE}
begin
  Writeln('Hi Console!');
  Readln;
end.
```

Here's the simple compilation:

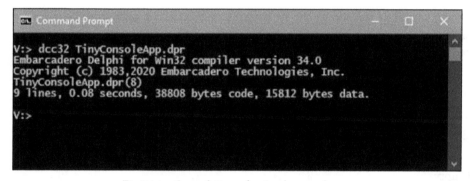

Fig 2.14 – Compilation of a simple console app

Now that we've got our feet wet at the command line, let's go deeper.

One approach to automated command-line builds is to first set up these options the way that you need using the IDE, possibly in a separate build configuration, and then build your project and copy the commands that were called by the IDE from the **Build** page of the **Messages** window to your script. From there, you can cut out what you don't need from the script and include other actions, such as checking for and logging the results of unit tests, running audits, and other tools your team utilizes.

> **NOTE**
> The default layout in Delphi doesn't initially show the **Messages** window. Remember, you can manually show it by selecting **View** | **Tool Windows** | **Messages** from the menu.

A helpful hint to know before writing your batch files is to use environment variables as shortcuts in your paths. Looking at the IDE's generated commands can make your eyes glaze over pretty quickly as you scan down the long path names because the IDE doesn't use shortcuts; you can—and should! This will not only shorten your paths and make your scripts more readable but they'll also work on someone else's system where Delphi may have been installed in a different location or when you upgrade to a newer version of Delphi and only need to change one or two lines at the top of your script.

To help with this, every version of Delphi ships with a batch file called `rsvars.bat` in the `bin` folder of the Delphi installation. Running this at the top of your script sets up `BDS`, `BDSINCLUDE`, `BDSCOMMONDIR`, and other useful environment variables pointing to folders you'll need.

The following example scripts use the techniques covered in this section to generate executables for an open source project, `DelphiVersions`, of which a portion was extracted for this book and is available here, along with a commented version of the build script:

`https://github.com/PacktPublishing/Fearless-Cross-Platform-Development-with-Delphi/tree/master/Chapter02/01_BuildScript`

The scripts start off by making sure that the destination folders exist, then clean the temporary build files, and finally, compile the projects. The first one is a simple console app:

```
call "%ProgramFiles(x86)%\embarcadero\studio\21.0\bin\rsvars.
bat"
if not exist .\Win32\Release mkdir .\Win32\Release
del .\Win32\Release\*.rsm
del .\Win32\Release\*.dcu
del .\Win32\Release\*.exe
dcc32.exe -$O+ -$W- --no-config -B -TX.exe -DRELEASE -E.\Win32\
Release -NU.\Win32\Release -NSSystem; -U"%BDS%\lib\Win32\
release" -V -VN -VR DelphiVersionsConsole.dpr
```

The next one builds a resource with an embedded style for both **Debug** and **Release** versions of a 64-bit Windows VCL application:

```
call "%ProgramFiles(x86)%\embarcadero\studio\21.0\bin\rsvars.
bat"
if not exist .\Win64\Debug   md .\Win64\Debug
if not exist .\Win64\Release md .\Win64\Release
del /s *.res
del /s *.rsm
del /s *.dcu
del /s *.exe
cgrc.exe -c65001 -v DelphiVersionsVCL.rc -foDelphiVersionsVCL.
res
dcc64.exe -$O- -$W+ --no-config -B -TX.exe -DDEBUG -E.\Win64\
Debug   -NU.\Win64\Debug 1-NSSystem;Vcl;Vcl.Shell;WinAPI
-R"%BDS%\lib\Win64\Release" -U"%BDS%\lib\Win64\debug" -V -VN
-VR DelphiVersionsVCL.dpr
dcc64.exe -$O+ -$W- --no-config -B -TX.exe -DRELEASE -E.\Win64\
Release -NU.\Win64\Release -NSSystem;Vcl;Vcl.Shell;WinAPI
-R"%BDS%\lib\Win64\Release" -U"%BDS%\lib\Win64\release" -V -VN
-VR DelphiVersionsVCL.dpr
```

Finally, we build two 64-bit FireMonkey apps—one for Windows and one for Mac:

```
call "%ProgramFiles(x86)%\embarcadero\studio\21.0\bin\rsvars.
bat"
if not exist .\Win64\Release md .\Win64\Release
if not exist .\OSX64\Release, md .\OSX64\Release
del /s *.dcu
del /s *.rsm
del /s *.exe
del .\OSX64\Release\*.o
dcc64.exe -$O+ -$W- --no-config -B -TX.exe -DRELEASE -E.\Win64\
Release -R"%BDS%\lib\Win64\release" -U"%BDS%\lib\Win64\release"
-K00400000 DelphiVersionsFM.dpr
dccosx64.exe -$O+ --no-config -B -DRELEASE -E.\OSX64\Release
-R"%BDS%\lib\OSX64\release";"%BDS%\redist\OSX64" -U"%BDS%\
lib\OSX64\release" -NO.\OSX64\Release -NU.\OSX64\Release
-NSSystem -O"%BDS%\lib\OSX64\release";"%BDS%\redist\OSX64"
--syslibroot:"%USERPROFILE%\Documents\Embarcadero\Studio\SDKs\
MacOSX10.14.sdk" DelphiVersionsFM.dpr
```

You can download the original project by following the link to **Open source project: DelphiVersions** in the *Further reading* section at the end of this chapter.

Summary

In this chapter, we covered all the project types and target platforms Delphi 10.4 supports—and some starting templates to jump-start the development of a cross-platform application. With powerful build configurations, you can apply sets of options quickly and launch external tools for certain conditions to save steps after the compilation process. Additionally, you learned how to easily manage a group of projects. With built-in version control, you seldom need to leave the IDE. But if you have unattended build processes that run on a regular basis, you learned that Delphi even supports that paradigm with command-line parameters for the unattended compilation of any project type.

In the next chapter, we'll build on these concepts by reviewing the language syntax of Delphi and noting several important additions over the years that have laid the foundation for building modern cross-platform applications.

Questions

1.   What project types support the Linux platform?

2.   What are the two main differences between the **Tabbed** and **Tabbed with Navigation** multi-device application templates?

3.   Which project type can provide callable functions for applications written in other programming languages?

4.   How do you customize the **File | New** list?

5.   How do you automatically build an installer whenever your **Release** compilation process succeeds?

6.   Where can you see the list of changes made to your files?

7.   What's the filename of command-line compiler for the Android platform?

Further reading

- The **Projects** window: `http://docwiki.embarcadero.com/RADStudio/Sydney/en/Projects_Window`

- Extensions of files generated: `http://docwiki.embarcadero.com/RADStudio/Sydney/en/File_Extensions_of_Files_Generated_by_RAD_Studio`

- Delphi's History Manager: `http://docwiki.embarcadero.com/RADStudio/Sydney/en/History_Manager`

- Version control support in Delphi: `http://docwiki.embarcadero.com/RADStudio/Sydney/en/Version_Control_Systems_in_the_IDE`

- Using SVN in Delphi: `http://docwiki.embarcadero.com/RADStudio/Sydney/en/Subversion_Integration_in_the_IDE`

- Using Git in Delphi: `http://docwiki.embarcadero.com/RADStudio/Sydney/en/Git_Integration_in_the_IDE`

- Using Mercurial in Delphi: `http://docwiki.embarcadero.com/RADStudio/Sydney/en/Mercurial_Integration_in_the_IDE`

- Command-line switches and options for starting the IDE: `http://docwiki.embarcadero.com/RADStudio/Sydney/en/IDE_Command_Line_Switches_and_Options`

- Command-line utilities index: `http://docwiki.embarcadero.com/RADStudio/Sydney/en/Command-Line_Utilities_Index`

- The Microsoft SDK Resource Compiler: `http://docwiki.embarcadero.com/RADStudio/Sydney/en/RC.EXE,_the_Microsoft_SDK_Resource_Compiler`

- Inno setup: `https://jrsoftware.org/isinfo.php`

- Open source project: DelphiVersions: `https://github.com/corneliusdavid/DelphiVersions`

3
A Modern-Day Language

This powerful development environment and toolset we've been discussing in this first section of the book would not be where it is today if it didn't have the language behind it to support the needs of the modern-day programmer.

This chapter will look back at where Delphi started, how it got here, and what the latest additions are that make it well-suited for any development task needed today. Read the following sections to understand the solid footing upon which Delphi stands:

- Remembering Delphi's Pascal roots
- Growing a language
- Learning about the latest enhancements

Technical requirements

You will need any edition of Delphi 10.4 Sydney to work with all the language features discussed in this chapter.

This chapter's sample code projects can be downloaded from GitHub at the following link:

```
https://github.com/PacktPublishing/Fearless-Cross-Platform-
Development-with-Delphi/tree/master/Chapter03
```

Remembering Delphi's Pascal roots

Some have pointed to the fact that Delphi—being based on Pascal, which was designed as a teaching language—is too simple to be a contender in today's complex programming environments. But nothing could be further from the truth. The extensions added to the language and the associated runtime library make it just as powerful as any other high-level, compiled programming language. In fact, the readability of this language and its strongly typed, structured design lends itself well to code that is easy to maintain and upgrade. It's object-oriented, compiles quickly, uses libraries and packages for modularity, and has a variety of frameworks to provide platform and device flexibility.

The Pascal syntax is often used in pseudocode for its universal readability. Instead of curly braces to define blocks (as in C# or JavaScript), Delphi uses `begin-end`. Instead of `-lt` or `-gt`, as in PowerShell, it uses less-than and greater-than symbols, < and >. Instead of the `DIM` statement to declare variables (as in BASIC), it uses the `var` keyword to start a variable declaration section. Instead of being declarative, like SQL, it's procedurally oriented.

Let's quickly review the basic structure of a Delphi program.

Reviewing the syntax

To see a working sample project, try opening one that comes with Delphi 10.4. Under the **Develop** section of Delphi's default **Welcome** page, click the **Open a sample project…** link and drill down through the **Object Pascal**, **VCL**, and **CardPanel** folders, and then open the **CardPanel** project.

Initially, the main form unit is displayed, but right-click on the project and select **View Source** to switch to the project file, `CardPanel.dpr`, where we see the `program` keyword, the `uses` clause listing the form unit, a compiler directive to link in the project's version information, compiled into a resource, and the `begin-end` block with a small bit of code to start the main application loop.

Back in the form unit, uCardPanel.pas, it starts with the unit keyword and has both an interface and implementation section, each with a uses clause. The interface section is used to tell other units or programs what types, variables, and methods are available from this unit. A TCardPanelForm type is declared as a descendant of TForm and a global variable of that type is declared, which is then created by the main program.

The implementation section is where the methods are implemented, and any identifiers declared there can only be used there. Again, you'll notice a compiler directive to link in a resource, only this time, it's the form's layout definition that is compiled into a resource instead of the project properties, as was seen in the .dpr file.

The const, types, and var sub-sections (to declare constants, types, and variables, respectively) can appear in any order and multiple times in both sections, as long as identifiers are declared before they are referenced.

Statements can be grouped into methods and named to break up your code. A procedure is a method that does not return a value. A function returns one value. Both can have zero or more parameters and declare constants, types, and variables local to that method.

Units can optionally include initialization and finalization sections. These sections do not need to use begin-end blocks. If the finalization section is used, the initialization section must be included, even if it's empty. These sections can be useful to implement startup code when the application starts or clean-up code when the application closes (unless the Halt procedure terminates it abruptly).

> **Note**
>
> The code in the initialization section is called in the order in which the unit appears in the project's uses clause—or in other units that are used by the project. Likewise, the code in the finalization section is called in the *reverse* order that they were loaded.

We won't go into great detail about the various control structures, conditional expressions, and so forth, but as we quickly talk about some of the highlights, you can see many of them demonstrated in the sample GitHub project, https://github.com/PacktPublishing/Fearless-Cross-Platform-Development-with-Delphi/tree/master/Chapter03/01_BasicSyntaxDemo.

The code shows typed constants and initialized variables (which are similar but have some important differences; see the *Declared constants* link in the *Further reading* section at the end of this chapter), nested `repeat-until` loops, a `case` statement with an `else` clause, and a `try-finally` block inside a `try-except` block for robust error-handling.

The code also demonstrates the following three types of comments. Comments are excluded in the compilation process, so it's safe to embed passwords and developer-level notes as they will never be found in the final distributable application or library:

- `//` **double slash**: These affect only the text after the double slash on the current line.
- `{ `**braces**` }`: These comment out a block of text; they cannot be nested—in other words, you cannot put braces around code that includes braced comments as the first end brace encountered will mark the end of the comment block.
- `(* `**parenthesis-star**` *)`: These also comment out a block of text and will also comment out other comments.

> **Tip**
> Use only the first two comment types in your regular programming. If and when you need to comment out a large block of code temporarily, use the parenthesis-star type as it supersedes the other two.

Now that we've covered the basic syntax, let's talk about what makes the language modern.

Growing a language

Delphi is not a stagnant language—it continues to grow and evolve. Sometimes, these changes are evolutionary, while sometimes they add optional functionality. Once in a while, they cause disruption when upgrading your code, but it has surprisingly good backward compatibility.

These extensions started with the earliest versions of Turbo Pascal, the precursor to Delphi back in the 1980s.

Adding objects to Pascal

In 1989, Turbo Pascal 5.5 was released. It was this historical upgrade that took Borland's extension of Pascal from a structured language to the object-oriented arena. A completely separate manual was provided just to educate developers about the concepts of object-oriented programming:

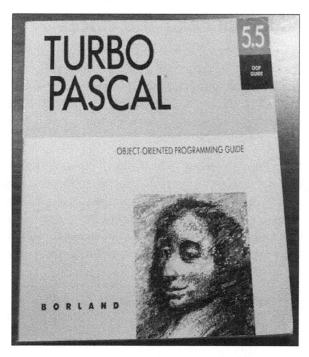

Figure 3.1 – Turbo Pascal's object-oriented guide

This laid the groundwork for class inheritance, encapsulation of data and functionality in objects that limit access to private fields through published properties, and polymorphism allowing descendant classes to override inherited methods. The first version of Delphi, which brought the Pascal language to mainstream Windows development, leveraged object-oriented class hierarchies with the powerful components prevalent in many systems today. You used these constructs in the previous chapter to build a new component, extending the capabilities of base components to provide a custom new functionality quickly and easily. These are important building blocks of the powerful frameworks we use today.

When you start a new GUI application, Delphi creates a class descendant of TForm. We're accustomed to placing controls on the form, adding event handlers and private functions, and rapidly building full-featured Windows VCL applications. This productivity, resulting from a well-conceived object-oriented class hierarchy, lends itself equally well to the rich set of database components in Delphi and FireDAC, which enable developing multi-tier database applications, as well as in the FireMonkey framework, allowing us to build cross-platform applications.

These topics could fill books on their own, but let's move on. Before we highlight the most recent enhancements, the following sections briefly describe several other important additions to the language.

Promising functionality with interfaces

Interfaces have been in Delphi since version 3, and you may think they are only needed when dealing with Microsoft's COM technology, but interfaces are quite useful constructs that provide code contracts, ensuring functionality in the classes that use them. For example, an interface can define the methods necessary for a design pattern, and then you can add that interface to one of your classes to ensure it fulfills, at least in the methods declared, the functionality of that design pattern.

Interfaces also allow dependency injection, which aids in building flexible unit tests. For example, a class can be designed to accept an interface reference instead of a specific class type. In the actual application, you would pass in the class for your real-world scenario, such as a database connection, while in a unit test, you could pass in a *mock* database, such as a class that gives random or hardcoded data instead of pulling from an actual database. This allows you to test your classes in isolation from other parts of your application.

There's an excellent paper and YouTube video that covers unit testing in detail, which can be found in the link titled *Unit testing and isolation frameworks* in the *Further reading* section at the end of this chapter.

Another language feature that arrived with Delphi 3 is variants.

Handling unknown data types

Sometimes, you don't have the privilege of knowing what data type you need to work with at compile time. The `variant` type was added to the language to support this. When a variant variable is created, its initial value is marked as `Unassigned`. After giving it a non-null value, you can query its type and perform operations on it. Variants can be simple data types (integer, string, and so on) but not structured ones (records, sets, files, pointers, and so on).

> **Note**
>
> Since a variant's type is not known at compile time, runtime execution will be slower as extra processing is required to know how to handle whatever value may be assigned. You should only use variants when required by external functions with which you need to interact.

A type of variant, `OleVariant`, is used to support Microsoft's **Object Linking and Embedding** technology, or **OLE**, which enables the passing of data between external applications. This allows you to, for example, read data directly from Microsoft Excel, or merge data into a Microsoft Word document.

Some language features, such as interfaces and variants, were necessary to keep compatibility with other technologies. Others added convenience or better programming methodologies to the language. The next section covers one of these.

Supporting nested types

Nested types are a great way to keep related constructs together. This can also avoid naming conflicts without using long class names. Declaring classes within a class can be easily demonstrated in a short program that adds people to a list. Normally, the people would be pulled from a database or web service, but here, we'll just add them manually each time:

```
private
  type
    TPerson = class
    private
      FFirstName: string;
      FLastName: string;
      FDOB: TDate;
    public
      constructor Create(NewFN, NewLN: string; NewDOB:
        TDate);
      function Age: Integer;
      property FirstName: string read FFirstName write
        FFirstName;
      property LastName: string read FLastName write
        FLastName;
      property DateOfBirth: TDate read FDateOfBirth write
        FDOB;
    end;
```

From the preceding code block, we see that the sub-class declares a TPerson type that gets added to TListBox, and then when the OnClick event occurs, the class is extracted and displayed on the screen. It is declared inside the private section of the form's class because it won't need to be used outside of the form, in this example.

The nested type's constructor simply assigns the parameters to local fields of the new object:

```
constructor TfrmPeopleList.TPerson.Create(
                    NewFN, NewLN string; NewDOB: TDate);
```

```
begin
  FFirstName := NewFN;
  FLastName  := NewLN;
  FDateOfBirth := NewDOB;
end;
```

The **Add** button creates a local `TPerson` object, assigns the values from the edit fields on screen, then adds it to the list, as follows:

```
procedure TfrmPeopleList.btnAddClick(Sender: TObject);
var
  NewPerson: TPerson;
begin
  if (Length(edtNewPersonFirstName.Text) > 0) and
     (Length(edtNewPersonLastName.Text) > 0) then begin
    NewPerson := TPerson.Create(edtNewPersonFirstName.Text,
                                edtNewPersonLastName.Text,
                                edtNewDOB.Date);
    AddPersonToList(NewPerson);
    edtNewPersonFirstName.Text := EmptyStr;
    edtNewPersonLastName.Text  := EmptyStr;
  end else
    ShowMessage('Please fill in both names.');
end;
```

When the user clicks on a row in `ListBox`, the associated `TPerson` object is extracted and shown:

```
procedure TfrmPeopleList.lbPeopleClick(Sender: TObject);
var
  APerson: TPerson;
begin
  if lbPeople.ItemIndex > -1 then begin
    APerson := lbPeople.Items.Objects[lbPeople.ItemIndex]
                 as TPerson;
    lblPersonName.Caption := APerson.FirstName + ' ' +
                             APerson.LastName;
    lblPersonDOB.Caption  := FormatDateTime('yyyy-mm-dd',
```

```
                              APerson.DateOfBirth);
    end;
  end;
```

You can find this code at https://github.com/PacktPublishing/Fear-less-Cross-Platform-Development-with-Delphi/tree/master/Chapter03/02_NestedTypesDemo.

We will likely be using nested types further in the book, so look for more elaborate examples. For now, there are still some important topics to cover, not least of which is Unicode.

Migrating to Unicode

Delphi has used UnicodeString as the default string type since Delphi 2009 (the type itself was available in earlier versions). It is the universally recognized standard for encoding all the world's characters and symbols. Each character in UnicodeString can take up to 4 bytes, and like all other string types in Delphi, indexing is 1-based.

Delphi has support for other string types as well, including AnsiString, UTF8String, RawByteString, and the original single-byte character type, ShortString, with a maximum of 255 characters. These all serve different purposes and there's a good, detailed white-paper, *Delphi Unicode Migration for Mere Mortals: Stories and Advice from the Front Lines*, referenced in the *Further reading* section at the end of this chapter, which covers some of these as it delves into the history and reasons behind the move from the simple string types of Delphi's early days.

This change was difficult for some applications but had long-term benefits for the language that outweighed the inconvenience to developers. I can't word it better than Cary Jensen as he introduces the subject:

> *"Embarcadero instantly enabled RAD Studio developers to build world class applications that treat both the graphical interfaces and the data they help manipulate in a globally-conscious manner, removing substantial barriers to building and deploying applications in an increasingly global marketplace."*

Along with the shift to Unicode, there were a couple more important language features to introduce before cross-platform support was really ready to take off. One of the most powerful new language concepts, introduced over 10 years ago now, is **generics**.

Applying strong type checking generically

Have you ever stuffed objects into a `TStringList` object? When you use them, you have to type-cast the extracted object to your custom type, much like we did when we extracted `TPerson` objects from `TListBox` in `NestedTypesDemo` earlier in this chapter (see the *Supporting nested types* section):

```
APerson := lbPeople.Items.Objects[lbPeople.ItemIndex] as
TPerson;
```

With generic types (also known as parameterized types, or just generics), you can create a list that knows what type each object is, which not only eliminates the need for type-casting but also provides syntax checking at compile time. Let's rework `NestedTypesDemo` to manage a generic list and provide more functionality. We'll do this with the following project:

https://github.com/PacktPublishing/Fearless-Cross-Platform-Development-with-Delphi/tree/master/Chapter03/03_GenericPeople

First, we'll separate the list of `TPerson` objects out from `TListBox` into a generic `TList`. The generic type declaration adds a parameter to `TList`, which may look a little strange at first:

```
FPeopleList: TList<TPerson>;
```

This tells the compiler that the `TList` object we are declaring will be holding objects of a specific type, `TPerson`.

Instantiating the object is similar:

```
procedure TfrmPeopleList.FormCreate(Sender: TObject);
begin
  FPeopleList := TList<TPerson>.Create;
end;
```

`AddPersonToList` is simplified:

```
procedure TfrmPeopleList.AddPersonToList(ANewPerson: TPerson);
begin
  FPeopleList.Add(ANewPerson);
end;
```

This procedure doesn't add the new `TPerson` to our ListBox as in our previous implementation because the user interface is no longer holding the list of `TPerson` objects. Best practices separate the display of the data from the management of the data. Instead, a new `ListPeople` procedure will copy the data from our `TList` to the ListBox whenever it needs to update the display:

```
procedure TfrmPeopleList.ListPeople;
var
  APerson: TPerson;
begin
  lbPeople.Items.Clear;
  for APerson in FPeopleList do
    lbPeople.Items.Add(Format('%s %s, Age: %d',
              [APerson.FirstName, APerson.LastName, APerson.
                Age]));
end;
```

Having the data separated gives us more flexibility. For example, we could show the list in `TListView` or `TMemo` format instead of `TListBox`. Or, we could print the list without it ever showing on the screen. We'll keep our `ListBox` and simply add sorting.

Generic collections provide a nifty way to implement sorting using another generic class: `TComparer`. This class's `Compare` function returns a positive integer if the first parameter is greater than the second, a negative integer if the first parameter is less than the second, or zero if they are equal. To use `TComparer`, the generic version of `TList` has an overloaded constructor that accepts an `IComparer<T>` parameter that allows you to specify the comparer function for sorting. It's a little complicated until you use it a few times.

The `TComparer` declaration is shown here:

```
TLastNameComparer = class(TComparer<TPerson>)
  function Compare(const Left, Right: TPerson): Integer;
    override;
end;
```

The implementation of our `TLastNameComparer.Compare` function is made quite simple with `Compare` functions from the `SysUtils` unit, which return the same values we need:

```
function TfrmPeopleList.TLastNameComparer.Compare(
                    const Left, Right: TPerson): Integer;
begin
  Result := CompareText(Left.LastName, Right.LastName);
end;
```

Now, we can modify the `ListPeople` procedure and add in last name sorting. Since our *master* list of `TPerson` objects did not implement sorting, we have to create a temporary one:

```
procedure TfrmPeopleList.ListPeople;
var
  APerson: TPerson;
  SortedList: TList<TPerson>;
begin
  lbPeople.Items.Clear;

  SortedList := TList<TPerson>.Create(TLastNameComparer.
    Create);
  for APerson in FPeopleList do
    SortedList.Add(APerson);

  SortedList.Sort;
  for APerson in SortedList do
    lbPeople.Items.Add(Format('%s %s, Age: %d',
              [APerson.FirstName, APerson.LastName, APerson.
                Age]));
end;
```

What if we wanted to sort by `First Name` or `Date of Birth` instead? We could create multiple `TComparer` types and determine which one to use in the `ListPeople` procedure. Doing that would encourage refactoring out the actual sort and display of the `TPerson` objects:

```
procedure TfrmPeopleList.ListSortedPeople(const Comparer:
IComparer<TPerson>);
```

```
var
  APerson: TPerson;
  SortedList: TList<TPerson>;
begin
  SortedList := TList<TPerson>.Create(Comparer);

  lbPeople.Items.Clear;
  for APerson in FPeopleList do
    SortedList.Add(APerson);

  SortedList.Sort;
  for APerson in SortedList do
    lbPeople.Items.Add(Format('%s %s, Age: %d',
                [APerson.FirstName, APerson.LastName, APerson.
                   Age]));
end;


procedure TfrmPeopleList.ListPeople;
begin
  case cmbPersonSort.ItemIndex of
    0: ListSortedPeople(TFirstNameComparer.Create);
    1: ListSortedPeople(TLastNameComparer.Create);
    2: ListSortedPeople(TBirthDateNameComparer.Create);
  end;
end;
```

The ListPeople procedure adds two sort options, which requires the creation of additional TComparer classes:

```
TFirstNameComparer = class(TComparer<TPerson>)
  function Compare(const Left, Right: TPerson): Integer;
override;
  end;
TBirthDateNameComparer = class(TComparer<TPerson>)
  function Compare(const Left, Right: TPerson): Integer;
override;
  end;
```

Implementing each of their `Compare` functions looks like this:

```
function TfrmPeopleList.TFirstNameComparer.Compare(const Left,
Right: TPerson): Integer;
begin
  Result := CompareText(Left.FirstName, Right.FirstName);
end;
function TfrmPeopleList.TBirthDateNameComparer.Compare(const
Left, Right: TPerson): Integer;
begin
  Result := CompareDate(Left.DateOfBirth, Right.DateOfBirth);
end;
```

That looks a little cluttered. Let's simplify this using anonymous methods.

Adding anonymous methods for cleaner code

An anonymous method is a shortcut way of passing a method as a parameter. You declare the method right within the parameter of a method call instead of declaring a separate, named method and then passing that method name as the parameter.

Continuing with the `GenericPeople` project of the last section, we can completely eliminate the three `TComparer` classes and implement the sort functions directly in the `ListPeople` procedure using anonymous methods. Refer to the following code:

```
procedure TfrmPeopleList.ListPeople;
begin
  case cmbPersonSort.ItemIndex of
    0: ListSortedPeople(TComparer<TPerson>.Construct(
        function (const Left, Right: TPerson): Integer
        begin
          Result := CompareText(Left.FirstName, Right.
            FirstName);
        end));
    1: ListSortedPeople(TComparer<TPerson>.Construct(
        function (const Left, Right: TPerson): Integer
        begin
          Result := CompareText(Left.LastName, Right.
            LastName);
        end));
```

```
    2:  ListSortedPeople(TComparer<TPerson>.Construct(
            function (const Left, Right: TPerson): Integer
            begin
                Result := CompareDate(Left.DateOfBirth, Right.
                    DateOfBirth);
            end));
    end;
end;
```

I suppose that some would say this also looks a little cluttered, but for procedures that only need to be called in one place, declaring them in place like this keeps the code together.

A great use case for anonymous methods is when synchronizing messages or data in a multi-threaded application. Let's say you have written a program that generates a large number of random numbers and some calculations take a long time. You want to display the numbers, update a progress bar, and have a way to stop the process cleanly. Take a look at a simple demo to simulate a threaded process:

https://github.com/PacktPublishing/Fearless-Cross-Platform-Development-with-Delphi/tree/master/Chapter03/04_ProcessSimulator

Since the VCL is not thread-safe, other threads must call `Synchronize` to call methods or set properties of the main thread. This simple demo does this twice. In this first example, it calls the main thread's public procedure, `ShowNumber`, to add the newly generated random number to the list:

```
procedure TfrmLongProcessMain.TProcessThread.
ShowRandNum(RandNum: Double);
begin
  TThread.Synchronize(TThread.CurrentThread,
    procedure
    begin
      frmLongProcessMain.ShowNumber(RandNum);
    end);
end;
```

The second synchronized call updates two `count` properties on the main form, which update a progress bar when called by `TTimer`'s `OnTimer` event:

```
procedure TfrmLongProcessMain.TProcessThread.UpdateStatus(const
```

```
Count, MaxCount: Integer);
begin
  TThread.Synchronize(TThread.CurrentThread,
    procedure
    begin
      frmLongProcessMain.Count := Count;
      frmLongProcessMain.MaxCount := MaxCount;
    end);
end;
```

We will implement anonymous methods several times when writing mobile applications in later chapters of this book.

Let's now look at one of the more esoteric language features: custom attributes.

Adding metadata to your classes with attributes

There are two subjects that are often discussed along with the topic of attributes in Delphi: **Aspect-Oriented Programming (AOP)** and **Runtime Type Information (RTTI)**. AOP is the idea of adding additional structure or support code that your core application classes will need but don't really fit the pure class hierarchy in which they were designed, functionality that cuts across many classes (you might see the term *cross-cutting* in your extended reading on this subject). For example, you might need to add logging to several unrelated classes in your application, or you will want to load and save configuration settings that are divided up into multiple classes. Adding behavior in a standard way without loading these different classes with additional code to support this functionality is what AOP promotes.

To implement this feature in our classes, we need code that can understand our code—in other words, code that can get information about our classes at runtime and perform the additional support functionality we need that we don't want to embed in those classes themselves. That's where RTTI comes in. RTTI has been a part of Delphi all along, but it was greatly extended in Delphi 2010 to enable greater support for, among other things, attributes.

To use RTTI, you need to create a TRttiContext. Then, you use that context to get information about classes, fields, properties, methods, and just about any Delphi type. For example, let's say you wanted to see information for components on a form in a VCL application. We could activate the OnClick event of each component and then dump property information for each component clicked to a ListBox:

```
procedure TfrmRttiMain.ComponentClick(Sender: TObject);
```

```
var
  LContext: TRttiContext;
  LRttiType: TRttiType;
  LProperty: TRttiProperty;
begin
  lbRttiInfo.Items.Clear;

  LContext := TRttiContext.Create;
  try
    LRttiType := LContext.GetType(Sender.ClassType);
    for LProperty in LRttiType.GetProperties do
      lbRttiInfo.Items.Add(LProperty.Name + ': ' +
                           LProperty.PropertyType.Name);
  finally
    LContext.Free;
  end;
end;
```

You can get the value of each property, but to display it, you need to convert each value into a string. PropertyType has an enumerated TypeKind property to help you do this. You'll need to add a TValue variable and then perform different types of conversion depending on the setting of TypeKind:

```
function TfrmRttiMain.GetPropertyValueAsString(Sender: TObject;
LProperty: TRttiProperty): string;
var
  LValue: TValue;
begin
  LValue := LProperty.GetValue(Sender);
  case LProperty.PropertyType.TypeKind of
    tkString, tkLString, tkUString:
      Result := LProperty.GetValue(Sender).AsString;
    tkInt64, tkInteger:
      Result := IntToStr(LProperty.GetValue(Sender).AsInteger);
    tkFloat:
      Result := FloatToStr(LProperty.GetValue(Sender).
        AsExtended);
    tkEnumeration:
```

```
        Result := GetEnumName(LProperty.PropertyType.Handle,
                              LValue.AsOrdinal);
    else
      Result := 'Unsupported type: ' + LProperty.PropertyType.
        Name;
    end;
  end;
```

With this added function, the `for` loop would change as follows:

```
    for LProperty in LRttiType.GetProperties do
      LPropEntry := LProperty.Name + ': ' +
                    LProperty.PropertyType.Name + ' = ' +
                    GetPropertyValueAsString(Sender, LProperty);
      lbRttiInfo.Items.Add(LPropEntry);
    end;
```

The completed project can be found on Github:

```
https://github.com/PacktPublishing/Fearless-Cross-Platform-De-
velopment-with-Delphi/tree/master/Chapter03/05_RttiExplorer
```

Armed with RTTI, AOP is accomplished through the use of attributes. An attribute in Delphi is a class you write that inherits from `TCustomAttribute`. To apply an attribute class to a type, put the name of the attribute class in square brackets just before the type declaration. This annotates or marks that type as *having* that attribute. Here's a simple attribute class:

```
    MySimpleAttribute = class(TCustomAttribute);
```

Now, you can mark up a type with it. When using the attribute, the `Attribute` part of the name is assumed, so it can be left off:

```
    [MySimple]
    TMyClass = class
    end;
```

Attributes can be applied to properties, fields, methods, entire classes, records, and even standalone procedures and functions. Getting these attributes is as simple as any other piece of information you can get from `RttiContext`. In our next example, let's create a custom class, apply attributes to it, then show what our code can do at runtime.

The purpose of this class is to hold some configuration settings that we want saved to a INI file and loaded when the program starts. We don't want the settings class to worry about where or how it is saved. Furthermore, we may want to change how the settings are saved at a later date (such as saving them to a database instead).

Here's the simple class for our settings, showing MyName and MyFavNum as the only properties:

```
TMySettings = class
private
  FMyName: string;
  FMyFavNum: Integer;
published
  property MyFavNum: Integer read FMyFavNum write FMyFavNum;
  property MyName: string read FMyName write FMyName;
end;
```

We want to save these to an INI file with a typical NAME=VALUE format:

```
[TMySettings]
MyFavNum=10
MyName=Fred
```

The class name will become the **section** and the property names will be the **key names**.

To use attributes to save these settings to a file, we need to define an attribute class and then methods that know how to use them. The first attribute we'll create will be one that marks a class as one we want to save to an INI file. I'll call it IniSaveClassAttribute and it will contain the name of the class it is working with in a property:

```
IniSaveClassAttribute = class(TCustomAttribute)
private
  FTheClassName: string;
published
  property TheClassName: string read FTheClassName write
    FTheClassName;
end;
```

Now, we can use this class to annotate the settings class:

```
[IniSaveClass]
TMySettings = class
  // ...
```

Now that we have an attribute in place, we need methods that know what to do with it. I'll create a class with some static functions for this purpose and call it TIniSave:

```
TIniSave = class
private
  class procedure SetValue(AData: string; var AValue:
    TValue);
  class function GetValue(var AValue: TValue): string;
  class function GetClassName(MyObj: TRttiObject; const
    MyObjName: string): IniSaveClassAttribute;
public
  class procedure Load(FileName: string; MyObj: TObject);
  class procedure Save(FileName: string; MyObj: TObject);
end;
```

Let's look at the Load procedure. We'll be passing in the INI filename and the object with the properties into which we'll store the values we read from the INI file:

1. First, we need to declare several variables:

    ```
    class procedure TIniSave.Load(FileName: string; MyObj:
    TObject);
    var
      LRttiContext : TRttiContext;
      LMyObjType : TRttiType;
      LMyProp: TRttiProperty;
      LIniClass: IniSaveClassAttribute;
      LData: string;
      LValue: TValue;
      LIniFile: TIniFile;
    ```

2. Next, we need to get the RTTI context and type info and create TIniFileObject:

    ```
    begin
      LRttiContext := TRttiContext.Create;
    ```

```
   try
     LMyObjType := LRttiContext.GetType(MyObj.ClassInfo);
     LIniFile := TIniFile.Create(FileName);
     try
       LIniClass := GetClassName(LMyObjType, MyObj.
         ClassName);
       // ...
```

3. Next, we need to get the class name because we'll use that as [Section] for the INI values. Currently, this is the only attribute used on this class, but the GetClassName function checks to make sure that IniClassAttribute is actually applied:

```
class function TIniSave.GetClassName(MyObj: TRttiObject;
                 const MyObjName: string):
                   IniSaveClassAttribute;
var
  Attr: TCustomAttribute;
begin
  Result := nil;
  for Attr in MyObj.GetAttributes do
    if Attr is IniSaveClassAttribute then begin
      Result := IniSaveClassAttribute.Create;
      Result.TheClassName := MyObjName;
      Break;
    end;
end;
```

4. Finally, back in the Load procedure, we'll go through all the properties, and if there is a value read from the INI file for that property, we'll set its property value to what was read:

```
for LMyProp in LMyObjType.GetProperties do begin
  LData := LIniFile.ReadString(LIniClass....
    TheClassName, LMyProp.Name,LDefaultData
  if Length(LData) > 0 then begin
    LValue := LMyProp.GetValue(MyObj);
    SetValue(LData, LValue);
    if LMyProp.IsWritable then
```

```
            LMyProp.SetValue(MyObj, LValue);
        end;
    // ...
```

We won't cover the implementation of GetValue and SetValue in this book as it
converts between string and various property types and can get a little tedious. See the
link for the full source at the end of this section.

Now that we have functions to load an object from an INI file using its property names as
the value names, here's how it's applied:

```
var
  LMySettings: TMySettings;
begin
  LMySettings := TMySettings.Create;
  TIniSave.Load(ChangeFileExt(ParamStr(0), '.ini'),
    LMySettings);
  Writeln('My name is: ' + LMySettings.MyName);
  Writeln('My favorite number is: ' +
        IntToStr(LMySettings.MyFavNum));
  // ...
```

In this book, we've only shown the details of the Load procedure. The Save procedure and
the implementation of the attribute classes we will discuss are in the uIniSave.pas unit:

```
https://github.com/PacktPublishing/Fearless-Cross-Platform-De-
velopment-with-Delphi/blob/master/Chapter03/06_SimpleAttri-
butes/uIniSave.pas
```

You may have made the observation that this example of using an attribute didn't really
enable the class to be saved to an INI file; it was all the work of RTTI in the TIniSave
class—we didn't really *need* to use an attribute. That's right. However, now that this class *is*
using RTTI to read through the class properties, we can add more attributes for additional
control and flexibility in expanding this settings class and for other settings classes we
might create—and thus make the case for using attributes a little stronger.

The first addition we'll make is to set a default value. If the INI file does not exist, all the initial values will be blank. Let's create an attribute to allow a default value to be set:

```
IniDefaultAttribute = class(TCustomAttribute)
private
  FDefVal: string;
public
  constructor Create(const NewDefaultValue: string);
published
  property DefaultValue: string read FDefVal write FDefVal;
end;
```

Notice that this attribute class has a constructor that takes a parameter. This parameter will set the class's field with the default value to be used for the property on which this attribute is applied, should that value be blank in the INI file when read.

Here's how it's used in our modified settings class:

```
[IniSaveClass]
TMySettings = class
private
  FMyName: string;
  FMyFavNum: Integer;
published
  [IniDefault('42')]
  property MyFavNumber: Integer read FMyFavNum write
    FMyFavNum;
  [IniDefault('Zeek')]
  property MyName: string read FMyName write FMyName;
  // ...
```

The Load procedure of TIniSave needs to be modified to check for this new attribute as it goes through the properties:

```
for LMyProp in LMyObjType.GetProperties do begin
  LDefaultData := GetDefaultAttributeValue(LMyProp);
  LData := LIniFile.ReadString(LIniClass.TheClassName,
                        LMyProp.Name, LDefaultData);
end;
```

The new `GetDefaultAttributeValue` function finds `IniDefaultAttribute` and returns the default value:

```
class function TIniSave.GetDefaultAttributeValue(MyObj:
TRttiObject): string;
var
  Attr: TCustomAttribute;
begin
  Result := EmptyStr;

  for Attr in MyObj.GetAttributes do
    if Attr is IniDefaultAttribute then begin
      Result := IniDefaultAttribute(Attr).DefaultValue;
      Break;
    end;
end;
```

This example shows that attributes can have parameters.

The last attribute example will mark properties that we don't want to be saved in the INI file. Why would we have a setting in our class that isn't saved? Sometimes, settings need to be modified from their raw value before saving. For example, a password should not be saved in raw, unencrypted form in a plain-text INI file. One way to approach this is to write encrypt and decrypt functions and use two different properties, one that is the raw password used in the application and the other that is the saved and encrypted version of the password.

I have found another use: `TDateTime` values. If you let the `TIniFile` class use its default save mechanism on `TDateTime` fields, then when the computer's date/time format is changed or if the settings file is copied to a different computer with a different date/time format, reading `TDateTime` values will cause an error. To avoid this problem, I always save `TDateTime` values as strings in a specific format and read and parse the string back to a date/time when loading it. Therefore, I would like to save the string version of a date/time but not the actual date/time property itself. The purpose of this attribute, then, will be to skip or ignore a property; I'll call it `IniIgnoreAttribute`:

```
IniIgnoreAttribute = class(TCustomAttribute);
```

This is the simplest type of attribute where there are no properties, no functions—nothing! All it is is an annotation to a property so that when we're going through the properties in the Load or Save procedures and find one with this attribute, we know to skip it:

```
for LMyProp in LMyObjType.GetProperties do begin
  LIniPropIgnore := GetPropIgnoreAttribute(LMyProp);
  if not Assigned(LIniPropIgnore) then begin
    // not ignored, get the value ...
```

The implementation of the GetPropIgnoreAttribute function is pretty simple:

```
class function TIniSave.GetPropIgnoreAttribute(MyObj:
TRttiObject): IniIgnoreAttribute;
var
  Attr: TCustomAttribute;
begin
  Result := nil;

  for Attr in MyObj.GetAttributes do
    if Attr is IniIgnoreAttribute then begin
      Result := IniIgnoreAttribute(Attr);
      Break;
    end;
end;
```

Finally, we have the addition to our class:

```
published
  // ...
  [IniIgnore]
  property MyBirthDate: TDateTime read GetMyDOB write
    SetMyDOB;
  property MyBirthDateStr: string read FMyDOBStr write
    FMyDOBStr;
  // ...
```

Now, the property, `MyBirthDate`, will not be loaded from or saved to the INI file but will instead use the getter and setter methods to translate the string property, `MyBirthDateStr`, which *is* in the INI file (because it is not annotated with the `IniIgnore` attribute).

Details of the getter and setter methods for `MyBirthDate` will not be listed here for brevity. The full, commented project, which includes some random data that fills the properties each time it is run, and many other details of this example project can be found in Packt's GitHub repository:

```
https://github.com/PacktPublishing/Fearless-Cross-Platform-De-
velopment-with-Delphi/tree/master/Chapter03/06_SimpleAttri-
butes
```

With such a strong language foundation, cross-platform computing was ready to take off! And in 2011, with Delphi XE2, compilers for the Mac and Windows 64-bit were released, along with the FireMonkey cross-platform GUI framework. Over the next several years, there were practically no changes to the Delphi language itself as the focus was on updating compilers and the GUI framework for mobile platforms, adding web services support, adding tools and features to the IDE, and improving memory management all around.

Finally, starting with Delphi 10.3 Rio in 2018, we got a few more small language enhancements.

Learning about the latest enhancements

The classic Pascal language rules have been followed in Delphi very closely—and that's a good thing, in my opinion. I remember stumbling through BASIC or JavaScript many times looking for the source of an error only to discover that a slight typo changed the name of a variable deep in a subroutine and subsequently lost the value I had been expecting. In Delphi, a mistake like that wouldn't even compile because identifiers have to be declared in a `var` or `const` section at the top of a method, or in a separate section of the class or unit. Some have decried this as old-fashioned and inconvenient, but with some IDE keyboard shortcuts and quick-setting bookmarks, any arguments for productivity loss have been removed.

Simplifying variable declaration

With Delphi 10.3 Rio, the declaration of local variables has been given a flexibility boost. You still have to declare them, but they no longer have to be declared in a separate `var` section above the code. These **inline variables** can also use **type inference**, where it infers the type of the variable from the value assigned if it's all done in one statement.

The most common example for this is using an integer loop variable in a `for` loop:

```
for var i := 1 to 10 do
   // ...
```

Here, the variable, i, is declared right in the `for` loop declaration and its type, `integer`, is inferred from the assignment of 1 at the start of the loop.

The first code example in this chapter, `BasicSyntaxDemo`, has been modified using three inline and type-inferred variables:

```
https://github.com/PacktPublishing/Fearless-Cross-Platform-De-
velopment-with-Delphi/tree/master/Chapter03/01_BasicSyntaxDemo
```

Those were the only language features that were added to with that version. In Delphi 10.4 Sydney, an interesting addition was made for the old `record` type.

Controlling initialization and the finalization of records

The record type in Delphi encapsulates fields of any type into a single type that is allocated as soon as a variable of that type is declared, and deallocated when it goes out of scope (unlike classes, which have to be manually created and freed). Delphi has managed the memory for a record's fields, such as strings, for a long time. Its default initialization code does not clear out memory for a new record variable, which could lead to unintentional behavior if you forget to do it yourself and start using the field values.

With the latest release of Delphi, you can write code to augment the built-in **initialize** and **finalize** operations. You do this by implementing two operator methods, named, aptly enough, `Initialize` and `Finalize`.

Here's a simple record with a few fields and these two methods:

```
TMyManagedFlags = record
   FlagHeader: string;
   Flag1: Boolean;
   Flag2: Boolean;
   Flag3: Integer;
   class operator Initialize(out Dest: TMyManagedFlags);
   class operator Finalize(var Dest: TMyManagedFlags);
end;
```

In the implementation, you simply write whatever code needs to be part of the record initialization. In our sample console app, I'll initialize the field values and write a message to the console:

```
class operator TMyManagedFlags.Initialize (out Dest:
TMyManagedFlags);
begin
  Dest.FlagHeader := 'My Custom Managed Flag Record';
  Dest.Flag1 := True;
  Dest.Flag2 := False;
  Dest.Flag3 := 100;
  Writeln(,Flags initialized');
end;


class operator TMyManagedFlags.Finalize(var Dest:
TMyManagedFlags);
begin
  Writeln('Flags finalized');
end;
```

The demo app has another, nearly identical record structure, but without the Initialize and Finalize methods as a comparison. A variable of each type is declared, and a procedure, ShowFlags, is called to list all the values:

```
procedure ShowFlags;
const
  FALSE_TRUE_STRS: array[Boolean] of string = ('False',
'True');
var
  uflags: TMyUnmanagedFlags;
  mflags: TMyManagedFlags;
begin
  Writeln('Here are the unmanaged flags...');
  Writeln(,  Header: , + uflags.FlagHeader);
  Writeln(,  Flag1 = , + FALSE_TRUE_STRS[uflags.Flag1]);
  Writeln(,  Flag2 = , + FALSE_TRUE_STRS[uflags.Flag2]);
  Writeln(,  Flag3 = , + uflags.Flag3.ToString);
```

```
  Writeln('Here are the managed flags...');
  Writeln('  Header: ' + mflags.FlagHeader);
  Writeln('  Flag1 = ' + FALSE_TRUE_STRS[mflags.Flag1]);
  Writeln(,  Flag2 = , + FALSE_TRUE_STRS[mflags.Flag2]);
  Writeln(,  Flag3 = , + mflags.Flag3.ToString);
end;
```

The body of the program simply writes a line to the console, calls `ShowFlags`, then writes another line to the console:

```
begin
  Writeln('Flag demo starting');
  ShowFlags;
  Writeln('Flag demo done');
  Readln;
end.
```

As might be expected, the output is as follows:

```
Flag demo starting
Flags initialized
Here are the unmanaged flags...
  Header:
  Flag1 = False
  Flag2 = False
  Flag3 = 0
Here are the managed flags...
  Header: My Custom Managed Flag Record
  Flag1 = True
  Flag2 = False
  Flag3 = 100
Flags finalized
Flag demo done
```

Notice the messages, Flags initialized and Flags finalized, before the first record is listed and after the second record is listed? That's telling us that before and after the body of the ShowFlags procedure, the mflags variable is being created and destroyed.

This may seem like a small thing, but if you use records, there must have been (or will be) some time when you've forgotten to initialize the fields of a record and, even though this simple example happened to initialize the values (empty string, false Booleans, zero integers), it doesn't always and it will bite you when you least expect it.

So, what do they call this new feature? **Custom managed records**.

Summary

This was a quick look back at the evolution of Delphi's extensions to Object Pascal, reviewing the basic syntax and briefly looking at interfaces, variants, nested types, and Unicode. Generics and anonymous methods were explained, RTTI and attributes were demonstrated, and finally, inline variables, type inference, and custom managed records were covered to complete the enhancements in the most recent versions. We didn't touch on all areas of what makes Delphi such a great language (such as operator overloading or class helpers) but hopefully, you learned some additional aspects of the language that will solidify your understanding of powerful modern-day programming concepts and increase your productivity with advanced techniques.

With a firm grasp of the IDE, an understanding of the toolset, knowledge of Delphi's project types, and making fine use of the language, you are now ready to go on to the next chapter and start your deep dive into cross-platform development.

Questions

After reading this, you should be able to answer the following questions:

1. In what scenario will the finalization section of a unit not get called?

2. What code comment style can comment out all other comments?

3. Can a nested type declare its own constant?

4. How long has Unicode been the default string type in Delphi?

5. What value should a `TComparer<T>.Compare` function return if its left parameter is less than its right parameter?

6. Why do you have to call `Synchronize` to access properties of the main form from a thread?

7. When do you need to get the RTTI context?

8. What version of Delphi introduced inline variables?

Further reading

- Delphi language overview: `http://docwiki.embarcadero.com/RADStudio/Sydney/en/Language_Overview`

- Programs and units: `http://docwiki.embarcadero.com/RADStudio/Sydney/en/Programs_and_Units_(Delphi)`

- Camel case: `https://en.wikipedia.org/wiki/Camel_case`

- Fundamental syntactic elements: `http://docwiki.embarcadero.com/RADStudio/Sydney/en/Fundamental_Syntactic_Elements_Index`

- Declared constants: `http://docwiki.embarcadero.com/RADStudio/Sydney/en/Declared_Constants`

- Data types: `http://docwiki.embarcadero.com/RADStudio/Sydney/en/About_Data_Types_(Delphi)`

- Object-oriented programming: `https://en.wikipedia.org/wiki/Object-oriented_programming`

- Unit testing and isolation frameworks: `https://www.embarcadero.com/rad-in-action/delphi-unit-testing`

- Marco Cantu's Essential Pascal – variants: `https://www.marcocantu.com/epascal/English/ch10var.htm`

- Nested type declarations: `http://docwiki.embarcadero.com/RADStudio/Sydney/en/Nested_Type_Declarations`

- Delphi Unicode Migration for Mere Mortals: Stories and Advice from the Front Lines: `http://www.embarcadero.com/images/dm/technical-papers/delphi-unicode-migration.pdf`

- Effectively using generics: `https://youtu.be/_DKx2_F3M6g`

- Dr. Bob examines anonymous methods: `http://www.drbob42.com/examines/examinA5.htm`

- What is aspect-oriented programming? `https://docs.jboss.org/aop/1.0/aspect-framework/userguide/en/html/what.html`

- Using attributes and `TCustomAttribute` descendants: `https://robstechcorner.blogspot.com/2009/09/using-attributes-and-tcustomattribute.html`

- Using Delphi attributes to unify source, test, and documentation: `https://marc.durdin.net/2014/05/using-delphi-attributes-to-unify-source-test-and-documentation/`

- Custom managed records new in Delphi 10.4 Sydney: `https://community.idera.com/developer-tools/b/blog/posts/custom-managed-records-coming-to-delphi-10-4`

Section 2: Cross-Platform Power

This section starts off by showing how to switch from the Windows-restricted VCL to the cross-platform FireMonkey visual framework, enabling you to build user-friendly business applications with modern features that span multiple platforms without rewriting it for each one. Then we create libraries, packages, and even components for other platforms, demonstrate how much more useful LiveBindings is than data-aware controls, and show how to stylize your apps and customize them for your unique look and feel. We end this section with a fun exploration of 3D, including a complete sample game you can play!

This section comprises the following chapters:

- *Chapter 4, Multiple Platforms, One Code Base*
- *Chapter 5, Libraries, Packages, and Components*
- *Chapter 6, All About LiveBindings*
- *Chapter 7, FireMonkey Styles*
- *Chapter 8, Exploring the World of 3D*

4
Multiple Platforms, One Code Base

We are now ready to dive into the real focus of this book: multiple platforms! We're going to wade through a lot of new territory in this chapter, introducing FireMonkey, setting up other platforms for deployment, and the best part, seeing your application running on your phone or tablet!

We'll take you through these concepts in the following sections:

- Moving to FireMonkey from the VCL
- Preparing other platforms
- Working with various screen sizes
- Writing code to support multiple platforms

Technical requirements

Besides the typical requirements of this book for running Delphi 10.4 Sydney on a Windows 10 64-bit computer, to run all the examples listed in this chapter and the ones in the rest of this book, you will also need access to one or more of the following:

- An Apple Mac running a minimum of macOS 10.13 High Sierra accessible via IP address from your Windows development machine

- A 64-bit iOS device (iPod Touch, iPad, or iPhone) running a minimum of iOS 11 connected to a Mac via USB

- A 32-bit or 64-bit Android device running a minimum of Android Marshmallow (6) connected to your Windows machine via USB

> **Note**
>
> The operating system version of the mobile device you want to test with can be a troublesome area when supporting multiple platforms with Delphi. If all you do is write Android apps with Android tools and Apple apps with Apple tools, you should be fine. But development tools that are native to neither must carefully test and debug the libraries that come from these platforms for changes or incompatibilities with the tool. Each version of Delphi lists the versions of macOS, iOS, and Android and lists the device types that have been tested and are supported. Even then, changes may occur between release cycles and your device may be upgraded suddenly to something not yet supported. This can be a point of great frustration if you're pushing a deadline. Indeed, this happened to me while writing this chapter. My iPhone upgraded from 13.7 to 14.0 and Delphi 10.4.1 was unable to get all the SDK files it needed for debugging. My Google Pixel phone was at Android 11 but Delphi 10.4 only supported up through Android 10, so would not recognize it until I manually copied over the Android 11 (API 30) SDK using Android Studio's tools and modified the paths within Delphi. Visit the Delphi forums for advice on getting past these hurdles.

The example projects for this chapter can be found on GitHub at `https://github.com/PacktPublishing/Fearless-Cross-Platform-Development-with-Delphi/tree/master/Chapter04`.

Moving to FireMonkey from the VCL

For many years, the only GUI development platform that came with Delphi was the **VCL**, or **Visual Component Library**. The VCL is powerful and mature but is so entrenched with Windows-specific API calls that it is not practical to turn it into something else. Instead, VGScene, a cross-platform library designed by Eugene Kryukov of KSDev, was acquired in 2011, rebranded as FireMonkey, and added to Delphi XE2. And now, many iterations later, it is the foundation for all cross-platform GUI applications in Delphi.

All Mac, iOS, and Android GUI development in Delphi uses the FireMonkey framework. Windows has the option to use either FireMonkey or the VCL. So, if you're familiar with the VCL, the first step toward understanding cross-platform GUI development is to switch your Windows application development to use FireMonkey. This first section will lead you through that step.

Starting a new Windows FireMonkey project

Starting a new FireMonkey project is pretty simple. We covered this briefly in the first section of *Chapter 2, Delphi Project Management*. Just select **File | New | Multi-Device Application** from the Delphi menu and select your template type. We'll just use the **Blank Application** template as our starting point for this first example as it most closely resembles the starting point for a new VCL application.

Once it is created, expand **Target Platforms** to see that Delphi has configured seven platform types for you with **Windows 32-bit** initially activated:

Figure 4.1 – New multi-device application targets

We'll just leave the target at the default of **Windows 32-bit** for now and switch to Unit1. Let's place a few common controls on the form:

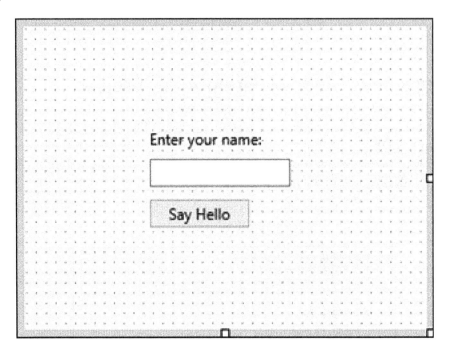

Figure 4.2 – FireMonkey form with simple controls

Name the TEdit control edtName, name the TButton control btnSayHello, then double-click the button to create the OnClick event handler:

```
procedure TForm1.btnSayHelloClick(Sender: TObject);
begin
  TDialogService.ShowMessage('Hello, ' + edtName.Text);
end;
```

(TDialogService is found in the FMX.DialogService unit, which you have to add manually under a uses clause.)

It looks fairly familiar, doesn't it? Save and run the project, type in a name in the box, and click **Say Hello** to see the name show up in a dialog message. You can download this simple project from GitHub:

```
https://github.com/PacktPublishing/Fearless-Cross-Platform-De-
velopment-with-Delphi/tree/master/Chapter04/01_HelloFireMonkey
```

FireMonkey is designed to have a lot of similarities to the VCL to aid experienced Delphi developers to quickly embrace this new framework. However, there are some notable differences. Here are several of them:

- VCL controls are positioned with `Top` and `Left` properties and sized with `Height` and `Width` whereas FireMonkey controls use a `Position` property with `X` and `Y` nested properties for positioning and a `Size` property with nested `Height` and `Width` properties for sizing.

- The text of a FireMonkey `TLabel` is in its `Text` property, not the `Caption` property as in VCL.

- Font properties in many FireMonkey controls are found under the `TextSettings` property. Furthermore, the sizes are given in **DIPs (Device-Independent Pixels)** and initialized at 96 per logical inch whereas VCL fonts use points and are set at 72 per logical inch, making the same font size in FireMonkey (compared with its VCL counterpart) smaller.

- FireMonkey's `TCheckBox` component changes a couple of things. First, the `Checked` property is now `IsChecked`. Second, to capture the click event and take action on the value of the `IsChecked` property that was just changed, use the `OnChange` event instead of the `OnClick` event like you did in VCL.

- FireMonkey's color constants are prefixed with `cla` instead of `cl`.

- There is no separate type of button to hold glyphs; instead, the FireMonkey `TButton` has an `Images` property that can point to a `TImageList` component holding multiple images and `ImageIndex` to select which one is showing.

- Global constants that are platform-independent, such as message dialog types, open options, modal results, and virtual key codes, have been moved to the `System.UITypes` unit and out of the VCL-specific unit, `Vcl.Controls`.

- VCL applications will use units such as `Vcl.Forms`, `Vcl.Dialogs`, `Vcl.Graphics`, `Vcl.Controls`, `Vcl.StdCtrls`, `Vcl.ComCtrls`, and `Vcl.ExtCtrls` but with FireMonkey, they'll use `FMX.Forms`, `FMX.Dialogs`, `FMX.Graphics`, `FMX.Controls`, `FMX.StdCtrls`, `FMX.Edit`, and `FMX.Controls.Presentation`.

- There are *no* data-aware components (for example, `TDBEdit`) in FireMonkey. Instead, you are encouraged to use **LiveBindings**, which was introduced at the same time FireMonkey came out in Delphi XE2 but is also available for VCL applications (we will cover LiveBindings in great detail in *Chapter 6, All About LiveBindings*).

I would suggest that you open the FireMonkey example projects that come with Delphi and play around with the controls and properties, trying out different ways of interacting with them and comparing your experience with the nuances of this new framework. Click on the **Open a Sample Project** link from the default welcome screen and in the file open dialog, drill down from **Samples** through **Object Pascal | Multi-Device Samples**, and select from the projects in **User Interface**. We will be covering several aspects of these later in this book but there's no better way to get familiar with them than to just dig in and start experimenting.

A much harder task than starting a new FireMonkey application is migrating an existing one written with the VCL framework. In some cases, it may not be practical.

Migrating an existing Windows VCL application to use FireMonkey

Based on the simple example project shown earlier and the list of differences just mentioned, you might think converting a VCL application over to use the FireMonkey framework won't be that difficult—just change the units, adjust the fonts, use different events and properties with checkboxes, and adjust for the other differences we discussed. But there is no easy way to convert a form using Delphi alone—you can't simply open a VCL form and save it as a FireMonkey form. Only in the case of porting it to a Mac desktop application might it make sense anyway as mobile devices have much smaller screens, which would require a re-design to make them usable.

> Note
> There is a third-party commercial product that makes a valiant attempt to do this that could save you many hours of work for large applications. It's called *Mida Converter* and has different capabilities depending on the price you pay. Its *Basic* edition is free for registered users of Delphi XE2 and XE3. Read more about it with the link in the *Further reading* section at the end of this chapter. If you want to make your desktop apps work on Macs, this shortcut may work for you.

The only pure-Delphi way of migrating your old VCL applications to FireMonkey is by starting a new blank multi-device application, then creating each form manually and placing new FireMonkey-equivalent controls, constantly referring to the VCL forms in the old application. This will be tedious but most of the code will be able to be ported over—unless you married the user interface close to the data and business logic, used data-aware controls, called low-level Windows APIs, or have a plethora of custom components. Most organizations feel it's better to just rewrite their application from scratch (and shed any technical debt built up with the old code!).

As a test of this process, I took the example project, `RttiExplorer`, used in *Chapter 3*, *A Modern Day Language*, and migrated it to use the FireMonkey framework. It took about an hour to recreate this one form and port the code over. Here are the issues I encountered:

- The panel in FireMonkey doesn't have a caption or text property, so I had to add `TLabel` to the top panel.

- Labels don't have an `OnClick` event as the VCL does.

- The FireMonkey `Align` property has many more options than in the VCL—it wasn't a problem but I had to look carefully through the list and learn about the new ones in order to choose the right one.

- `TLabel`'s `AutoSize` property is `False` by default in FireMonkey—and it works differently.

- There is no `TLabeledEdit` component in FireMonkey.

- There is no `TRadioGroup`; I had to place a `TGroupBox` and then add individual `TRadioButtons` inside to replicate the functionality from the VCL version.

- Positioning components of a group box is independent of the bounds of the group box; however, a component placed inside a group box and dragged outside of it still shows in the **Structure** pane as being in the GroupBox's container.

- The FireMonkey version of the application runs slower on Windows than the VCL version on the same platform.

You can see the migrated project on GitHub at the following link:

```
https://github.com/PacktPublishing/Fearless-Cross-Platform-De-
velopment-with-Delphi/tree/master/Chapter04/02_RttiExplorerFM
```

Whether you start afresh or migrate your code, you'll end up with an application whose GUI can work across platforms, no longer tied directly to Windows. The `RTTI Explorer` project was a simple one to migrate but, as it was originally designed, doesn't lend itself well to small screens. If all you're targeting are desktop platforms (for example, Windows and macOS), that's fine but it will be a common problem when you evaluate the future of legacy desktop applications and need to factor in phone-sized screen real estate.

So, once you have your FireMonkey application ready for cross-platform deployment, how do you actually get it onto other platforms? It takes a little preparation.

Preparing other platforms

You're probably very eager to see your new cross-platform code running on your phone or tablet. But there is a little setup to get it deployed to other platforms. Fortunately, Delphi provides tools to greatly simplify the process.

When I say "other platforms," that means anything other than what Delphi is, a Windows 32-bit application. So, even Window 64-bit is technically another platform. However, since Windows supports both 32-bit and 64-bit applications running side by side, deploying to Windows 64-bit is really straightforward but it is a good thing to keep in mind. There is no preparatory step for Windows 64-bit.

> **Note**
>
> If you have the Enterprise or Architect versions of Delphi, you'll also have the ability to write Linux applications. Out of the box, Delphi only supports console applications on Linux as Delphi's current implementation of FireMonkey does not support Linux but KSDev, the original producers of what later became FireMonkey, have ported the code to Linux and made this available under the name **FMXLinux**. You can download this through the GetIt Package Manager. We'll cover deploying apps to the Linux platform in *Chapter 12, Console-Based Server Apps and Services*.

The two main platforms outside of Windows that we will discuss in this chapter are the Mac and Android platforms. iOS apps require a Mac to deploy, so we'll cover that in conjunction with the discussion of setting up the Mac platform. Let's start there.

Preparing a Mac for cross-platform development

Debugging and deploying Delphi applications on Apple platforms (Mac or iOS) requires installing the **Platform Assistant Server** (**PAServer**) on a Mac accessible over a local network that is connected to your development PC. Delphi will then communicate with the PAServer, sending your packaged application and receiving application execution information back when it's running. Some developers use a Mac and run Windows in a virtual machine—and that works quite well with the Windows session able to communicate with the local Mac session.

`PAServer21.0.pkg` is the installer for the Mac and can be found in the `PAServer` folder in your Delphi directory. Simply copy it over the network to the Mac and run it. Installing the PAServer is a simple step-by-step process leading you through a common wizard interface. Click **Continue** through the steps until the last page where it lets you know it's installed and where to find it:

Install PAServer 21.0
The installation was completed successfully.
Introduction
License
Destination Select
Installation Type
Installation
Summary
Installation is complete for these two applications:
PAServer-21.0
PAServer Manager
Go to your /Applications folder and double-click the application you want to run
Go Back
Close

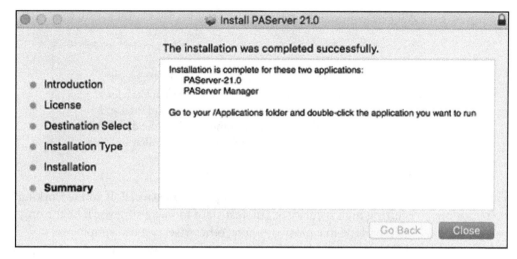

Figure 4.3 – PAServer for the Mac, installation complete

Now, navigate to the `Applications` folder on the Mac and look for it:

PAServer-21.0

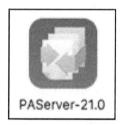

Figure 4.4 – PAServer application icon

Launching the PAServer starts a command-line application on the Mac:

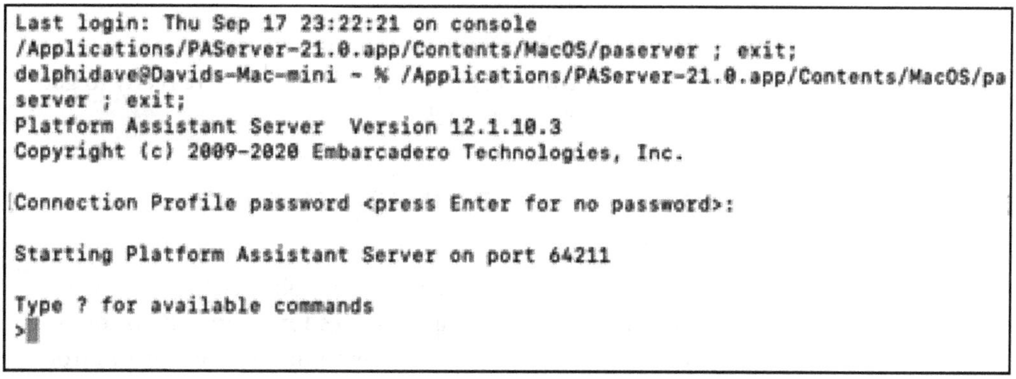

Last login: Thu Sep 17 23:22:21 on console
/Applications/PAServer-21.0.app/Contents/MacOS/paserver ; exit;
delphidave@Davids-Mac-mini ~ % /Applications/PAServer-21.0.app/Contents/MacOS/pa
server ; exit;
Platform Assistant Server Version 12.1.10.3
Copyright (c) 2009-2020 Embarcadero Technologies, Inc.
Connection Profile password <press Enter for no password>:
Starting Platform Assistant Server on port 64211
Type ? for available commands
>

Figure 4.5 – PAServer running on a Mac

This brings up the PAServer command interface.

> **Note**
>
> On some systems, a terminal window opens in addition to the PAServer window. macOS is based on the NeXT operating system (which has a lot of history with BSD Unix) and so the terminal window's command shell and utilities are similar to what you'd see in Unix or Linux. In fact, the default shell for many versions of macOS is the Bash shell, which is very familiar on Unix and Linux.

When you start the PAServer, it asks for a **connection profile password**. If you're working in an environment requiring high security or you just want to make sure you'll be the only one accessing this PAServer, type in a password here; otherwise, you can simply press *Enter* for no password. The PAServer then shows the port currently configured and drops to a prompt. You can type ? for a list of available commands:

```
Type ? for available commands
>?
q - stop the server
c - print all clients
p - print port number
i - print available IP addresses
s - print scratch directory
g - generate login passfile
v - toggle verbose mode
r - reset, terminate all child processes
>
```

Figure 4.6 – PAServer commands

From here, you can get information about the currently running PAServer and its connected clients or terminate processes running on it. That's it—now we're ready to get our first application connected and running on a Mac!

Running your first cross-platform application

Create a new multi-device application based on the **Master-Detail** template. It runs without modification immediately and gives us an easy starting point. Get familiar with it by selecting the **Windows 32-bit** target and running it. It uses **Live-Binding** to connect details in labels and a memo on the right side of the screen with the selected person from ListView on the left side of the screen. Now that you know what it looks like and how it should behave on Windows, let's run it on the Mac.

For Delphi to know where to send your Mac app, it has to connect to the PAServer we set up in the previous section. Since you can have multiple PAServer instances running on various machines in your network, Delphi needs a way to identify them and allow you to switch between them. This is done with **connection profiles**. To manage these, open **Tools | Options** from the Delphi menu and select the **Deployment | Connection Profile Manager** screen. Here, you can add, copy, rename, and test connection profiles—even share them with other developers on your team.

> **Tip**
>
> You can also simply right-click and select **Properties** on your project's target platform in the **Project Manager** to select and edit a single connection profile.

Our first target platform will be **macOSX 64-bit** and will connect to the PAServer we installed in the previous section. From the **Connection Profile Manager**, click **Add…** and the following dialog shows:

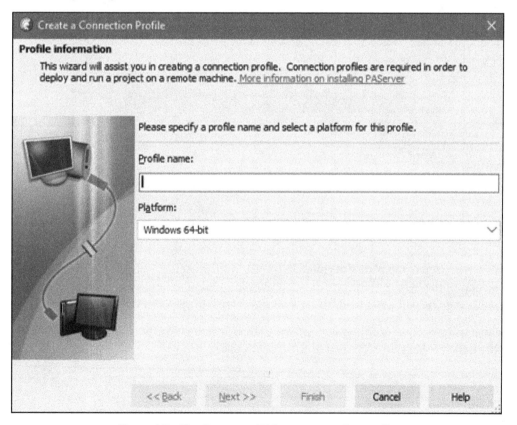

Figure 4.7 – Creating a new PAServer connection profile

Give a descriptive name for the PAServer to which you're connecting (I named mine `mac mini`) and change the selected platform to **macOS 64-bit**, then click **Next**. Enter the remote machine's name or IP address, set the port (if it's different from the default), and optionally, enter a password if you configured your PAServer with one. Once enough information has been entered, the **Test Connection** button will become enabled. Click it to verify the connection.

Now that Delphi knows where to send your app, it needs to know what type of device is on the other end of that platform connection. Under the same **Deployment** section of Delphi's **Options** screen, drop down to **SDK Manager** and click **Add**:

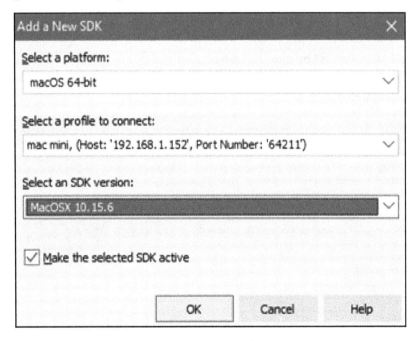

Figure 4.8 – Add a New SDK for Mac OS X

In this screen, select the platform (**macOS 64-bit** in this case), the profile just created, and the version of the SDK, or software development kit, that matches the Mac on the other side. Clicking **OK** will download the files necessary for this SDK (if not already installed).

Now that a connection profile exists and the SDK is installed for the macOS 64-bit platform, the profile shows up next to the macOS platform:

> macOS 64-bit - MacOSX 10.15.6 - mac mini profile

Figure 4.9 – Mac target platform using the new connection profile and SDK

With the PAServer running on a Mac in your network and a connection profile successfully configured and tested in Delphi for your selected macOS 64-bit target, there's only one thing left to do: run it!

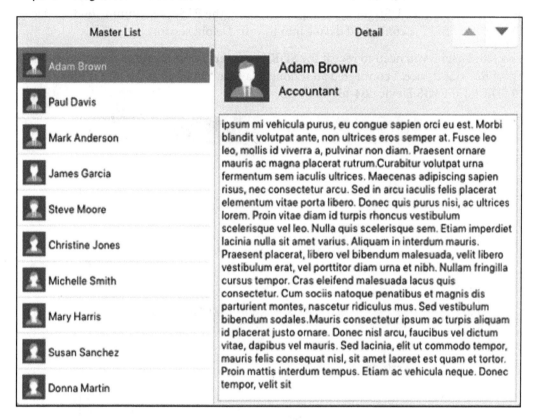

Figure 4.10 – The ClientList app running on a Mac

It doesn't launch as quickly as a Windows application on your local computer—and you may wonder whether it will even start—but after a few moments, you'll be rewarded.

I'll call this app `ClientList`; the code for it is on GitHub and we'll be using it several more times in this chapter:

```
https://github.com/PacktPublishing/Fearless-Cross-Platform-De-
velopment-with-Delphi/tree/master/Chapter04/03_ClientList
```

Now, let's hook up an iPhone.

Preparing for iOS development and deployment

Now that the PAServer is successfully running on your Mac, Delphi can use it to deploy both Mac and iOS apps. If you haven't already, connect your iOS device (iPhone, iPad, or iPod) to your Mac via a USB (or lightning) cable. Now, the PAServer running on the Mac will see it and send the connected device info to your Delphi session.

So, back in Delphi, you need to revisit the **SDK Manager** on the **Options** screen and add support for your device. I connected an iPhone to a Mac Mini and added the **iPhoneOS 14.0** SDK for the **iOS Device 64-bit** platform:

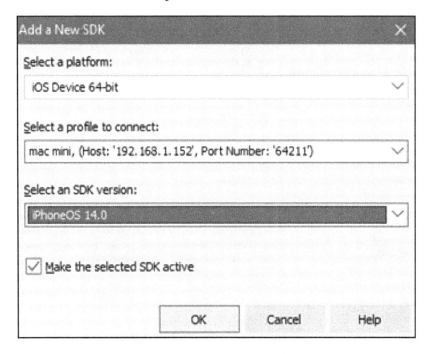

Figure 4.11 – Adding a new SDK for iOS

Again, you'll note the new profile is now showing next to the iOS platform for the project:

Figure 4.12 – iOS target platform and SDK

It also shows up in Delphi's title bar:

Figure 4.13 – iPhone device selected

Now that Delphi knows the type of iOS device to build and where to send it, the last part of getting your app prepared involves putting several pieces of information into your app so that the Xcode utilities know how to sign it and your iOS device accepts it and runs it. But there's a problem: the PAServer doesn't know how to tell Xcode to sign code that comes straight out of Delphi without an Apple Developer account. So, follow the link at the end of this chapter to sign up for one—it'll cost $99 per year but allows you to distribute apps on the Apple App Store and prevents your apps' digital certificates from expiring.

> **Note**
>
> It is possible to get Delphi apps onto iOS devices without an Apple Developer account but it's convoluted. You have to create a simple app in Xcode and name it exactly what your Delphi app is named, deploy it, then once the iOS device is prepped to receive an app by that name, and has allocated the security clearance for it, copy the "bundle" information found in the Xcode project to the **Version Info** fields of your Delphi project and run it through the PAServer. Your iOS device will then see your app as an update to the one that is already there. The process requires some navigating through Xcode and is outside the scope of this book. (But for the record, I have done it!)

Setting up an Apple Developer account isn't terribly difficult. Just follow the instructions at the *Apple Developer Program* link in the *Further reading* section at the end of this chapter and soon you'll be able to log in and manage your account. This will allow you to create identities for the apps you will build and certificates that will be used to sign them.

You only need one or two certificates to sign all the apps you will build, so let's set that up first. There are several different certificate types; the one we're interested in for developing and testing Mac and iOS apps is **Apple Development** (you could get more specific and get one just for **iOS App Development** if you choose—read about the various types and select the one that works for you and your team; you can create additional ones as needed). You'll use the *Keychain Access* app on your Mac to create a certificate request file via its **Certificate Assistant**, and save it to disk:

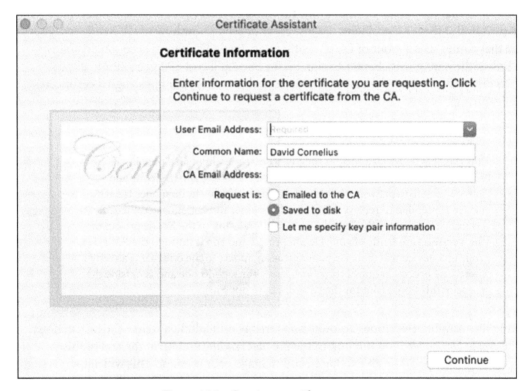

Figure 4.14 – Creating a certificate request

Then, from your Apple Developer account, create a new certificate, select your certificate type, then upload the certificate request file you just created. Your signing certificate is created immediately, allowing you to download your new signing certificate to your Mac and add it to the certificates in your Keychain Access list:

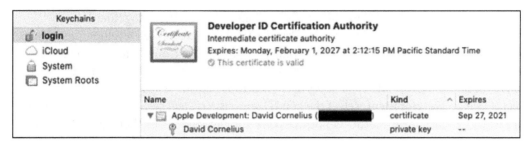

Figure 4.15 – New certificate added to Keychain Access

You now need to create **app identities** for each app you deploy. In your Apple Developer account, go to the **Identifiers** list and create a new identifier. As with certificates, there are several types of identifiers; we'll just use the standard **App ID** type for now.

Fill out the **Description** and **Bundle ID** settings for your app (I recommend an **Explicit** ID for now; we'll cover **Wildcard** IDs much later). Follow the recommended naming convention when setting up the bundle ID; it must be unique among all app IDs, so prefix it with the name of your organization and append a unique, descriptive app name. For the sample `ClientList` app created in the previous section, the app identifier I created is `book.fearless-crossplatform-delphi-dev.ClientList`:

Figure 4.16 – Creating a new app ID

Now that we have an app ID and a signing certificate, we hook those together into a **provisioning profile**. This is done under **Profiles** (still in your Apple Developer account). As might be expected, there are several profile types to choose from. Later, when you're ready to deploy an app to the App Store or to unknown devices within your organization, you'll choose the appropriate type under **Distribution**. For our testing and development, select **iOS App Development**, select the app ID you just created, select the certificate you just created, and lastly, select the iOS device connected to your Mac. Then, give your new profile a name:

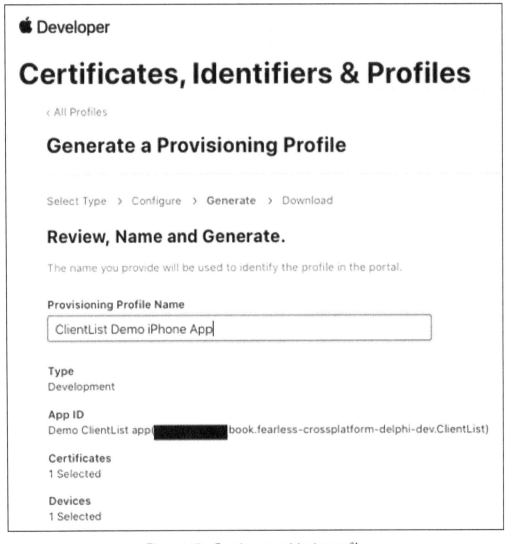

Figure 4.17 – Creating a provisioning profile

With a completed provisioning profile now present in your Apple Developer account, enter the app ID into the **CFBundleIdentifier** field of the **Version Info** screen for the iOS target platform. If your application name exactly matches the last part of the app ID, you can replace that part with $(ModuleName):

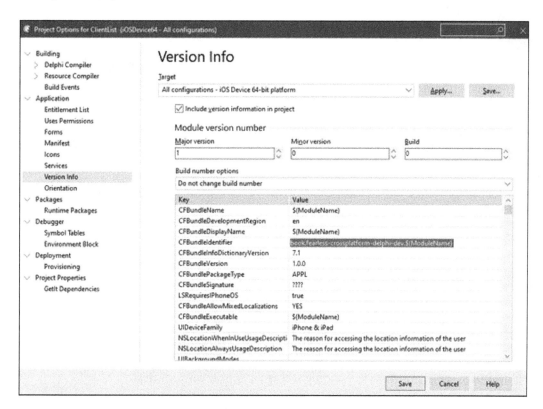

Figure 4.18 – Adding the app ID into the Delphi project's version info CFBundleIdentifier field

Finally, in the project's **Deployment | Provisioning** options, select the **Build type**, **Provision Profile**, and **Developer Certificate** settings associated with this app in your Apple Developer account (usually, **Auto** will suffice):

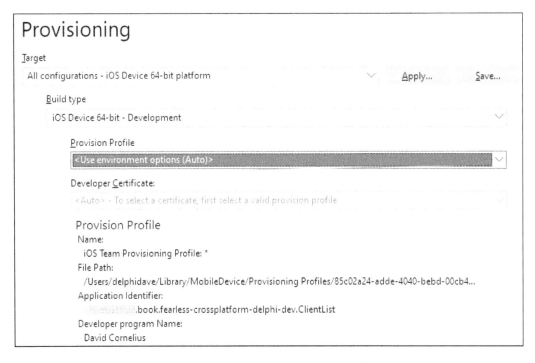

Figure 4.19 – Setting the Provisioning information for the project

At last, launch the application and soon you'll see it appear on your device:

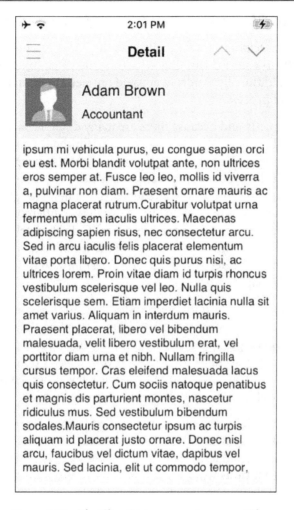

Figure 4.20 – The ClientList app running on an iPhone

These steps can be repeated for other devices (create additional device identities in your Apple account for testing) and other applications (create a provisioning profile for every application you will deploy).

Now that we've got the Mac and iOS platforms under our belt, let's move on to Android.

Preparing your PC to deploy to an Android device

Android development in Delphi is not done through the PAServer. Instead, Android devices connect directly to your PC running Windows, then Delphi deploys and debugs it over the USB cable. Additionally, Android apps do not need to be signed, saving several steps in the development process.

Because they connect directly and because there are many different manufacturers of Android devices, not only do you have to install an SDK for the version of Android you want to support (similar to the Mac and iOS platforms), you also have to install a device driver to support each device type that you want to test with. For example, I have an old ZTE phone for testing, an Amazon Kindle Fire tablet, and a new Google Pixel phone. In order to test each of these, I have to download and install two different device drivers, one for each of my non-Google devices, onto my Windows PC. Fortunately, the drivers are fairly easy to find and it's a one-time install. Many of them can be found by following links for your device manufacturer under *Android device driver links* in the *Further reading* section at the end of this chapter.

Once the driver is installed, you need to set up your device for USB debugging. On most Android devices, this is done by going into **Settings** and enabling the setting under **Developer options**. If **Developer options** is not visible, select **About** and click the *build number* seven times (this may seem a little strange if you've never done this before but that's what you have to do!). You'll likely want to enable a few other developer options, such as **Stay awake while charging**, so look through the ones available on your device.

If you selected Android development when you installed Delphi, chances are the Android SDK and **NDK** (**native development kit**) are already installed. If not, or if you need to use a different version of the Android tools that were originally installed, you'll need to revisit the **SDK Manager** in the **Options** window. See the *Adding an Android SDK* link in the *Further reading* section at the end of this chapter for details.

Now, everything should be configured and in place for Delphi to see your device. With the ClientList app we were working with previously, select the **Android 32-bit** target platform in the project. If your Android device is configured and connected properly, it should show up next to the target platform:

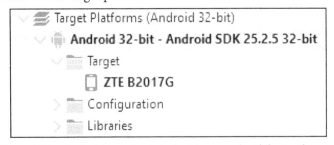

Figure 4.21 – Android target selected with an Android device detected

Running the app is as simple as other Delphi apps and automatically deploys and launches it on your Android device:

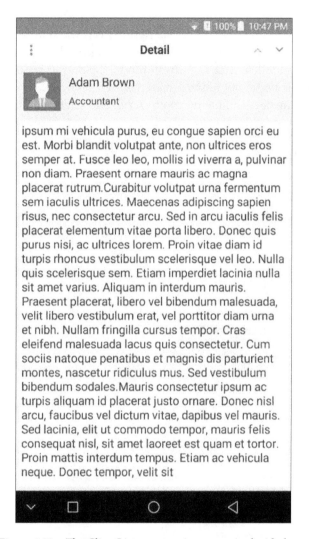

Figure 4.22 – The ClientList app running on an Android phone

Well, there it is! A single code project deployed to four very different platforms: Windows, Mac, iPad, and an Android phone. Now, let's dig under the hood and see what it took to make this happen—and some tricks you'll take to other projects.

Working with various screen sizes

The first thing you probably noticed when working with the `ClientList` app we used in the previous section is that Delphi's design-time view shows a master list of people on the left and a **Detail** view of a single person on the right. This is the standard desktop interface and was how both the Windows and Mac desktop applications looked. However, when we deploy the app to a mobile device in portrait mode, the master list disappears and all you see is a button at the upper left of the screen that, when clicked, reveals the master list in a slide-out list over the detail, allowing you to scroll and select a different person, after which it hides again.

How does the app look and behave differently on a smaller screen?

Exploring target views

Before I answer that, look at the right side of the toolbar at top of the design view for the `ClientList` project. There's a combo box labeled **View**, and if you drop down the list, you'll see the various combinations of target devices and screen sizes:

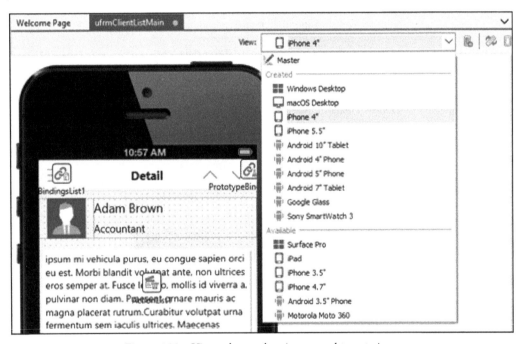

Figure 4.23 – View selector showing several target views

The first view that is automatically selected when you load a FireMonkey project is the Master view. This is the base view upon which all others inherit from. The one selected above shows what the app would look like on a 4" iPhone.

Look in the upper-left corner of the detail toolbar, in either an Android or iPhone view, at the `MasterButton` component, a **TSpeedButton** that has the `StyleLookup` property set to `detailstoolbutton`. This is the button that shows the Master list on a small screen when in portrait mode. On an Android, it's three vertically aligned dots; on an iPhone, it's three horizontal lines.

Having different views allows you to arrange components differently and set properties differently for the different devices that will use those views. You can add and remove components only on the Master view, not any of the other views; however, some properties will be available on some views and not on others depending on the capabilities of the device that supports that view. Also, different views have different styles that correspond to that device's native look and feel. `MasterButton`'s property is the same in all views—the view's style is what dictates how it appears. We will cover these concepts in greater detail later in the book.

Let's get back to the question of how a particular view is selected. Views are tied to Delphi devices, defined in **Tools | Options | User Interface| Form Designer | Device Manager**. These devices have properties that define the size and orientation of the form (you can create custom devices from this screen for your use in Delphi if you need to). When your application runs, the FireMonkey framework goes through the defined views for the current platform and picks the closest match based on the device size and class.

In addition to customizing the layout and properties of components on forms at design-time, you can also write code that executes only on specific devices.

Writing code to support multiple platforms

While the bulk of your code will be the same regardless of the platform upon which it's running, there will be some cases where you want to do things differently or provide different options if the application is running on a specific device. These unique application characteristics cannot always be implemented by simply customizing properties on a view but must be specifically handled by the code you write. And sometimes that code will only run on devices with very particular hardware architectures. To handle these cases, use **conditional compilation**.

In previous chapters, you've seen specially formatted comments that tell Delphi to link in resources (for example, `{$R *.fmx}`) or to create a console app (for example, `{$APPTYPE CONSOLE}`). This same type of syntax is used to include or exclude specific lines of code based on defined constants. These are not *code* identifiers, which are defined in a `const` section, but *compilation* identifiers defined during the compilation process depending on the target platform.

For example, if you're compiling a Windows application, the MSWINDOWS identifier is defined. If your target is Windows 32-bit, then WIN32 is also defined, whereas if your target is Windows 64-bit, then WIN64 is also defined.

The Mac OS X platform defines MACOS and MACOS64. Here are several more common ones:

- Android platforms: ANDROID, ANDROID32, ANDROID64

- iOS platforms: IOS, IOS32, IOS64

- Linux platforms: LINUX, LINUX32, LINUX64

There are also compilation identifiers for various CPU architectures, for example, CPU386, CPU64BITS, and CPUARM. Some are useful for libraries that span several versions of Delphi and need to know whether certain language features are supported, such as UNICODE.

To use these identifiers, surround lines of code with conditional directives such as {$IF}/{$IFDEF}/{$IFNDEF}, {$ELSE}/{$ELSEIF}, and {$ENDIF}.

Check out this FormCreate procedure to see them in action:

```
procedure TfrmHello.FormCreate(Sender: TObject);
begin
  {$IFDEF MSWINDOWS}
  lblHello.Text := 'Hello Microsoft!';
  {$ELSEIF DEFINED(MACOS)}
  lblHello.Text := 'Hello Apple!';
  {$ELSEIF DEFINED(ANDROID)}
  lblHello.Text := 'Hello Android device!';
  {$ELSE}
  lblHello.Text := 'Hello Unknown device!';
  {$ENDIF}
end;
```

This was taken from the sample project on GitHub at the following link:

https://github.com/PacktPublishing/Fearless-Cross-Platform-Development-with-Delphi/tree/master/Chapter04/04_HelloDevice

There are several other compiler identifiers available—see the *Conditional compilation* link in the *Further reading* section at the end of this chapter. You can create your own. When we talked about **build configurations** in *Chapter 2, Delphi Project Management*, you learned that Delphi's default **Debug** and **Release** configurations define DEBUG and RELEASE, respectively. You can customize the list of defined identifiers in the configurations or in your code using the {$DEFINE} directive. One reason you might do this is if you want to build multiple versions of your application for different customers or with different features.

Conditional compilation is another powerful management tool for the professional cross-platform developer.

Summary

In this chapter, we've migrated your skills as a Windows VCL developer to the FireMonkey framework, walked you through preparing devices from other platforms for development and testing, and finally, saw your Delphi apps running on them. We also showed you how different devices can utilize custom form views and how to execute different code based on conditional compilation.

Before we look deeper into advanced aspects of multi-device development, the next chapter talks about strategies for sharing code with other projects or other members of your team with libraries, packages, and components and how that is done with cross-platform goals in place.

Questions

1. What FireMonkey project template most closely resembles a new VCL project?

2. List two ways a FireMonkey checkbox differs from a VCL checkbox.

3. What can you use in FireMonkey to mimic the VCL's data-aware capability?

4. What GUI framework is available on Linux that is similar to FireMonkey?

5. Where can you find PAServer installers for other platforms?

6. What does a provisioning profile for iOS consist of?

7. Besides the SDK, what must be in place for testing and debugging Android apps?

8. How does a FireMonkey application know which view to use when it runs?

9. What compiler identifier would you use to compile code only on 64-bit platforms?

Further reading

- The FireMonkey platform: `https://en.wikipedia.org/wiki/FireMonkey`

- FireMonkey versus VCL: `http://delphi.org/2016/10/firemonkey-vs-vcl/`

- Mida Converter: `http://midaconverter.com/`

- FMX Linux: `https://www.fmxlinux.com/`

- *PAServer, the Platform Assistant Server Application*: `http://docwiki.embarcadero.com/RADStudio/Sydney/en/PAServer,_the_Platform_Assistant_Server_Application`

- *iOS Mobile Application Development*: `http://docwiki.embarcadero.com/RADStudio/Sydney/en/IOS_Mobile_Application_Development`

- Apple Developer Program: `https://developer.apple.com/programs/`

- Apple app signing certificates: `https://help.apple.com/xcode/mac/current/#/dev3a05256b8`

- Android device driver links: `https://developer.android.com/studio/run/oem-usb`

- *Configure Your System to Detect Your Android Device*: `http://docwiki.embarcadero.com/RADStudio/Sydney/en/Configuring_Your_System_to_Detect_Your_Android_Device`

- *Adding an Android SDK*: `http://docwiki.embarcadero.com/RADStudio/Sydney/en/Adding_an_Android_SDK`

- *Using FireMonkey Views*: `http://docwiki.embarcadero.com/RADStudio/Sydney/en/Using_FireMonkey_Views`

- *Adding a Customized View to the View Selector*: `http://docwiki.embarcadero.com/RADStudio/Sydney/en/Adding_a_Customized_View_to_the_View_Selector`

- Conditional compilation: `http://docwiki.embarcadero.com/RADStudio/Sydney/en/Conditional_compilation_(Delphi)`

5
Libraries, Packages, and Components

There's a joke listed as one of Murphy's computer laws that goes like this:

Any given program will expand to fill all the available memory.

Over the life of an application, customers demand more, operating systems get updated, new devices are supported, and so on. It is inevitable that useful software will grow in size and complexity over time.

Today's programs share code, call external services, and dynamically load resources in a variety of ways. Delphi has supported these techniques for a long time, but a refresher is in order, not only because it's more important than ever but also because there are some cross-platform considerations.

We'll start with the basics by simply pulling out procedures and functions into a **Dynamically Loaded Library** (**DLL**) that can be called from any language. Then, we'll move to working with packages and show you how to install your own custom components into Delphi's Tool Palette. Finally, we'll show how your components in the Delphi IDE can be made to link project code from other platforms. Here are the sections for these topics covered in this chapter:

- Sharing code in libraries
- Modularizing applications with packages
- Building components for multiple platforms

Technical requirements

In addition to having Delphi 10.4 Sydney Professional installed in a Windows 10 64-bit environment, you should also have access to a Mac and/or a mobile device to build the sample applications discussed in this chapter for other platforms. For details, refer to the Technical Requirements of the previous chapter.

> **NOTE**
> If you have Delphi Enterprise or Architect and want to try the examples in this book on the Linux platform, you will also need 64-bit Ubuntu 14.04, 16.04, or 18.04, or Red Hat Enterprise 7. If you don't have a separate machine with Linux, you can install the Linux subsystem right within Windows 10 and it works well as a test environment for Delphi server applications.

All the code for this chapter can be found on GitHub at `https://github.com/PacktPublishing/Fearless-Cross-Platform-Development-with-Delphi/blob/master/Chapter05`.

Sharing code in libraries

Let's begin with a simple Windows FireMonkey application that hides a string. Create a blank multi-device application, and add a label for a prompt, an edit box for input, a button to take action on the entered text in the edit box, and another label to display the hidden text. Your form should look something like this:

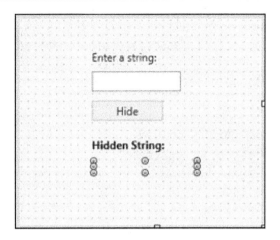

Figure 5.1 – Hide string demo FireMonkey main form

Add a function that takes a string and returns another with the original string hidden in some way. Here's a simple example:

```
function TfrmHideStrMain.HideString(const MyString: string):
string;
begin
  // manipulate the string to hide its original contents
  for var i := 1 to Length(MyString) do
    Result := Result + Chr(Random(26) + Ord('A')) +
      MyString[i];
end;
```

Now, call that function from the button's `OnClick` event and display the results in the result string (which I've called `lblHidden`):

```
procedure TfrmHideStrMain.btnHideStringClick(Sender: TObject);
begin
  lblHidden.Text := HideString(edtInput.Text);
end;
```

Compile and run this program as a Windows 32-bit application just as a quick test.

Now, let's take the `HideString` function and extract it to a library. To do that, create a **Dynamic Library** project, copy the `HideString` function, and add an `exports` clause:

```
library HideStringLib;
function HideString(const MyString: string): string; stdcall;
begin
  // manipulate the string to hide its original contents
  for var i := 1 to Length(MyString) do
    Result := Result + Chr(Random(26) + Ord('A')) +
      MyString[i]; end;


exports HideString;


begin
end.
```

Then, modify the main project code by taking out the declaration to the function in the form and the body of the function in the `implementation` section and replace it with just the declaration and a reference to the DLL in which it exists:

```
function HideString(const MyString: string): string;
  stdcall; external 'HideStringLib.dll' name 'HideString';
```

Compile the library project, making sure that the DLL will end up in the same folder as the executable for the main demo app, then run the demo app to check that it still works. The code for both projects is on GitHub at the following link:

https://github.com/PacktPublishing/Fearless-Cross-Platform-Development-with-Delphi/tree/master/Chapter05/01_HideStringDLL

This is the quickest and simplest way to extract and use functions in a DLL—but it's not the best way. There are a couple of things to discuss here. But first, a few tips about writing DLLs for Windows.

Things to keep in mind for the Windows platform

There are a couple of things to consider when writing DLLs for the Windows platform. First of all, when you created the Dynamic Library project, did you see the big comment inserted near the top of the project file that warned about using the Delphi string type? Strings are easy to work with in Delphi, but they are not directly compatible with other languages. Since DLLs should be written in a language-agnostic way, you should avoid using the string type in DLLs. If your DLL will only be used with your own applications, you can make dealing with strings simpler by adding the SharMem unit as the very first entry in the uses clause of both the project and the library (which I did in this example), plus ship the BORLANDMM.DLL library with your project.

Another language-agnostic tip to keep in mind only needs to be considered if you're writing 32-bit libraries (which implies Windows only since you cannot build 32-bit Mac or Linux projects). The parameter-passing mechanism in 32-bit Delphi applications is different than other languages. By including the stdcall keyword, it forces parameters to be internally used in the same order as other languages. In 64-bit compilations, there is no difference, so the keyword is ignored in order to avoid having to use {$IFDEF} compiler directives. It's a common practice to simply always include stdcall at the end of your Delphi library procedures and functions to ensure compatibility.

> **Tip**
> You can also add the delayed keyword to the end of the declaration for Windows-only libraries. If you do, the library won't be loaded until the function is actually called.

This hardcoded reference to the DLL file in code is not the ideal way to call functions in a library. It's what we call "brittle" because it breaks easily and inelegantly if the file is missing. So, while it takes more work, there's a much better way to link external code—since these are *dynamic* libraries, they should be loaded dynamically, right?

Loading libraries dynamically versus statically

The method of loading the library function shown previously is called *static* loading. It should only be used for quick prototyping or when you can be assured that the file will always be there. For example, the Winapi.Windows unit that comes with Delphi relies on several Windows DLLs to be in place and statically links many Windows platform **Runtime Library (RTL)** functions available in Delphi. This is acceptable because if those files are missing, you have some serious problems with Windows, and many other applications that rely on these DLLs would also break!

Your application should ensure that the DLLs you supply are where you expect them to be, and if not, display a nice message to the user, rather than just crash with an ugly error. To load DLLs dynamically, you need to call the LoadLibrary function, which returns a non-zero handle to the loaded library if it was successful. On Windows platforms, that function is in the WinAPI.Windows unit, so you need to add that to the uses clause:

```
uses
  WinAPI.Windows, // ...
```

Continuing with our HideString demo app, we need to declare what type of function we'll be expecting in the library once it's loaded and create variables to hold both the handle to the library and a pointer to the function it contains. Add these in the implementation section of the demo project's main form:

```
type
  THideStringFunc = function(const MyString: string): string;
    stdcall;
var
  DllHandle: HModule;
  HideString: THideStringFunc;
```

Now, we need to load the library before we use it and free it when we're done. You can do that in several different ways, such as in some initialization code, a separate unit if you have multiple libraries to load, the form's OnCreate and OnDestroy event handlers, or as I've done in this example, using the OnActivate and OnDeactivate event handlers:

```
procedure TfrmHideStrMain.FormActivate(Sender: TObject);
begin
  DllHandle := LoadLibrary('HideStringLib.dll');
  if DLLHandle = 0 then begin
    lblHidden.Text := 'Could not find function library';
    btnHideString.Enabled := False;
  end else
    @HideString := GetProcAddress(DllHandle, 'HideString');
end;


procedure TfrmHideStrMain.FormDeactivate(Sender: TObject);
begin
```

```
  FreeLibrary(DllHandle);
end;
```

In this code, I check to see whether the result of the call to `LoadLibrary` is `0`, and if it is, a label on the form gets an error message and I disable the button that would otherwise call the function in the library. If it is successful, the `HideString` function variable gets the address of the function in the library that matches the exported name.

Finally, once this is all set up and the library is successfully loaded, the button's event handler can safely call the function as before. As a test, compile and run the demo app before compiling the library so that you can see the nice error message that tells you the library cannot be found. Then, compile the library and re-run the demo app to see that it now works as expected. This version of the app is on GitHub at the following link:

```
https://github.com/PacktPublishing/Fearless-Cross-
Platform-Development-with-Delphi/tree/master/Chapter05/02_
HideStringDynamicDLL
```

Without touching the library, the application has been improved to dynamically load it and run the same function—or give an error message and exit cleanly if the library does not exist. You can see how this could be easily extended to search a folder for DLL files, pass them into the `LoadLibrary` function, and enable or disable certain features in your application depending on which ones exist. Additionally, you could replace `HideStringLib.DLL` with a better string obfuscation routine—as long as the function declaration does not change or is backward-compatible.

We will use this method of using libraries from here on.

> **Note**
> Although Delphi's identifiers are case-insensitive, most other languages' identifiers are not. DLLs fall into the latter category, so when calling `GetProcAddress` and looking up the name of an exported function, the case is important. In other words, a library function declared and exported as `HideString` will not be found with `GetProcAddress('HIDESTRING')`.

Before we move on, we should cover libraries on other platforms.

Things to keep in mind for non-Windows platforms

Delphi can create dynamic libraries on other desktop platforms as well (Mac and Linux). Here's a list of the platforms and the filenames they will generate:

- **Windows**: `ProjectName.dll`
- **Mac OS X**: `libProjectName.dylib`
- **Linux**: `libProjectName.so`

Notice that on the Mac and Linux platforms, the base filename is prefixed with `lib` and all three platforms have different extensions. This means that if your application needs to call dynamic libraries on multiple platforms, it will need to construct the right filename based on the platform it's running on using compiler directives.

Here are a few more coding differences for writing libraries for non-Windows platforms:

- **Mac and Linux**: The `ShareMem` unit is neither required nor available.
- **Mac**: All exported functions and procedures in the library should start with an underscore (_) if you want them to be compatible with other languages.
- **Mac**: The `exports` clause cannot be in the main project file (`.dpr`) but must be in a unit.

Let's take our previous example and extend it to also work on a Mac. There are several changes to make. The first is that we only want to use the `WinAPI.Windows` unit on the Windows platform, so use compiler directives to restrict that unit's inclusion in the `uses` clause to only the Windows platform:

```
uses
  {$IFDEF MSWINDOWS} Winapi.Windows, {$ENDIF}
  System.SysUtils;
```

The second is to create a constant that references a platform-specific library name by once again using compiler directives:

```
const
  {$IFDEF MSWINDOWS}          LIB_NAME = 'HideStringLib.dll';
  {$ELSEIF DEFINED(MACOS)}    LIB_NAME = 'libHideStringLib.dylib';
  {$ELSEIF DEFINED(LINUX)}    LIB_NAME = 'HideStringLib.so';
  {$ENDIF}
```

Then, use that constant instead of the hardcoded string:

```
begin
  // initialize the DLL handle
  DllHandle := LoadLibrary(LIB_NAME);
  // ...
```

If your projects use the ShareMem unit as this example does, be sure to put that unit reference in platform-specific compiler directives. Here's the library project modification:

```
library HideStringLib;
{$IFDEF MSWINDOWS}
uses
  ShareMem;
{$ENDIF}
// ...
```

Here's the demo application project modification:

```
program HideStringDynLibDemo;
uses
  {$IFDEF MSWINDOWS} ShareMem, {$ENDIF}
// ...
```

Finally, since the exports clause cannot be in the main library file on a Mac, move the function to its own, new unit:

```
unit uHideStringFunc;

interface

function HideString(const MyString: string): string; stdcall;

implementation

function HideString(const MyString: string): string; stdcall;
begin
  // manipulate the string to hide its original contents
  for var i := 1 to Length(MyString) do
```

```
    Result := Result + Chr(Random(26) + Ord('A')) +
        MyString[i];
end;


exports HideString;
end.
```

With the Windows 32-bit platform still active, run the project to see that it still works. Then, add the **macOS 64-bit** platform to the project, compile, and run it through the pre-established **Platform Assistant Server (PAServer)** to deploy and run it on your Mac:

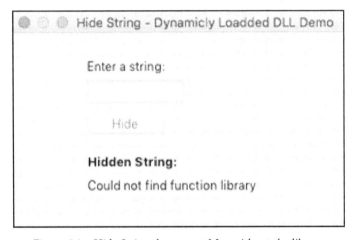

Figure 5.2 – Hide String demo on a Mac without the library

But wait—why can't it find the function library? Delphi doesn't know that your compiled application will load a library, so it only copied the demo app that you told it to run—and nothing more. We need to modify the deployment properties of the demo app to include the compiled library when it sends the package over to the PAServer on the Mac.

To do that, make sure the demo app is the active project, then click **Project | Deployment** from the Delphi menu. A new tab in the IDE opens up with the files to deploy for the selected project:

Local Path	Local Name ▲	Type	Confi...	Platforms	Remote Path	Remote Name
☑ $(BDS)\bin\	delphi_PROJECTICNS.icns	ProjectOSXResou...	Debug	[OSX64]	Contents\Resources\	HideStringDynLibDemo.icns
☑ OSX64\Debug\	HideStringDynLibDemo	ProjectOutput	Debug	[OSX64]	Contents\MacOS\	HideStringDynLibDemo
☑ OSX64\Debug\	HideStringDynLibDemo.dSYM	ProjectOSXDebug	Debug	[OSX64]	..\$(PROJECTNAME).ap...	HideStringDynLibDemo
☑ OSX64\Debug\	HideStringDynLibDemo.entitle...	ProjectOSXEntitle...	Debug	[OSX64]	..\	HideStringDynLibDemo.entitl
☑ OSX64\Debug\	HideStringDynLibDemo.info.plist	ProjectOSXInfoPL...	Debug	[OSX64]	Contents\	Info.plist

Figure 5.3 – Project Deployment files

To add the dynamic library to the package deployment, click the second toolbar button from the left to **Add Files**, then select the compiled dynamic library file for the Mac—libHideStringLib.dylib, in our example.

There's just one last step before this will work. The default folder for the dynamic library isn't always correct. Make sure the remote path for the library file you just added is set to Contents\MacOS\ so that the project will find the file when running on the Mac—you don't even have to type it in as that column's value has a drop-down list of the most frequent paths and you can simply select it. At last, compile and debug your app, click the **Hide** button, and see your string obfuscated. The cross-platform version of this dynamic library project is at the following link:

https://github.com/PacktPublishing/Fearless-Cross-Platform-Development-with-Delphi/tree/master/Chapter05/03_HideStringCrossPlatDLL

Now, let's find out what advantages there are to putting these functions into a package instead of a dynamic library.

Putting code into packages

Runtime packages are another way to share code and are used far more frequently than dynamic libraries as these are available on *all* platforms and are the basis of components—which we will cover in the next section. Packages are simpler to write than dynamic libraries because they don't have the parameter-passing issues to worry about, and functions and objects can be called and passed around just as if the code were part of the project. But they can only be used by Delphi packages and applications written in the same version of Delphi that is used to compile the package.

> **Note**
> You also need to know how to write packages to use RAD Server (Enterprise or Architect edition).

Let's turn our HideString library into a package. This actually involves taking out some of the scaffolding code we put in to support a dynamic library. In fact, when we're done, it simply looks like a collection of used units we'd link directly in the main application—which is exactly what it is. Therefore, to make sure the demo application is actually using the compiled package instead of linking in the source units directly, make sure the package code is in a separate folder from the application code when you set up this set of example projects.

Create a new **Package** project and add a unit for the `HideString` function:

```
package HideStringPkg;
{$R *.res}
requires
  rtl;
contains
  uHideStringFunc in 'uHideStringFunc.pas';
end.
```

The `requires` section lists other packages that may be needed. The `contains` sections lists the source units that comprise this package.

The `uHideStringFunc` unit is virtually the same as the one from the dynamic library project used in the last section; the only difference is that there's no longer an `exports` clause and we don't need the `cdecl` keyword:

```
unit uHideStringFunc;
interface
function HideString(const MyString: string): string;

implementation
function HideString(const MyString: string): string;
begin
  // manipulate the string to hide its original contents
  for var i := 1 to Length(MyString) do
    Result := Result + Chr(Random(26) + Ord('A')) +
      MyString[i];
end;
end.
```

Copy the `HideString` demo application from the first example project in this chapter to a different folder from the package project just created and remove the reference to the static library.

Before we build and run this application, here's a short but important note on how your packages get distributed with your application. By default, Delphi links all the packages together into one large executable. This is very convenient, but there are some situations where either this is not feasible due to some library constraint or you want a suite of applications to share packaged code.

To change this so that Delphi compiles just a bare-bones executable and references library code in shared packages instead, go to the project's options and click on **Packages | Runtime Packages**, then uncheck the **Link with runtime packages** checkbox. When you do that, the line below it, **Runtime packages**, is enabled and you can edit the list of packages upon which the project depends:

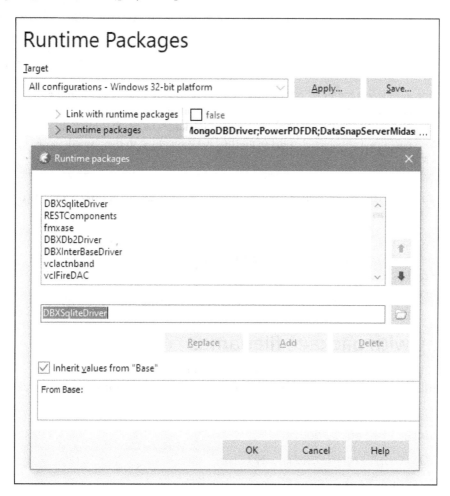

Figure 5.4 – Editing runtime packages for a project

The default list of packages includes everything you might need. You will never need all of them because they include both VCL and FireMonkey, which are incompatible, so be sure and delete ones not used in your project. Also, be sure to select the right target platform before editing the package list—initially, the selected platform is the one you're currently working with (for example, **Debug** or **Release**) and you probably want the package list to be the same for both configurations of a platform.

Our demo application only needs one package, so delete all of the ones listed (the `rtl` package is included by the Delphi compiler whether it's listed or not) and simply add `HideStringPkg`.

Before you build the demo application, you need to build the package so that Delphi will find the package when linking the demo app. So, back in the package project, make sure its project options send the DCP and BPL (package output) files to the folder where the demo executable will be built.

> **Tip**
> When a project is built without runtime packages, Delphi needs to link in the **Delphi Compiled Unit (DCU)** files. But when a project is built with runtime packages, Delphi needs the **Delphi Compiled Package (DCP)** files instead, which contains a concatenated list of all the DCUs included in the named package.

Now, you can build the package and then build and run the demo application that uses the package. Compare your results with the set of projects on GitHub:

```
https://github.com/PacktPublishing/Fearless-Cross-
Platform-Development-with-Delphi/tree/master/Chapter05/04_
HideStringPackage
```

Before we move on, we need to talk about package file naming.

Working with package filenames

If you build the package for each of the platforms, you'll notice that the filenames follow a similar pattern as was shown previously with dynamic libraries—with some key differences:

- **Windows**: `ProjectName.bpl`
- **Mac OS X**: `bplProjectName.dylib`
- **iOS/Android**: `libProjectName.a`
- **Linux**: `bplProjectName.so`

Notice that both iOS and Android use the same naming scheme and they both prefix the project name with `lib`, while the other platforms have `bpl` as part of the filename (from **Borland Package Library**).

Take a look in Delphi's `bin` folder at all the BPLs there. Notice how they all have `270` on the end? You can easily number your packages in a similar manner without changing the package name. Go into **Project Options** and look at the **Description** page of the package project you just created. Enter `270` for **LIB suffix** and save the options. Notice the package name now has `270` appended, and when you build the Windows package, the filename is `HideStringPkg270.bpl`. Similarly, **LIB prefix** will prepend text to the package name and **LIB version** will append the version text to the end of the base filename with a period and before the extension is added.

Applications and other packages find and use packages based on the package name without these name extensions, so in our example projects, we didn't have to modify the application's runtime package list to have it still load the `HideStringPkg` package, even though its filename changed. This simplifies project management and helps identify and separate different versions of packages, allowing multiple versions of packages to exist in system paths without conflict. Another benefit is that you may want to modify a package to fit in with other files in a specific scenario, such as prepending `mod_` to packages that get deployed as Apache web modules.

> **Tip**
> It is highly advised that you use some sort of unique file naming mechanism for your packages (the **LIB prefix** and **LIB suffix** features are very convenient!) even if you don't use multiple versions of Delphi. Hunting down the source of why a unit seemingly isn't getting updated only to find out you have another version of the same package name that is getting picked up earlier in the system path is very frustrating, not to mention time-consuming.

Now that you have seen how simple it is to create a package and that there's nothing more to a package than a few units put together, you will be delighted to know that it's just a small step from there to make it a component that can be installed into Delphi's Tool Palette.

Turning a package into a component

Let's continue working with the `HideString` example and turn it into a component that gets installed on Delphi's Tool Palette. First, we'll copy and rename the demo project and package over to a new set of folders. Then, we'll use the **New Component** wizard to create an empty design-time package that gets registered in Delphi as a component and brings in our runtime package. Finally, we'll add some properties and expand the `HideString` function to be a little more useful and make it easy to configure. Each step is detailed in the following sections.

For reference through these steps, here are the project filenames I will use (everything will be in the same folder for simplicity):

- `HideStringComponentDemo.dpr`: FireMonkey demo application (renamed)

- `HideStringRT.dpk`: Runtime package (renamed from `HideStringPkg`)

- `HideStringDT.dpk`: Design-time package (new)

Let's dive in!

Creating our first component

A component is simply a runtime package that has been installed by a design-time package. What might surprise you even more is that a package can function as both a runtime package *and* a design-time package! In fact, that's the default setting when creating a new package (however, to get the component to actually show up on the Tool Palette requires a little additional code, as we will soon see).

Open up **Project Options** of the `HideStringRT` package project and go to the **Description** page. Near the bottom are some radio buttons for **Usage options**. There are three options:

- **Designtime only**: The package's only purpose is to add functionality within the Delphi IDE, by registering components, providing property editors for components, or augmenting IDE functionality with special Delphi hook functions.

- **Runtime only**: The package contains code meant for distributing with other packages and/or applications.

- **Designtime and runtime**: The package can act as both a design-time and a runtime package.

Up to this point, we've built only runtime packages; even though the package may be marked as both design-time and runtime, we don't have any design-time code in place yet. So, just for good measure (and partially to reduce possible confusion later), change this project to **Runtime only**.

Let's build a design-time component to install our runtime HideStringRT package:

1. From the menu, select **Component | New Component...** (if your package project is active, you'll be prompted for the framework—select **FireMonkey for Delphi**). Click **Next**.

2. Select **TComponent** from the **Ancestor Component** list and click **Next**.

3. Enter THideString for **Class Name**, leave **Palette Page** at **Samples**, select the destination folder in our project structure, and enter uHideStringComponent. pas for **Unit name**, leaving **Search path** blank, and then click **Next**.

4. Select **Install to a New Package** and click **Next**.

5. Enter HideStringDT for **Package Name** (which becomes the component project filename) in the same folder, add a description, and click on **Finish**.

It will confirm saving the unit file you selected, then compile and install the new component. After reading the message that the component has been registered, it will load the new project it just created.

> **Note**
>
> This is a non-visual component because it inherits from TComponent, the most basic component from which all others descend. You can create a visual control by selecting a different ancestor component in *step 2*, but that is beyond the scope of this book.

So, what did this **New Component** wizard do for us? It created a package containing a unit with enough code to register itself in the Delphi Tool Palette—and that's about it. It doesn't yet link in the actual HideString runtime package we have. The important part of this step is in the new unit: a class descending from TComponent and a Register procedure:

```
procedure Register;
begin
  RegisterComponents('Samples', [THideString]);
end;
```

This tells the IDE what class to register and what category to list it under in the Tool Palette. The class name becomes the component name you see in the Tool Palette (viewable when designing a FireMonkey form or data module):

Figure 5.5 – Our new component in Delphi's Tool Palette

You can also see the new entry by selecting **Component | Install Packages** from Delphi's menu and scrolling down to the description you entered (if you left the **Description** field blank when creating the component, it'll just list the filename). You can edit the `HideStringDT` project's description, rebuild the project, and then come back and see it updated on this list.

As a final preparatory step to using this component, open up the design-time project options and set **Usage options** to **Designtime only**.

> **Tip**
>
> A good practice to be in is to set each package's usage type to its specific purpose, therefore marking `HideStringRT` as **Runtime only** and `HideStringDT` as **Designtime only**. While I've seen some packages combine design-time and runtime code in the same package, I would not advise it as it adds unnecessary bloat to the distributed application since it carries Tool Palette registration code (and sometimes custom property editor code), which is only relevant to the Delphi environment.

Now, let's make this new component actually do something.

Adding in the code to a component

Linking in code to a design-time package is quite simple if you already have a runtime package prepared. Open the new design-time project source, `HideStringDT.dpr`, and add the name of the runtime package, `HideStringRT`, we created earlier under the `requires` clause:

```
package HideStringDT;
{$R *.res}
```

```
requires
  rtl,
  HideStringRT;
contains
  uHideStringComponent in 'uHideStringComponent.pas';
end.
```

Here, we have the new component package referencing our runtime package (with the default compiler directives removed temporarily for brevity). But we're not done—we need to make a way to actually use the code in that package.

Open the contained unit, uHideStringComponent.pas, which registers the component, and add a new public Execute function in the class:

```
type
  THideString = class(TComponent)
  public
    function Execute(const OrigString: string): string;
  end;
```

Add the uHideStringFunc unit, included in the runtime package, to the uses clause in the implementation section and fill in the body of the new Execute function:

```
uses
  uHideStringFunc;

function THideString.Execute(const OrigString: string): string;
begin
  Result := HideString(OrigString);
end;
```

Our new component now calls the HideString function in the runtime package we wrote earlier. So, let's test it!

In the demo application's main form, remove the uses clause under implementation, then switch to **Design** mode, find our new component in the Tool Palette, and drop it on the form. (Later, when you save the form unit, this new component will notify Delphi of any units it needs and Delphi will automatically add them to the interface's uses clause if they are missing—try it, and watch uHideStringComponent get added.)

Finally, modify the `OnClick` event handler for the button to call the `Execute` function of the component instead of the `HideString` function directly:

```
procedure TfrmHideStrMain.btnHideStringClick(Sender: TObject);
begin
  lblHidden.Text := HideString1.Execute(edtInput.Text);
end;
```

Build everything and run the demo app to see it working.

The component, as it is, now works, but if we stopped here, there would've been no point in going to the trouble of creating a component—we could've simply shipped the runtime package and documented this function. But now that we have a component, we should make it more useful with design-time properties.

Adding design-time properties to a component

Before adding properties to the component, we need to make a few additions to the runtime code. Let's add options to use numbers instead of letters and also to optionally reverse the characters. Modify the declaration of the `HideString` function in the `uHideStringFunc.pas` unit by adding two optional parameters, like this:

```
function HideString(const MyString: string;
                    const UseNumbers: Boolean = False;
                    const ReverseStr: Boolean = False): string;
```

By providing default values to the two new parameters, we've kept the code backward-compatible as they can be left out. This is useful if there are existing applications using the runtime package that call `HideString` with only one parameter.

Now modify the `HideString` implementation to support these two new parameters:

```
function HideString(const MyString: string;
                    const UseNumbers: Boolean = False;
                    const ReverseStr: Boolean = False): string;
begin
  if ReverseStr then
    for var i := Length(MyString) downto 1 do
      Result := Result + RandomChar(UseNumbers) + MyString[i]
  else
    for var i := 1 to Length(MyString) do
```

```
        Result := Result + RandomChar(UseNumbers) + MyString[i];
end;
```

Add the supporting function:

```
function RandomChar(const ReturnDigit: Boolean = False): Char;
begin
  if ReturnDigit then
    Result := Chr(Random(10) + Ord('0'))
  else
    Result := Chr(Random(26) + Ord('A'));
end;
```

With the runtime package updated and the HideString function ready to accept additional parameters, add properties to the component in the uHideStringComponent unit to support them:

```
type
  THideString = class(TComponent)
  private
    FReverse: Boolean;
    FUseDigits: Boolean;
  public
    function Execute(const OrigString: string): string;
  published
    property Reverse: Boolean read FReverse write FReverse;
    property UseDigits: Boolean read FUseDigits write
      FUseDigits;
  end;
```

With the properties and their corresponding fields in place, the Execute function can be modified to call the HideString function's new parameters:

```
function THideString.Execute(const OrigString: string): string;
begin
  Result := HideString(OrigString, FUseDigits, FReverse);
end;
```

Recompile the package, `HideStringDT`, which refreshes the component. Switch to the **Design** view of the form in the demo app and look at the `HideString` component's properties in the Object Inspector. See the new properties?

```
Object Inspector

HideString1  THideString

Properties   Events
   LiveBindings De   LiveBindings Designer
   Name              HideString1
   Reverse           [ ] False
   Tag               0
   UseDigits         [ ] False
```

Figure 5.6 – The updated HideString component's new properties in the Object Inspector

Play around with the demo project, checking the boxes on or off and seeing how the `HideString` function's results show up differently. You could even put a couple of checkboxes on the form to manipulate those properties at runtime. The example on GitHub does just this:

```
https://github.com/PacktPublishing/Fearless-Cross-
Platform-Development-with-Delphi/tree/master/Chapter05/05_
HideStringComponent
```

We now have a fully functional component that uses properties visible in the **Object Inspector** and calls a function in a runtime package that is linked to an application. But so far, it's limited to Windows 32-bit applications. We need to broaden our horizons.

Adding cross-platform support to components

Since the Delphi IDE is a Windows 32-bit application, and since packages can only be loaded by packages or applications in the same version of Delphi and on the same platform, component packages are, of course, limited to the Windows 32-bit platform. But the IDE launches compilers and linkers to support eight different platforms. Certainly, there's a way for components to instruct the IDE to bring in the right code?

Yes, there is. That is done by adding an **attribute** to the component class, `THideString`, in the `uHideStringComponent` unit:

```
type
  [ComponentPlatforms(pidWin32 or pidWin64 or pidOSX64)]
  THideString = class(TComponent)
  // ...
```

The attribute parameter is a bitwise `or` of platform identifiers, found in the `System.Classes` unit. Here, I've added support for both Windows platforms and the Mac. Windows 32-bit was already supported, but once the `ComponentPlatforms` attribute is added, all platforms that need to be supported must be listed. There are preset combinations of these bitwise identifiers for convenience. For example, `pfidWindows` adds support for both 32-bit and 64-bit versions of Windows, and `pidAllPlatforms` adds all supported platforms.

Recompile the package to update the definition in Delphi, then try running it on other platforms.

Back in your sample application, you can test which platforms are supported for a component by activating a target platform, pulling up a form in design mode and looking in the Tool Palette. If the component is disabled, it's not available for that platform; if you can add it to the form, it is.

> **Note**
> This idea of adding metadata to the design-time component is simply the programmer telling the IDE that there is a runtime package available for the platforms listed—there is no mechanism to check whether it will actually work. It is up to the programmer to ensure that it will be available when it's time to build the project.

After modifying the demo app by adding a couple of checkboxes to control the component, I ran it on a Mac:

Figure 5.7 – HideString demo app with checkboxes to set component properties, running on a Mac

Now, you can create components, install them in Delphi's 32-bit Windows IDE, and tell them to link 64-bit code for other platforms!

Summary

You now know how to separate your code for working with teams or developers with other tools by creating dynamic libraries. You also learned how to modularize your code into runtime packages and then bundle them into components for quick placement on forms. Throughout this chapter, tips and examples for supporting other platforms were shown, giving you the flexibility to move forward in a way that best suits your workflow. Organized libraries of code structured in ways to support a variety of scenarios will not only increase the value and life of projects you develop but also add awareness to your Delphi programming expertise.

Now let's dive into LiveBindings, a feature of Delphi that allows you to visually hook up components at design time—basically, a way to make FireMonkey components data-aware.

Questions

1. What are the default file extensions for dynamic libraries on the three supported platforms?

2. Where must the exports clause be in a dynamic library created for the Mac?

3. How do you add the Delphi compiler version to the end of a package without changing its name?

4. What are the differences between design-time and runtime packages?

5. How does a 32-bit design-time package tell Delphi to link a 64-bit runtime package for a cross-platform application?

Further reading

- Murphy's computer laws: `http://murphys-laws.com/murphy/murphy-computer.html`

- Requirements for supported target platforms: `http://docwiki.embarcadero.com/RADStudio/Sydney/en/Installation_Notes#Requirements_for_Supported_Target_Platforms`

- Libraries and packages: `http://docwiki.embarcadero.com/RADStudio/Sydney/en/Libraries_and_Packages_(Delphi)`

- Sharing libraries on Mac: `http://docwiki.embarcadero.com/RADStudio/Rio/en/Shared_Libraries_for_macOS`

- Procedure and function calling conventions: `http://docwiki.embarcadero.com/RADStudio/Sydney/en/Procedures_and_Functions_(Delphi)`

- Loading packages: `http://docwiki.embarcadero.com/RADStudio/Sydney/en/Loading_Packages_in_an_Application`

- Design-time packages: `http://docwiki.embarcadero.com/RADStudio/Sydney/en/Design-time_Packages`

- Making components available at design time and runtime: `http://docwiki.embarcadero.com/RADStudio/Sydney/en/64-bit_Windows_Application_Development#Making_Your_Components_Available_at_Design_Time_and_Run_Time`

6
All About LiveBindings

Windows VCL programmers have often used data-aware controls to link data sources with a user interface, reducing the amount of code they need to write. These data sources can be redirected or disabled, fields can be combined on-the-fly by setting properties or hooking into event handlers, and large datasets can be hooked up to powerful grids, enabling complex views with minimal work. Data-aware controls enable prototypes that can easily be turned into production applications.

FireMonkey's components do not come with data-aware controls, but we are not left to code everything ourselves. Right alongside this new GUI framework, another framework was introduced (in Delphi XE2) – one that used expressions to connect datasets to objects in a powerful and flexible way. It's called **LiveBindings**, and it's not limited to just FireMonkey controls – it works with VCL as well!

The **binding expressions** can be as simple as connecting a label to a field in a database table, or as complex as defining custom expressions for powerful manipulations at runtime. We'll progress from the simple to the advanced by covering the following topics:

- Using the LiveBindings designer to get started quickly
- Creating magic with the LiveBindings Wizard
- Applying custom formatting and parsing to your bound data
- Coding your own LiveBindings methods

By the end of this chapter, you will be able to connect a wide array of components, datasets, and objects to each other and amaze your colleagues with how little code you have to maintain. Perhaps they'll be convinced to start using LiveBindings more as well!

Technical requirements

There are no special requirements for this chapter beyond what is needed to build Delphi 10.4 applications. All the examples will be built with the cross-platform FireMonkey framework, but since the concepts presented here are not platform-specific, nothing more than a Windows computer running Delphi is required. It is recommended that the sample InterBase database that comes with Delphi is installed so that you can work with data in some of the examples provided, all of which can be found on GitHub:

```
https://github.com/PacktPublishing/Fearless-Cross-Platform-
Development-with-Delphi/tree/master/Chapter06
```

Using the LiveBindings designer to get started quickly

Let's start by creating a **Multi-Device Application** and using the **Header/Footer with Navigation** template. Follow these simple steps:

1. Drop a couple of labels and an edit box onto the first tab of the TabControl so that the first label prompts the user to type some text (for example, their name) into the edit box; then, place the second label somewhere below the edit box. Change the names of each of these controls so that they can be easily identified:

Figure 6.1 – Hooking up the first controls – a label with an edit box

2. Now, right-click on the edit box you just placed and select **Bind visually…** from
    the pop-up menu. The **LiveBindings Designer** window will appear, showing all
    the components of the form in little rounded-corner boxes. If you took the time to
    name your controls, they should be easy to pick out. You can drag the boxes around;
    move all but the edit box and the label under it a little way away.

3. To bind these together, simply drag a line from the `Text` property in the edit box to the `Text` property of the label under it. Your LiveBindings Designer should look somewhat like this:

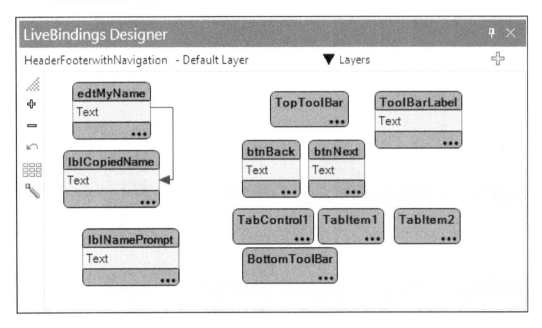

Figure 6.2 – LiveBindings Designer – first link

With your first two components hooked up, run the program, type something into the edit box, and hit *Tab*. As soon as your focus leaves the edit box, the LiveBindings expression will update the label with the new value from the edit box. This is a one-way link and is indicated as such by the arrow in the designer pointing from the edit box to the label.

Let's try another one. Right-click on TabControl1 and select **Next tab** from the pop-up menu. On this second tab, place a TrackBar and another label. Again, give them unique names to help identify them. In the LiveBindings Designer, hook the TrackBar's `Value` property to the new label's `Text` property. When you run the program and switch to the second tab, adjusting the TrackBar now displays its position in the new label.

Note

Notice that the numbers from the FireMonkey TrackBar are floating-point values, not integers like the VCL equivalent. This is true for many FireMonkey controls, giving you more control and enabling additional functionality we will explore later in the book.

This is pretty simple, isn't it? Of course, nothing ever stays simple for long. So, before we go any further, let's do organization inside the LiveBindings Designer.

Using layers to group LiveBindings elements

In the upper-right corner of the LiveBindings Designer is the word **Layers**. Follow these steps to group the elements:

1. Click the little down arrow just to its left to reveal a drop-down menu. Initially, there is only one layer, the **Default** layer.

2. Click the plus (+) sign to add another layer, then double-click on it to rename it.

3. Once you have a new layer defined, *Ctrl + click* various components in the Designer or drag a box around a group of them, then right-click on one of them and select **Add to layer <LayerName>**. They will be moved to that layer.

I created a layer for each of the two tabs we have so far, moved the first two labels and the edit box to the **Tab1** layer, and then moved the TrackBar and the third label to the **Tab2** layer. Now, I can click the little eye icon next to either of the two new layers to highlight only the elements in that layer:

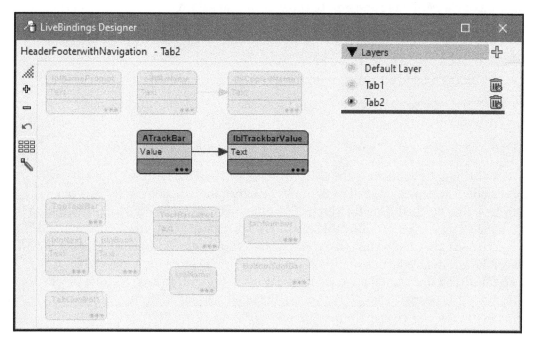

Figure 6.3 – LiveBindings layers with "Tab2" selected

This will be a tremendous help when you have a plethora of components on a form and you're wading through them in the LiveBindings Designer, looking for the ones you need to work with.

> **Tip**
> Start organizing early in the design process and always give descriptive names to your components.

Another way to keep the designer clean is by hiding elements you won't be binding to data. For example, the first label you added to the screen to prompt the user to enter a name is static text. To hide components like this from cluttering the LiveBindings Designer, select one or more component elements, right-click on one of them and select **Hide Element(s)**. To show them again or see the complete list of elements that you can selectively show or hide, right-click in an empty area of the designer and select **Show/ Hide elements…**.

Revealing embedded component properties

Not all properties of the components in the designer are initially shown. To demonstrate when you might need more and how to show them, add a third tab to TabControl and put the following components on it, along with some initial properties:

- A `TRectangle`, `Align = Client`

- A `TSpinBox` and a `TLabel` with the text `Margins` inside `TRectangle`

- A `TComboColorBox` and a `TLabel` with the text `Background` inside `TRectangle`

- A `TArcDial` and a `TLabel` with the text `Rotate Me!` inside `TRectangle`

We want to bind the SpinEdit to the `Margins` properties of the `Rectangle`, but when you click on that element in the LiveBindings Designer, they are not visible. This can easily be solved by clicking on the ellipsis (…) button in the bottom-right of the element's box in the Designer. A list of **Bindable Members** will appear. Scroll down until you find `Margins.Bottom`, `Margins.Left`, `Margins.Right`, and `Margins.Top`; check each of these to enable them and then click **OK**. With these properties now visible, repeatedly drag a line from the `Value` property of the SpinBox to each of those properties so that they'll all be updated together, setting the same margins all the way around. At runtime, you can now change the margins of `TRectangle` by clicking the arrow buttons of the SpinEdit control.

Changing the background color of TRectangle can be done in a similar manner: reveal the Fill.Color property of TRectangle and then hook up the Color field of TComboColorBox to it. Run it and try different background colors by selecting them from the color combo.

Finally, show the RotationAngle property of the label with Rotate Me! as the text and bind the ArcDial's Value property to it so that when it's running, turning the dial will rotate the text.

Your form and LiveBindings Designer windows may now look something like this:

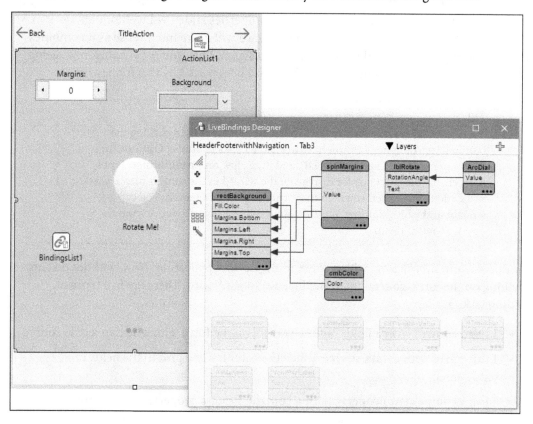

Figure 6.4 – Additional properties revealed and bound

You can find this sample project on GitHub at https://github.com/ PacktPublishing/Fearless-Cross-Platform-Development-with- Delphi/tree/master/Chapter06/01_DesignerIntro.

There are a few buttons on the left-hand side of the LiveBindings Designer that allow you to fit all the controls in the window as best it can, zoom the Designer window in or out, and have the designer rearrange all the elements for you. (These are also available in the pop-up menu when you right-click inside an empty area of the designer.) The last button is **LiveBindings Wizard**.

Let's explore the magic waiting for us there.

Creating magic with the LiveBindings Wizard

Now that you're getting the feel for connecting properties and how LiveBindings work in the designer, the actions of the LiveBindings Wizard will make more sense as it combines several actions you performed manually in the designer. Plus, you're probably now seeing how the data-aware controls in Windows VCL will be replaced by this framework.

> Tip
> There's a quicker way to get to the LiveBindings Wizard than clicking the button in the designer, but you have to enable it. Select **Tools | Options** from the main Delphi menu, expand **IDE**, and then click on **LiveBindings**. From here, enable the checkbox for displaying the wizard in the context menu. Once you've done this, when you're in the form designer of a project, you can go straight to the LiveBindings Wizard with a right-click and select it from the pop-up context menu.

For this section, we will start with a blank multi-device application. Now, without placing anything on the form, go straight to the LiveBindings Wizard. There are five types of binding tasks available:

- **Link a control with a field**: Binds a control on the form with a field in a data source
- **Link a grid with a data source**: Binds the columns in a grid to the fields in a data source
- **Link a component property with a control**: Binds a property of a component (visual or non-visual) with a control on the form
- **Link a component property with a field**: Binds a property of a component (visual or non-visual) with a field in a data source
- **Create a data source**: Creates a new data source

Based on which one of these you select, the tasks and options on the left will change. We'll go through each of these and show what the wizard can do for us.

Choose the first option and click **Next**. Since this is a blank form, select **TEdit** from the **New Control** list and click **Next**. Since there is no data source here, we need to create a new one. We have three options in the **New Source** list:

- **FireDAC**
- **TBindSourceDBX**
- **TPrototypeBindSource**

The first two bind controls to database sources, FireDAC and dbExpress, respectively. The third one allows us to prototype sample data from custom objects at design time, a very useful mechanism we'll explore in a bit. Right now, let's make "data-aware controls" with the LiveBindings Wizard using FireDAC, so select that and click **Next**.

Pulling in fields from a database

If you installed the InterBase database and sample data when you installed Delphi, then you have a good option readily available: select IB for **Driver**, Employee (IB) for **Connection Name**, Table for **Command Type**, and Employee (table) for **Command Text**. You can click **Test Command** to make sure your connection works. If you get a database login prompt during this process, the defaults for the sample InterBase database are "sysdba" and "masterkey" for **User Name** and **Password**, respectively. After clicking **Next** again, select **FIRST_NAME** from the **Field Name** list, click **Next**, and finally check both option boxes; that is, **Add data source navigator** and **Add control label**.

Now, click **Finish**. You will see that a mess of components have been added to the middle of the form. Spread them out to see what the wizard has created for you:

- FDConnectionEMPLOYEE: The FireDAC database connection component
- FDPhysIBDriverLink1: The FireDAC InterBase driver used by the connection
- FDTableEMPLOYEE: The FireDAC table component connected to the FireDAC connection, which allows access to the Employee InterBase table
- FDGUIxWaitCursor1: Allows the FireDAC database components to switch the mouse cursor to a wait cursor during long database operations in a cross-platform way
- BindSourceEMPLOYEE: The LiveBindings equivalent of Windows VCL's TDataSource component

- `EditFIRST_NAME`: A `TEdit` control showing the contents of the `FIRST_NAME` field

- `LabelFIRST_NAME`: A `TLabel` for the `FIRST_NAME` edit field (set as a child of the `TEdit` control so that it follows it around like the `TLabeledEdit` VCL component does)

> **Note**
>
> If your database uses a non-standard port or requires custom connection parameters, then you might need to drop a `TFDConnection` component and associated `TFDPhysDriverLink` descendant first and test the connection. Then, you'll need to start over in the LiveBindings Wizard and complete these steps, in which case you must select an existing driver and connection instead of creating them.

You can go back through the LiveBindings Wizard multiple times to add more controls to your form, selecting the existing source (for example, `BindSourceEMPLOYEE`) for subsequent passes. I added `LAST_NAME`, `HIRE_DATE`, `JOB_CODE`, and `SALARY` and ended up with this on my form:

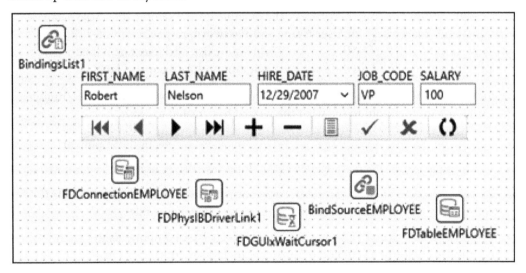

Figure 6.5 – FireMonkey "data-aware" controls created by the LiveBindings Wizard

If you look in the LiveBindings Designer, you'll notice that most of the arrows of the binding links are pointing in both directions now. This is because the data source can update the edit controls and vice versa – if their values change. These are two-way bindings, rather than the one-way bindings we saw in the *Using the LiveBindings designer to get started quickly* section. This makes sense as labels and background colors cannot be directly changed by a user (for example, you can't type on a label), so those are one-way links, whereas the controls, such as the edit boxes on this form, can.

So, what we have so far is a nice start, but it needs something more.

Adding more controls through the wizard

Go back to the LiveBindings Wizard from the form and select the second option, **Link a grid with a data source**. Click **Next**, select `TGrid` from the **New grid** tab, select the `BindSourceEMPLOYEE` member we created previously, and then click **Finish**. (It'll look best if you extend the height of the form and align the grid to the bottom.) Now, we have a grid hooked up to the data source and can see all the fields in the Employee table.

> Tip
> The grid's headers and data are set from the fields in the Employee table, and then linked through `BindSourceEMPLOYEE` the `TBindSourceDB` component. You can right-click the grid and select **Columns Editor** to modify the list of columns shown in the grid. If the dataset is active and you try to add a column, you'll get errors as it tries to display data in the column before it has a fieldname assigned. Click the **Add All Fields** button and then remove the columns you don't want. If you don't, you'll see that as soon as you add one grid column, you have to add all the ones you want to see. This is because the columns in the data source are only automatically displayed if there are no columns defined in the grid.

We linked a component property (`TLabel.Text`) to a control (`TEdit`) in the first LiveBindings Designer example. The third option in the form's LiveBindings Wizard, **Link a component property with a control**, will do that using the wizard. Here, you select an existing component on the form, along with a property of that component, and then either select an existing control or create a new one and bind the two. We'll skip this and move on to the next one.

What I'd like to do is add a big name banner at the top of the form. I created a
TRectangle aligned to the top, set its fill color to black, then placed a TLabel inside
it, and then aligned it with the client and with a large, bold, white font. To bind this
new label in the LiveBindings Wizard, select the fourth option, **Link a component
property with a field**, select the new label and its Text property, click **Next**, select the
BindSourceEMPLOYEE data source, click **Next**, select the FULL_NAME property, and
click **Finish**.

Now, when you run it, you might see something like this (I adjusted some labels and
formatting to make it look nice):

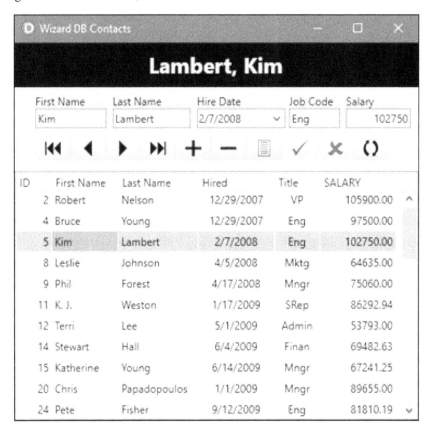

Figure 6.6 – Fully editable InterBase application built with LiveBindings Wizard

We have still not typed a single line of code! Of course, it's a pretty simple app, but we can
add a lot more to this through LiveBindings, as we'll soon discover.

You can find this project on GitHub at https://github.com/PacktPublishing/
Fearless-Cross-Platform-Development-with-Delphi/tree/master/
Chapter06/02_WizardBoundDB.

Now that you know how to hook up databases to the GUI, is there a way to bind a list of your own custom objects? Of course, there is! The drawback is that you can't see what your data looks like at design time like you can by activating a data table. But don't worry – LiveBindings provides a very handy solution for this.

Prototyping custom objects at design time

Starting again with the clean slate of a blank multi-device application, start up the LiveBindings Wizard and click on the first option, **Link a control with a field**, select **TEdit**, and click **Next**. The last item in the list of sources, `TPrototypeBindSource`, provides sample data for us to see at design time when we are working with our own custom objects. This is very cool!

After selecting the PrototypeBindSource and clicking **Next**, you will come to a blank screen where it prompts you to add a **Field Name**. You should add fields that will mimic your custom object as this data source will be replaced at runtime. You can add several types of fields and when you're in the **Add Field** dialog, it will show you what the sample data will look like, allowing you to click the **Next Value** button to see other sample values:

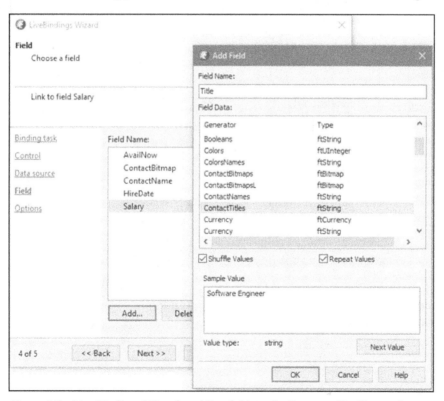

Figure 6.7 – Live Bindings Wizard – adding fields to the PrototypeBindSource dataset

Once you have a few fields selected, click **Next** and, optionally, add a data source navigator and control label, then click **Finish**. It actually only places a control and label for the last field you added to the PrototypeBindSource, but now, you can go through the Wizard again and add controls to the other fields you just created (or use the designer discussed in the first section, whichever is easier for you). You can change the text of the labels and arrange the controls on a form to make it look like a real-life application, and then add a grid and the big name banner at the top of the form just like we did with the InterBase app earlier.

And when you run it, you won't be surprised to see something like this:

Figure 6.8 – Prototype app built with LiveBindings Wizard, with sample data, a grid, and a name banner

You can find my version of this sample project on GitHub at `https://github.com/PacktPublishing/Fearless-Cross-Platform-Development-with-Delphi/tree/master/Chapter06/03_WizardBoundProto`.

Again, you can throw this working prototype together in a few minutes with no database and no code! But what do you do with the PrototypeBindSource when you're ready to go to production with the real data in your objects?

Swapping out prototype data for your own custom data

Inside the PrototypeBindSource is a public property that has been used in the previous examples by the wizard. This is actually a separate component in the Tool Palette that you can place on a form or data module manually. This is known as TDataGeneratorAdapter. At design time, the PrototypeBindSource's internal Adapter property gets its data from its internal DataGeneratorAdapter so that you can see what the data will look like while you're developing it. To use your own data, that Adapter property needs to be redirected so that it points to a different data source.

Your own data may have been populated from a parsed CSV file, filled in with the results of a web service call, and so on. In any case, let's say your data structure looks like this:

```
type
  TEmployee = class
  private
    FContactBitmap: TBitmap;
    FContactName: string;
    FTitle: string;
    FHireDate: TDate;
    FSalary: Integer;
    FAvailNow: Boolean;
  public
    constructor Create(const NewName: string;
        const NewTitle: string; const NewHireDate: TDate;
        const NewSalary: Integer; const NewAvail: Boolean);
    property ContactBitmap: TBitmap read FContactBitmap write
      FContactBitmap;
    property ContactName: string read FContactName write
      FContactName;
    property HireDate: TDate read FHireDate write FHireDate;
    property Salary: Integer read FSalary write FSalary;
    property AvailNow: Boolean read FAvailNow write FAvailNow;
  end;
```

This class would likely be created in a separate unit with a unique identifier field, unit tested with ways to load data, and so forth. For the quick demo we've been building, this class will simply be part of the main form's code. (In order to show images in the LiveBindings example, I loaded some local images into the project as project resources and provided names that matched my hand-created contacts. I then used those with my hardcoded data to create the user-defined object data. You probably won't need to do this in a real-life application, of course; the goal here is to quickly produce a variety of data types to show you how LiveBindings works.)

Here is the constructor for the class, filling the bitmap field with the objects from the project's built-in images:

```
constructor TEmployee.Create(const NewName: string;
         const NewTitle: string;   const NewHireDate: TDate;
         const NewSalary: Integer; const NewAvail: Boolean);
var
  NewBitmap: TBitmap;
  ResStream: TResourceStream;
begin
  ResStream := TResourceStream.Create(HINSTANCE, 'Bitmap_' +
    NewName, RT_RCDATA);
  try
    NewBitmap := TBitmap.Create;
    NewBitmap.LoadFromStream(ResStream);
  finally
    ResStream.Free;
  end;
  FContactName   := NewName;
  FTitle         := NewTitle;
  FContactBitmap := NewBitmap;
  FHireDate      := NewHireDate;
  FSalary        := NewSalary;
  FAvailNow      := NewAvail;
end;
```

At some point in your application, prior to showing the form containing controls with LiveBindings, your data will be filled. For our demo, we'll just populate an object list when the application is created so that our simulation mimics a real data source that has been loaded:

```
constructor TForm1.Create(AOwner: TComponent);
begin
  FEmployeeList := TObjectList<TEmployee>.Create;

  FEmployeeList.Add(TEmployee.Create('Adam', 'Manager',
    EncodeDate(2012, 1, 1), 50000, True));
  FEmployeeList.Add(TEmployee.Create('George', 'Driver',
    EncodeDate(2017, 7, 11), 75000, False));
  FEmployeeList.Add(TEmployee.Create('Brenda', 'Coder',
    EncodeDate(2014, 11, 5), 68000, True));
  FEmployeeList.Add(TEmployee.Create('Jack', 'Janitor',
    EncodeDate(2019, 5, 20), 35000, False));

  inherited;
end;
```

You may have noticed that instead of using the form's OnCreate method, we added an overriding constructor instead. Why? Because we embedded the data within the form and must make sure the data is ready before we bind it. This happens when the form is created. Therefore, the OnCreate method would be too late to start creating data.

Now, let's get back to the Adapter property of TPrototypeBindSource. To bind it to our custom data at runtime, we can use the PrototypeBindSource's event handler, OnCreateAdapter:

```
procedure TForm1.PrototypeBindSource1CreateAdapter(Sender:
TObject;
  var ABindSourceAdapter: TBindSourceAdapter);
begin
  ABindSourceAdapter := TListBindSourceAdapter<TEmployee>.
    Create(self, FEmployeeList, True);
end;
```

This creates a TListBindSourceAdapter of the TEmployee type and hooks up our global FEmployeeList object. Now that our BindSourceAdapter is pointing to our data, running this version of the application shows our hardcoded data instead of the randomly generated data we saw at design time, meaning our own data source is being used successfully:

Figure 6.9 – Custom data objects replacing the prototype data at runtime

> **Note**
> This section only discusses bringing up the LiveBindings Wizard from the form. If you enabled the LiveBindings Wizard context menu option, as described earlier, and then chose to select that after right-clicking on one of the controls, the wizard's options will be limited to those available for just that control.

What we've done thus far in this chapter has been fairly straightforward. We've only written one line of code dealing with LiveBindings and have been able to hook up controls in a variety of ways to both data and other controls, just by simply clicking and dragging or stepping through a series of options.

There's so much more just below the surface! While working through these demos, you've probably noticed a few things you'd like to change by manipulating or formatting the data.

Let's start digging into the properties of the LiveBindings links.

Applying custom formatting and parsing to your bound data

To continue studying the capabilities of LiveBindings, we can simply build on the example project we created in the previous section using the PrototypeBindSource. I copied the project into a new folder, added a few data items, and enhanced the look a little.

Before we start manipulating the properties of the bindings, we need to know how to get to them.

Getting to the BindingsList

You may have noticed that several fields have been added to your form's class with types such as `TLinkPropertyToField`, `TLinkControlToField`, and `TLinkGridToDataSource`. These have been added as you've added LiveBindings links and they're components with properties and events, just like other non-visual components you can place on a form or data module. However, you can't find them in the **Tool Palette** menu as they are managed from the BindingsList component that appeared on the form when you first started creating LiveBindings links.

To see and edit this list of bindings, either double-click on the BindingsList component or right-click on it and select **Binding Components...** from the pop-up menu. A floating window will appear, showing you the components that are used for binding controls, properties, and fields:

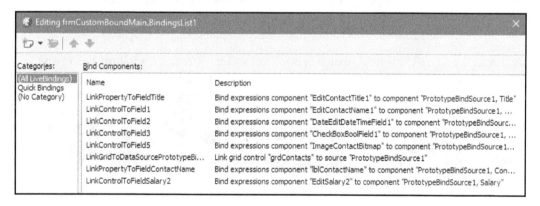

Figure 6.10 – Binding Components List

Clicking on one of the components in the Bindings List reveals its properties in the Object Inspector and allows you to have full control over what and how it binds data. You can manually create new ones and hook them up yourself rather than using the LiveBindings Wizard or Designer. You can categorize them (several were automatically set to `Quick Bindings` by the wizard). You can also change the direction of the binding. There are two other properties for these link component's binding controls and properties that look interesting: `CustomFormat` and `CustomParse`.

Customizing the display

The first thing we'll do is very simple: show the name banner all in uppercase. Click on its `TLinkPropertyToField` component (in the Bindings List or the Designer) and type this into the `CustomFormat` property in the Object Inspector:

```
UpperCase(%s)
```

`UpperCase` is one of the methods that can manipulate data in a LiveBindings expression, and `%s` is a placeholder for the current value. `%s` has an implicit `self` prefixed to it, representing the object that it belongs to. You could write `UpperCase(self.%s)` instead. The owner of the object is referenced with `Owner`. So, if you wanted to access a different field of the data source, you can. Let's see how that can be done.

I want to indicate in the banner whether the contact is available for a new position by adding to the name if they are available but leaving it as-is if not. We can add to the string in the `CustomFormat` property and use another method, `IfThen`, like this:

```
UpperCase(%s) + IfThen(Owner.AvailNow.Value, " (Yes!)", "")
```

This checks the `AvailNow` field (which is Boolean) and if it's `True`, it appends the `(Yes!)` string to the name; otherwise, it appends nothing. Running it reveals the following:

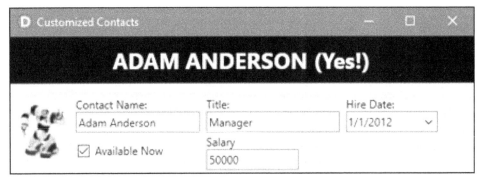

Figure 6.11 – Custom formatting applied to the name banner

To see what other methods are available for use in LiveBindings expressions, click the BindingList component and double-click the `Methods` property in the **Object Inspector**. This is just a viewable list – you can't use them from here. However, you can enable and disable one or many of them – if you disable one that is used in an expression, you'll get a runtime error.

> **Note**
>
> LiveBindings expression method names are case-sensitive. Expressions can use single or double quotes.

Let's make one more change, this time by adding a dollar sign (*$*) to the salary. In the `CustomFormat` property of the `TLinkControlToField` member that binds `TEdit` to the `Salary` field of the data source, type in the following:

```
'$' + %s
```

This is pretty simple, but when you run the application and try to change the amount to `55000`, for example, notice that the value that gets saved in the data source and redisplayed is incorrect:

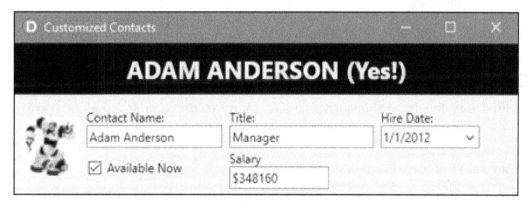

Figure 6.12 – Editing custom formatted data in the Salary field gives us strange results

The reason this happens is because the dollar sign that was added to the `CustomFormat` property of the binding link has become part of the value. When that whole value gets sent back to the data source, which is expecting an integer in this case, it interprets it incorrectly. What we need to do is parse the dollar sign out of the edited amount.

Parsing edited data

The property just below `CustomFormat` in the Object Inspector for the `TLinkControlToField` component is `CustomParse`. This is what we'll use to undo any formatting we made to allow the control to send the raw data back to the data source. It's as simple as stripping off the first character:

```
SubString(%s, 1, 15)
```

The `SubString` method, just like in Delphi, takes a string value, a starting index, and a character count to extract a substring. Unlike Delphi, the string is zero-based, so passing in `1` for the starting index references the second character. There's no built-in method to get the length of the string, so we just pass in a number higher than the maximum length allowed on the field, which is 15 digits.

This version of our demo can be found on GitHub at

```
https://github.com/PacktPublishing/Fearless-Cross-
Platform-Development-with-Delphi/tree/master/Chapter06/04_
FormattedContcts.
```

But what if you'd like to format your data using an approach that's not supported by one of the built-in methods? For this, you can create your own methods and register them, which means it's time to write some code!

Coding your own LiveBindings methods

For this last section on LiveBindings, we'll continue with the prior project, but I've copied it again to a new folder so that if you haven't compiled the package introduced in this section yet, you'll still be able to run the sample in the previous section.

What I'd like to do here is show the number of years' experience each of the contacts in this app has. Since the `Hire Date` field is a date and the date type is a double, all I will need is a way to get the current date; then, I can subtract the two and divide it by 365 to get the number of years difference. Surprisingly, there doesn't seem to be a way to do that. However, with a little knowledge of how these methods are created, it's not too difficult to create one ourselves.

LiveBindings methods are Delphi packages that are installed on the IDE to give us various design-time functionality and syntax checking functionalities; the units also need to be compiled into the application to perform the required functionality at runtime. In the `uses` clause of the sample project we've been building is the `Data.Bind.EngExt` unit, if you view the source to that, you'll see several `Make...` functions of the `IInvokable` type. Then, in the `initialization` section, there should be a call to `RegisterMethods` that registers them. This is what we'll do.

To start, create a new package project, set it to **Designtime only**, and add a new unit. I named my unit `Fearless.Bind` in the same fashion as Delphi's binding expression units. There won't be anything in the `interface` section, but the `uses` clause in the `implementation` section needs several units:

```
uses
  System.SysUtils,
  System.Bindings.Methods,  System.Bindings.EvalProtocol,
  System.Bindings.Consts,  System.TypInfo;
```

Create a function that will make a LiveBindings invokable function; this function will be sent an array of arguments and return a value:

```
function MakeNowMethod: IInvokable;
begin
  Result := MakeInvokable(function(Args: TArray<IValue>):
    IValue
    begin
      Result := TValueWrapper.Create(Now);
    end);
end;
```

For our purposes, we won't be using any arguments that have been passed in, so this function can skip checking and validating them and simply return the value we're after: the current date and time. This can be done by using the `Now` system function. The `TValueWrapper` class wraps whatever value we send it into an `IValue` in order to return it to the LiveBindings expression.

The only thing we need to do now is register it in the `initialization` section of the unit and unregister it in the `finalization` section:

```
const
  sNowMethod = 'Now';
```

```
    sNowDescription = 'Returns the current date/time';
initialization
  TBindingMethodsFactory.RegisterMethod(
    TMethodDescription.Create(MakeNowMethod, sNowMethod,
      sNowMethod, EmptyStr, True, sNowDescription, nil));
finalization
  TBindingMethodsFactory.UnRegisterMethod(sNowMethod);
end.
```

Now, compile and install it (right-click on the project and select **Install**) and that's it! Click on the `BindingList` component in the form of the demo project and double-click on the `Methods` property in the Object Inspector, then scroll down to view the new Now function.

> **Note**
>
> Even though this package is installed and shows up immediately in the list of bindings methods, trying to use it in a binding expression sometimes results in it not being found. Closing and reloading the project, or even restarting Delphi, fixes this.

Now, we can finally finish our quest and calculate the number of years' experience based on the Hire Date. Place a label on the form and bind its `Text` property to the `Hire Date` field – it should show exactly what the `TEdit` we already have on the form shows. Now, click the new `LinkPropertyToField` binding link we just created and type the following into the `CustomFormat` property:

```
"Experience: " + ToStr(Round((Now() - %s) / 365)) + " years"
```

Even though our Now function does not need any arguments, the syntax requires parenthesis; otherwise, you'll get an `Interface not supported` error. `%s` in this context is a `TDate` (double), which means we can do arithmetic on it. Subtracting that value (the `Hire Date` field) from the current date gives us the number of days difference, and dividing that by the number of days in a year gives us the number of years; we'll round that to the nearest year. The label needs a string in the `Text` property so that we can convert the result using `ToStr()`, prepend it with `Experience:`, and append it with `years` to complete the expression.

Before you run it, remember that this same code needs to be registered with your application (in addition to Delphi) because you want the LiveBindings expression to work at runtime as well. This is done by simply including the unit name that created and registered the new LiveBindings method (`Fearless.Bind`) in the `uses` clause of your application. I placed the new `Experience` label directly under the Hire Date edit. If that's where you placed yours, you should see the following:

Figure 6.13 – Custom LiveBindings Method showing experience in years

As an alternative to implementing the `Now` method, we could have done the calculations in the method. Here's the code for a method called `YearsSince` that takes the given date parameter and returns the number of years that have elapsed since that date:

```
function MakeYearsSinceMethod: IInvokable;
begin
  Result := MakeInvokable(function(Args: TArray<IValue>):
    IValue
  var
    InputValue: IValue;
    InputDate: TDate;
    YearsSince: Double;
  begin
```

```
    if Length(Args) <> 1 then
      raise EEvaluatorError.Create(Format
        (sUnexpectedArgCount, [1, Length(Args)]));
    InputValue := Args[0];

    if not (InputValue.GetType.Kind in [tkFloat]) then
      raise EEvaluatorError.Create('Argument to YearsSince
        must be TDate');

    InputDate := InputValue.GetValue.AsExtended;
    YearsSince := TTimeSpan.Subtract(Now, InputDate).
      TotalDays / 365.25;
    Result := TValueWrapper.Create(YearsSince);
  end);
end;
```

Note that this time, we need one parameter, so we validate there's only one and that it's a date type (double, also known as a float). With this binding method registered, the `experience-in-years` expression is simplified to the following:

```
"Experience: " + ToStr(Round(YearsSince(%s))) + " years"
```

Now, you can fill in the register and unregister calls, or simply download this project from GitHub at

```
https://github.com/PacktPublishing/Fearless-Cross-
Platform-Development-with-Delphi/tree/master/Chapter06/05_
ContactsWithCustomMethods.
```

There are many more examples that could be provided, but this is a good start. Look through Delphi's code and online tutorials to learn more about validating arguments and other nuances of creating LiveBindings methods. There are also examples to be found with a quick search on GitHub.

Summary

In this chapter, we've showed you how LiveBindings not only replaces the need for data-aware controls you've used in VCL applications but has greater potential and is easier to extend. There are multiple ways to use it, and it's really quite intuitive once you understand the basics.

First, we learned how to hook up controls to other controls and data sources, how you can use both the Designer and the Wizard to your advantage to speed up prototyping, and then how to transform your demos into production-ready applications with very little effort. Now, you know how to manipulate how date is displayed through LiveBindings custom formatting, and even how to extend the LiveBindings capabilities with custom methods.

This is not only a step forward in productivity due to it reducing your time to market with the ease of use LiveBindings provides – it also lowers the barriers to improving your interface in subtle ways that will make a big difference to your users.

In the next chapter, we will add to these user interface improvements in not-so-subtle ways by taking control of the styles that FireMonkey uses to govern the entire look and feel of your application.

Questions

1. Are there platform or GUI framework restrictions for using LiveBindings?

2. What are the two ways we can organize the elements in the LiveBindings Designer?

3. Which class type is the field of the `TPrototypeBindSource` component that provides random data visible at design time, and which has a component in the Tool Palette that can be used independently?

4. How do you access the properties of a `TLinkPropertyToField` or `TLinkControlToField` component?

5. Do you always have to put something in the `CustomParse` property of a binding link component if `CustomFormat` has something in it?

6. What needs to be included in your application's `uses` clause to provide a custom method's functionality that has been used in a binding expression?

Further reading

- LiveBindings in RAD Studio: `http://docwiki.embarcadero.com/RADStudio/Sydney/en/LiveBindings_in_RAD_Studio`

- Introduction to LiveBindings Wizard: `https://youtu.be/tpipscFNTGA`

- LiveBindings and Rapid Prototyping: `https://youtu.be/FQminjTLS0E`

- Quickly Bind Your Data Source to Different Types of Data with LiveBindings: `https://blogs.embarcadero.com/quickly-bind-your-data-source-to-different-types-of-data-with-livebindings-delphi-adapterbindsource-sample-app/`

- Developer Skill Sprint: LiveBindings From Forms to Code: `https://youtu.be/_sGeY_VWBMI`

- Using Custom Format and Parse Expressions in LiveBindings: `http://docwiki.embarcadero.com/RADStudio/Sydney/en/Using_Custom_Format_and_Parse_Expressions_in_LiveBindings`

7
FireMonkey Styles

Styles are fundamental building blocks of FireMonkey controls. All FireMonkey controls can be colored, bordered, and animated, and most can glow when focused. There are default styles for each platform to provide a native look and feel for various devices, but you can customize them to suit your needs or add a creative flair to set your application apart. You can share styles, package them with your applications, and install them in the Delphi IDE. You can also build cross-platform applications without editing a single FireMonkey style – but where's the fun in that?

This chapter will discuss the role FireMonkey styles play in your applications, how to control and use them, how to modify them, and ways to share them. Due to this, we will cover the following topics:

- Understanding and using FireMonkey styles
- Customizing FireMonkey styles with the Style Designer
- Styling your applications with ease

After reading this chapter, you'll have a firm grasp of how to use and manipulate FireMonkey styles to suit your team's needs.

Technical requirements

This chapter will show you how applications are rendered on different platforms and how styles make that happen; therefore, all the examples will be able to be run on Windows, Mac, Android, and iOS devices, and screenshots will be provided throughout to showcase these differences. At a minimum, a Windows computer running Delphi 10.4 will be required– it is up to you to decide which other platforms to use for personal education and testing.

Understanding and using FireMonkey styles

Every FireMonkey control has a default style for each platform. These come with Delphi and are hardcoded to give you a consistent, platform-specific base to work from that conforms to user interface guidelines. Back in *Chapter 4*, *Multiple Platforms, One Code Base*, we showed you how to look at different views of a form to preview how it would look on different devices. Instead of switching a view, you can keep the **Master** view active and simply change its **Style**.

To illustrate this further, open the sample project; that is, `MobileControls` (this can be found in your Delphi Samples folder under `Object Pascal\Multi-Device Samples\User Interface\Controls`). You can switch **Style** to Android and see how it will look without changing the view or running the application (or even having an Android device hooked up):

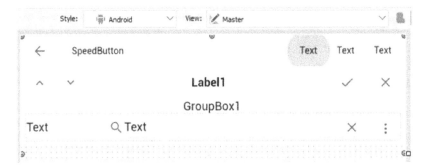

Figure 7.1 – The Toolbars tab of the MobileControls sample application in Android style

As we mentioned in that same chapter, running the application, unmodified, on various platforms will result in a different look on each. The following is a collage of screenshots from the MobileControls app running on four different platforms:

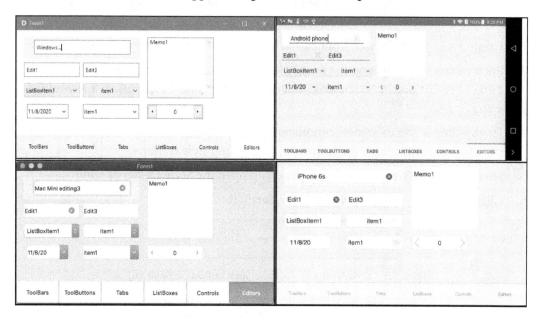

Figure 7.2 – MobileControls application with the "Editors" tab running on various platforms

This shows FireMonkey's default styles at work. The FireMonkey framework chooses the appropriate view for the device it's running on, then uses the default style for that view. Again, at design time, when you change the view, it selects the style for you (as it does at runtime); when you're on the Master view, you can choose which platform's style is active.

But what if you'd like to override the style from its default so that your users can uniquely color or shade the controls in your application? You've probably noticed several styles available in the GetIt Package Manager. Let's learn how to use them in our cross-platform apps.

Loading style sets

If you've used styles in a VCL application, you might have gone looking in a FireMonkey project's options for a way to add styles directly to the project definition. There is a way to do that, which we'll discuss later in this chapter. However, in this section, we'll use a component that's been loaded at design time.

In the following steps, we will use a pre-packaged FireMonkey style that's been loaded in a new app:

1. If you don't already have it, get the *CopperDark FMX Premium Style* from **GetIt Package Manager**.

2. Create a new **Multi-Device Application** and choose the **Blank Application** template.

3. Put a variety of controls on the form: TRadioButton, TCheckBox, TButton, TEdit, TListBox, TComboBox, TProgressBar, and TTrackBar. I also added a TGroupBox

4. Finally, drop a TStyleBook component onto the form. This is not a control but, as its name suggests, a book (or container) for holding style resources.

5. Double-click the StyleBook component to open the **Style Designer** window.

6. Click the **Open** button on the Style Designer's Toolbar:

Figure 7.3 – Style Designer Toolbar, Open button highlighted

7. Navigate to the platform-specific folder where your GetIt Package Manager downloads styles and select the .style file for the selected platform's *CopperDark* style. For example, the default path for Windows styles would be C:\Users\ Public\Documents\Embarcadero\Studio\21.0\Styles\Win, while the file would be CopperDark.Win.style.

8. Repeat *steps 6* and *7* with the different platform style files until all four platform's versions of the *CopperDark* style have been loaded (besides the Windows style, the other style files are CopperDark.MacOS.style, iOSCopperDark.style, and AndroidCopperDark.style).

9. Close the **Style Designer** window and click **Yes** when you're prompted to apply these changes.

10. Set the StyleBook property of the form to the StyleBook component.

As soon as the form's StyleBook property is assigned to a StyleBook component, all the controls that have been defined in that style for the current platform will show the styled controls. You can switch through the different platform styles and see how the controls look for each one. Running it looks like this:

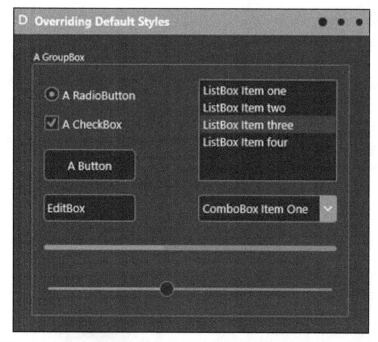

Figure 7.4 – FireMonkey app with the CopperDark style running on the Windows platform

At runtime, the form's StyleBook property tells the FireMonkey framework to use the style that has been loaded in the StyleBook component. This is how you establish a custom style for your application without any code!

This project can be found on Github at `https://github.com/PacktPublishing/Fearless-Cross-Platform-Development-with-Delphi/tree/master/Chapter07/07_01_DefaultStyleControls`.

Now, let's learn how to use multiple sets of styles.

Selecting between multiple StyleBooks

Let's say you want to give your users the option to select a style of their preference. To do that, we need to add another StyleBook to the project we've been working on. Let's name the first one we added `StyleBookDark` and the second one `StyleBookGreen`. Follow the steps provided in the previous section to add another style from the GetIt Package Manager; for example, *Emerald Crystal*. Change the form's `StyleBook` property to the new style book to see how the controls will look – this is how you can view your application with different styles at design time. Providing the option to switch styles at runtime for your users takes just a few lines of code.

Change the items of the ComboBox that's on the form to the following item strings:
`Default style`, `Dark style`, and `Green style`. In the `OnChange` event, add the
following code:

```
procedure TfrmStylesIntro.cmbStyleChange(Sender: TObject);
begin
   case cmbStyle.ItemIndex of
      1: StyleBook := StyleBookDark;
      2: StyleBook := StyleBookGreen;
   else
      StyleBook := nil;   // ItemIndex = 0
   end;
end;
```

By simply reassigning the `StyleBook` property of the form, the style of the controls
changes at runtime, just like it does at design time:

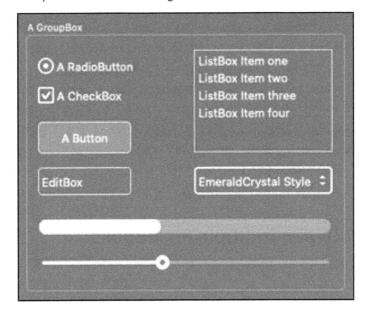

Figure 7.5 – Demo app using the Emerald Green style on a Mac computer

By setting the `StyleBook` property to `nil`, we have set the style back to its built-in default:

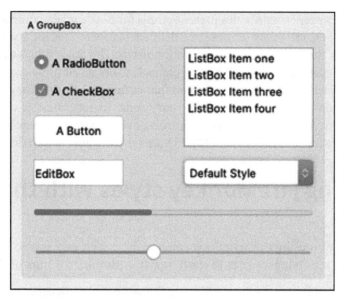

Figure 7.6 – Demo app using the default style on a Mac computer

There are a couple more aspects to working with styles in FireMonkey that you should be aware of before we talk about customizing a style. Most of the controls we've placed on the form in this demo app have one defined style per platform. However, three of them have more than one.

Accessing substyle definitions

If you select the TListBox component and then look at the StyleLookup property, you'll see it's a drop-down list containing two items: listboxstyle (the default) and transparentlistbox. If you switch the value to the latter, the ListBox becomes transparent.

TEdit also has more than one item for its StyleLookup property. You can experiment with its eight different substyles and see how they might be useful in various situations.

The most interesting one on this form is TButton. It has quite a few options for its StyleLookup property. Some of them look the same on different platforms, while others look quite different. For example, set the value to replytoolbuttonbordered and then look at the different platform's styles:

Figure 7.7 – The replytoolbuttonbordered StyleLookup of a TButton is rendered differently on Windows, MacOS, iOS, and Android

The `StyleLookup` property provides preset options for how a given control looks. The great thing is that there are many of these pre-defined, platform-specific styles and lookup options for many of the most common application needs. The downside is that it could be quite tedious to customize the entire set of controls to create all these component style definitions. Fortunately, you seldom need to do that as there are many that could serve as excellent starting points and then modify for your needs. So, regardless of whether you need to modify the style of just one or two controls or create an entire new look for all the controls in your application, the built-in Style Designer will allow you to do that.

Customizing FireMonkey styles with the Style Designer

With a good handle on using the StyleBook component and an easy way to switch sets of styles, let's dig into the Style Designer to learn how to customize the look of a control.

As we continue through this chapter, we'll reuse the same project because it already contains several controls that make it nice to see the effects of the style changes we'll be making. To "reset" the styles to their built-in defaults, you can simply delete the StyleBook components from the form. I will do that as well, but also copy the project to preserve the state of the previous demo. This section's source can be found on GitHub at `https://github.com/PacktPublishing/Fearless-Cross-Platform-Development-with-Delphi/tree/master/Chapter07/07_02_CustomStyleControls`.

If you right-click on a control, you will see two menu options for editing styles; here's what they do:

- `Edit Custom Style...`: This creates and/or edits a style definition for the selected control – and only that specifically named control – for the selected platform style.
- `Edit Default Style...`: This creates and/or edits a style definition for all the controls of the selected type for the selected platform style.

Let's illustrate the difference in our demo app by creating two more copies of the `TEdit` component. First, we'll make changes to the default font and color used by all `TEdit` controls and then customize the style of a specific `TEdit`.

Creating a default style for a control type

Right-click on the first `Tedit` and select **Edit Default Style...** from the pop-up menu to get back into the Style Designer. Here's where the **Structure** window becomes very useful. FireMonkey controls often have several style parts, and `TEdit` is a prime example. Expand `Editstyle` and select the `font` subcomponent, which reveals its properties in the **Object Inspector** window. Now, you can change the font's `Family`, `Size`, and `Style` to whatever you'd like. You could also select the `foreground` style component and change the color of the text. As you make these changes, a preview will appear on the right-hand side of the designer. Here's what I did with mine:

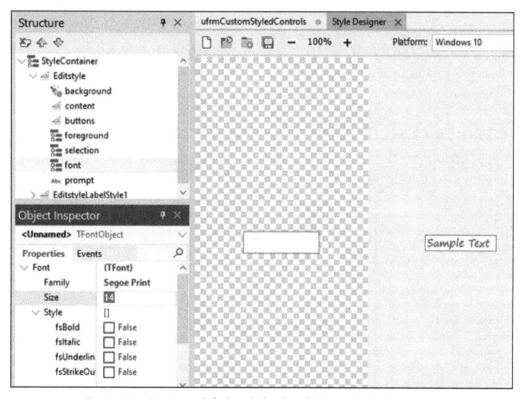

Figure 7.8 – Creating a default style for the Editbox in the Style Designer

Click the *Apply style* button on the far right of the Style Designer's toolbar or simply close the Style Designer and apply the changes. Then, back in the form, you will notice that all three `TEdit` components have the new style of text:

Figure 7.9 – TEdits at design time with a new default style

All the edit boxes now have the same style. Additionally, all future edit boxes we drop on this form will get this same style because it's now the default style for the `TEdit` controls on this form.

Creating a custom style for a specific control

Now, right-click on the third `TEdit` control and select **Edit Custom Style…**. You'll see the following in the **Structure** window:

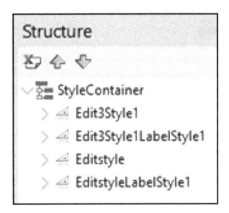

Figure 7.10 – The Structure window shows a custom TEdit style available for Edit3 in the Style Designer

I named my `TEdit` components `Edit1`, `Edit2`, and `Edit3`. The custom style I'm now editing, `Edit3Style1`, is only for the `Edit3` component and will not affect the other two Edit controls. After going through the same process we did previously but with a different font and color to distinguish it from the others, this is how my form looks:

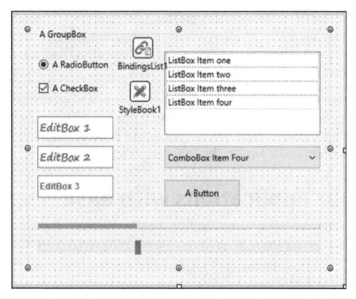

Figure 7.11 – Design time view of customized styles applied to our TEdit controls

If you deleted the previous StyleBook components from the form at the beginning of this section, you'll notice that one has reappeared. It contains both the default style for all the TEdit components on this form (though this may be added later) and the customized style for Edit3. As we mentioned earlier, the hardcoded default styles for all the controls on each of the platforms are part of the FireMonkey framework. As soon as you customize any of them, a copy is made and is stored in the properties of a StyleBook. If no StyleBook component is on the form, the only styles available are the built-in, default styles.

The styles we've just edited are only for the selected platform style. If you select the StyleBook and double-click its Styles property, there will only be one platform listed after Default. To edit the styles for other platforms, switch the style selector to the desired platform and then go through the aforementioned process to modify additional styles for those platforms. For example, if I create at least one style (either default or custom) for the macOS platform in addition to Windows, my StyleBook's Styles list would show this:

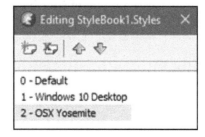

Figure 7.12 – Showing collections of custom styles in the StyleBook

The style collection lists the different platforms that have controls for custom styles. In this example, the Android and iOS platforms have not been customized and are not listed so they would continue using the built-in platform styles, regardless of what style is selected for the other platforms.

So far, we've just been dealing with StyleBook components placed on a single form. This has been great for simple demos and learning about styles, but it's not a good practice in real-world applications with multiple forms. We'll need a good strategy going forward to manage styles on a larger scale. The next section lists some ways to do that.

Styling your applications with ease

If your application contains many forms, putting these StyleBook components on each form will quickly discourage many programmers from using styles. Fortunately, there are three other ways to centrally manage styles in a Delphi project that will make this task easier. Let's take a look at them.

Quickly setting a single, application-wide style

The first and simplest way to style your entire application is to drop a `TStyleBook` component onto any form or data module in your application, load it up with the style of your choice, as we've done earlier in this chapter, and then set its `UseStyleManager` property to `True`. As soon as that component is created, it calls a global `TStyleManager` class, which styles every active form. If your form is not auto-created when the application starts up, no style will be applied until that form is created – at which point every form is immediately styled.

> **Note**
> If you have more than one `StyleBook` component in your application that has the `UseStyleManager` option checked, only the last one that you created will affect the style of the application.

This approach is simple and takes the least amount of code but has some limitations, as we'll soon see.

Customizing styles per form

If you'd like to give your users the option of selecting from a list of styles you've provided or assigning different styles to different forms, consider putting several StyleBook components on a shared data module. While only a form has a `StyleBook` property, the `TStyleBook` component itself is non-visual and can reside on either.

An example of how to do this can be found on GitHub at `https://github.com/ PacktPublishing/Fearless-Cross-Platform-Development-with- Delphi/tree/master/Chapter07/07_03_MultiFormControls`.

Follow these steps to learn how you can build this yourself:

1. Download four styles from **GetIt Package Manager**.

2. Create a new **Multi-Device Application** and use the **Blank Application** template.

3. Put a bunch of controls on the form to see what they'll look like when a style is applied.

4. Add three more forms just like this one, naming them appropriately and copying the controls to each of them.

5. Add the names of the units of the newly added forms to the `uses` clause in the main form's `implementation` section.

6. Add a `TActionList` to the main form and create an action for each of the other forms.

7. Create an `OnExecute` event for each of the `TAction lists` in the `ActionList` that calls the `Show` method of the corresponding form.

8. Create three buttons and assign a different `TAction` to each.

9. Add a data module to the project and add its unit name to the `uses` clause in the `implementation` section of every form in the project.

10. In the data module, add four `TStyleBook` components.

11. For each of the StyleBooks, double-click and add one of the downloaded styles for each platform you'd like to run this application (for example, `AndroidJet. style`, `iOSJet.style`, `Jet.style`, and `Jet.Win.style` to load the *Jet* style for all four platforms into one of the StyleBooks).

12. Assign a different StyleBook from the data module to each of the form's `StyleBook` properties.

Now, run the application. I used the *Jet*, *Material Patterns Blue*, *Coral Crystal*, and *Puerto Rico* styles. When I ran it on my Mac and opened all the forms, I saw the following:

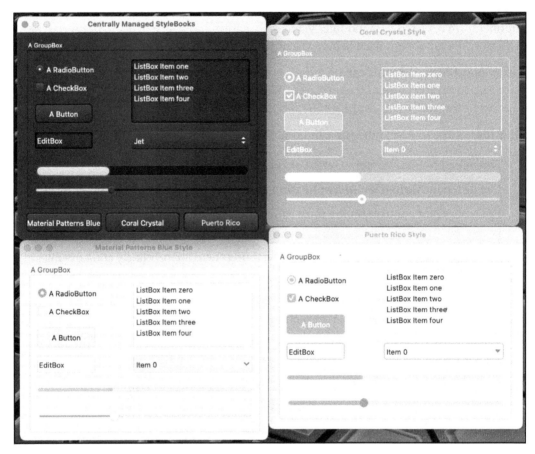

Figure 7.13 – An example of multiple forms loading custom styles from a shared data module

By having multiple StyleBooks and assigning the `StyleBook` property of each form, you've set the stage for independent and flexible styling in your application.

There's another way of styling your applications that is not only the most powerful and flexible option, but also takes the most amount of code.

Managing style resources with code

The third way to accomplish centralized style management in your Delphi projects is with the `TStyleManager` class in the `FMX.Styles` unit.

> **Note**
> This is what was used behind the scenes when you checked the
> `UseStyleManager` option on a StyleBook component in the first part of
> this section. But as we'll see, you have more control when managing the style
> using the methods of the `TStyleManager` class in your code.

With this approach, styles are loaded as resources into the application instead of properties
on a component. Style files can be loaded at runtime for greater flexibility and so that we
can update or replace them after deployment, but our next demo will add them to the
project using the resource manager. Follow these steps to set up a project in this manner:

1. Create a new **Multi-Device Application** and use the **Blank Application** template.

2. Put a `ListBox` and three buttons on the form and save the project to a new folder.

3. Download four new styles from **GetIt Package Manager** or select ones you already
    have; then, copy the `.style` files for as many platforms you want to support to
    a subfolder of the project directory. For example, I selected the *Calypso*, *Emerald
    Crystal*, *Vapor*, and *Wedgewood Light* styles and want to support all four platforms,
    so I've got 16 files in my version of the project:

Figure 7.14 – List of style files to support four platforms for four styles

4. Select **Project | Resources and Images...** from the main Delphi menu and load the style files, assigning descriptive resource names and setting the resource type to RCDATA for each:

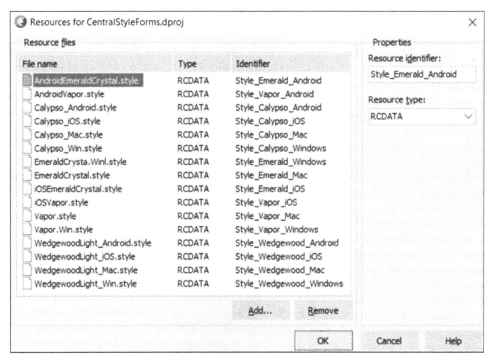

Figure 7.15 – Four sets style files for each of the platforms loaded into the project as embedded resources

5. Add a data module to the project and add its unit name to the uses clause in the implementation section of the main form.

6. In the data module, create a class with four static procedures; name these procedures based on the styles you're using:

```
type
  TStyleMgr = class
    class procedure LoadCalypsoStyle;
    class procedure LoadVaporStyle;
    class procedure LoadWedgewoodStyle;
    class procedure LoadEmeraldStyle;
  end;
```

7. Add the FMX.Styes unit to the uses clause in the implementation section of the data module.

8. In the body of each procedure, call `TStyleManager`.
    `TrySetStyleFromResource` for each resource, putting each call into
    platform-specific compiler directives so that the right style is loaded for the right
    platform. For example, here's my implementation of `LoadCalypsoStyle`:

```
class procedure TStyleMgr.LoadCalypsoStyle;
begin
  {$IFDEF MSWINDOWS}
  TStyleManager.TrySetStyleFromResource('Style_Calypso_
    Windows');
  {$ENDIF}
  {$IFDEF MACOS}
  TStyleManager.TrySetStyleFromResource('Style_Calypso_
    Mac');
  {$ENDIF}
  {$IFDEF IOS}
  TStyleManager.TrySetStyleFromResource('Style_Calypso_
    iOS');
  {$ENDIF}
  {$IFDEF ANDROID}
  TStyleManager.TrySetStyleFromResource('Style_Calypso_
    Android');
  {$ENDIF}
end;
```

9. Add the names of the four styles you added to the items of our `ListBox`; at this
    point, your main form might look something like this:

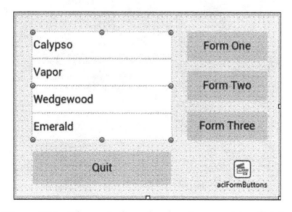

Figure 7.16 – The main form for showing a variety of styles

10. Create three additional forms for demonstration purposes and put a bunch of controls on them to show what they'll look like when a style is applied. Then, set up the buttons to launch these forms in a similar manner to what we did in the previous demo project.

11. In the OnChange event of our ListBox, call the corresponding static procedure from the data module to load the named style:

```
procedure TfrmMultiFormsMain.lbStylesChange(Sender:
TObject);
begin
  case lbStyles.ItemIndex of
    0: TStyleMgr.LoadCalypsoStyle;
    1: TStyleMgr.LoadVaporStyle;
    2: TStyleMgr.LoadWedgewoodStyle;
    3: TStyleMgr.LoadEmeraldStyle;
  end;
end;
```

This should be enough to run the program, show some forms, and then look at different styles by clicking on them in the ListBox component. Here's what it looks like on Windows with the Calypso style selected:

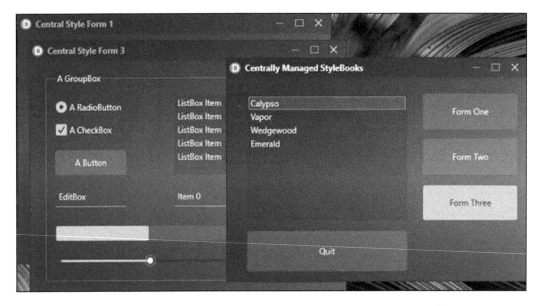

Figure 7.17 – Application using TStyleManager to set the style for all the forms

The TStyleManager class affects all the forms in the application – just like the first method we introduced at beginning of this section. You might worry that this removes the flexibility of independently setting the style of one form, but setting the StyleBook property on a form to a TStyleBook component with a loaded style overrides the global effect of the TStyleManager class for that form.

> **Note**
>
> This concept of overriding a global style with form-level StyleBook property settings also applies to the first approach of using the UseStyleManager property on a StyleBook, as described at the beginning of this section. However, mixing styling methods can lead to confusion and increase debugging time in large projects or teams.

You can test this out by adding a StyleBook to one of the forms, loading a style, setting the form's StyleBook property for the form, and then running the application and switching styles from the main form, as you did previously. You'll see that the other forms still follow the global style established by TStyleManager but that the form with the assigned StyleBook remains unchanged:

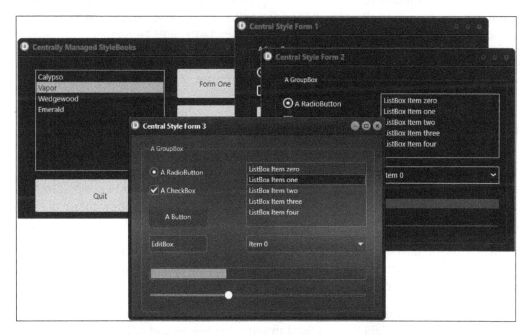

Figure 7.18 – Application using the TStyleManager set to load the Vapor style globally but with Form3 loading the Ubuntu Clear Fantasy style via a StyleBook

The source code for this project can be found on GitHub at `https://github.com/PacktPublishing/Fearless-Cross-Platform-Development-with-Delphi/tree/master/Chapter07/07_04_CentralStyle`.

> **Note**
>
> If a style loaded by a StyleBook is not supported on a particular platform, the global style set by the `TStyleManager` class will take effect. This can be seen in the example project we used here, if you download the project from GitHub and run it on iOS or Android. The preceding screenshot shows the application running on Windows with the Vapor style selected globally but *Form Three* using the *Ubuntu Clear Fantasy* style. Currently, that style is only available for Windows and MacOS, so when you run it on iOS or Android, the global style set by `TStyleManager`, which is Vapor in this example, would take effect.

Remember that style files can be loaded at runtime. We embedded them into the `TStyleBook` component and loaded them as resources to the project for the `TStyleManager` class to simplify the examples we used in this section, but there's nothing stopping you from bundling the `.style` files separately from your application to provide selection options, or even to allow for style updates after deployment. To load a style file at runtime using `TStyleManager`, simply use its `SetStyleFromFile` method.

Loading them at design time does make the code larger and adds an element of inflexibility, but the options are there for you and your team, if you decide to use them.

Summary

This chapter has shown you that using different styles in your cross-platform applications is not that much more involved than using them in VCL applications. Plus, there are a couple of different ways to load and use them in your application. If your needs require unique styling to be applied, then you now have a great start in terms of how to use the Style Designer to make the necessary customizations.

Utilizing styling options in your application can greatly enhance an otherwise standard interface. But there are other ways to help applications stand out from your competition. Some of them may be useful, while others might just be for fun.

In the next chapter, we will explore components that can move and rotate in a three-dimensional space, apply light sources for diffused color effects, and utilize other graphically intensive operations you might have thought were out of your reach.

Questions

1. What are two ways we can change the visual style of a FireMonkey form?

2. Is it possible to change the built-in platform styles that come with FireMonkey?

3. What is the difference between a default style and a custom style?

4. How do you select the various style components in the Style Designer?

5. What resource type should be assigned to embedded styles?

6. How do you override the global style established by the `TStyleManager` class?

Further reading

- Customizing FireMonkey Applications with Styles: `http://docwiki.embarcadero.com/RADStudio/Sydney/en/Customizing_FireMonkey_Applications_with_Styles`

- Mobile Tutorial: Using a Button Component with Different Styles: `http://docwiki.embarcadero.com/RADStudio/Sydney/en/Mobile_Tutorial:_Using_a_Button_Component_with_Different_Styles_(iOS_and_Android)`

- FireMonkey Style Designer: `http://docwiki.embarcadero.com/RADStudio/Sydney/en/FireMonkey_Style_Designer`

- Working with Custom and Native FireMonkey Styles: `http://docwiki.embarcadero.com/RADStudio/Sydney/en/Working_with_Native_and_Custom_FireMonkey_Styles`

8
Exploring the World of 3D

Adding 3D capabilities to your application may not be at the top of the priority list for many business applications, but the ease of use with which you can incorporate these graphical capabilities may open your mind to creative new ways to enhance your product's look and feel. Whether you're a game developer building a new world, a product designer wanting to incorporate 3D controls and effects, or you're just programming for fun, this chapter will get you on the road to crafting your next breed of visual interfaces with your favorite cross-platform development tool.

The beauty of using Delphi and the FireMonkey framework is that you don't have to learn a new language or **application programming interface (API)**—simply use a common set of controls that are similar to others you have been using in this book. The framework and controls hide the complexities of calling APIs to utilize the underlying **graphics processing unit (GPU)** engines.

We'll start simple and go over how to mix **2D** and 3D controls in the same form and place basic shapes in a viewport. Then, we go over adding more complex shapes and learn how to add colored material to cover them and a light source to enhance the 3D look. Next, we'll animate various objects and colors and show how to import a 3D model. Switching camera views and testing everything out in our demo app will complete the introduction to 3D. Taking these new concepts, we'll then build an actual game—and reveal a few more tricks of the trade.

This will all be explained in the following sections:

- Getting started with 3D in Delphi

- Adding basic and extruded shapes

- Adding color, lighting, and movement

- Importing 3D models

- Changing the camera

- Let's write a game!

There's a lot in this chapter but don't skip it—we'll have a lot of fun!

Technical requirements

This chapter will show how to build 3D applications that run on different platforms and how the FireMonkey framework gets you there. The examples will be able to be run on Windows, Mac, Android, and iOS devices. As always, a Windows computer running Delphi 10.4 will be the minimum requirement—it is up to you to decide which other platforms to use for personal education and testing.

In addition to standard Delphi requirements, the 3D components discussed in this chapter utilize the following advanced capabilities of the GPU engines expected to be available:

- DirectX on Windows

- OpenGL or Metal on Mac OS X

- **OpenGL for Embedded Systems** (**OpenGL ES**) on iOS and Android

> **Note**
>
> While OpenGL is still the default API used by FireMonkey on the Mac at the time of writing, support for the Metal engine was added in Delphi 10.4 Sydney and may someday be the default. Follow the *Boost Mac performance with Metal and Delphi 10.4* link in the *Further reading* section at the end of this chapter to learn why this is important and how simple it now is to activate it in your Mac applications. The source code for the examples and sample applications used in this chapter is available on GitHub at `https://github.com/PacktPublishing/Fearless-Cross-Platform-Development-with-Delphi/tree/master/Chapter08`.

Getting started with 3D in Delphi

Building a FireMonkey application with 3D capabilities requires a `TViewPort3D` component container for holding 3D objects. Starting a new 3D application gives you a form unit based on `TForm3D` that conveniently bundles the `TViewPort3D` container in for you. If you have an existing 2D FireMonkey application, you can add a `TViewPort3D` container onto it, in which you can put 3D objects.

There's no better way to learn than to dive in and start playing in this new arena. I would suggest creating a new multi-device application, selecting the **3D Application** template, and placing a few controls from the **3D Shapes** section of the palette onto the form. As you place each one, you'll notice the objects have four handles with which to resize and rotate the object in 3D space. Most also have a default color of red.

Another thing you'll notice is that all 3D controls are initially placed in the center of the viewport. `ViewPort3D` is measured in width (x axis), height (y axis), and depth (z axis). The center of `ViewPort3D` is the center of your 3D space; more specifically, the x, y, and z coordinates at the center are (0, 0, 0). This is a big difference from what you're used to in the **Visual Component Library** (**VCL**) or 2D FireMonkey forms where the (0, 0) position is in the upper-left corner. Additionally, the y axis increases as it goes down the page, yielding negative values for both x and y positions in the upper-left quadrant of `ViewPort`, and the z axis increases as you go further into the screen, so to pull something forward where it looks as though it's in front of the screen, set its `Position.Z` property to a negative. Finally, the position coordinates are floating-point values, not integers.

Here's a screenshot of a sample Delphi application we'll build in this chapter, with several types of 3D objects represented in the four quadrants of the viewport's space:

Figure 8.1 – FireMonkey 3D application with a variety of objects in the four quadrants of the viewport

There are many interesting (and unrelated) objects in this app, all showing different techniques we'll explore in this chapter. I suggest you try to create them on your own, but if you get stuck, the source for this can be found on GitHub at `https://github.com/PacktPublishing/Fearless-Cross-Platform-Development-with-Delphi/tree/master/Chapter08/01_Quadrants`.

Adding 2D controls to a 3D form

We'll get to the 3D objects on the form shortly but first, notice the right side of this application has what looks like a collapsed panel with an "arrow" button. By starting with `TForm3D` as this application does, the only way to get 2D controls on the form is by adding `TLayer3D`. When you do that, it covers over a portion of the underlying viewport. Since I wanted the full space of the viewport, what I did was aligned `TLayer3D` to the right of the form, made it very narrow, and added a button that, when clicked, expands the `TLayer3D` area that shows the controls I've placed there. I named the button `btnShowOptions`, and set its `StyleLookup` property to `'arrowlefttoolbutton'`. Here's the `OnClick` event handler:

```
procedure Tfrm3DQuadrants.btnShowOptionsClick(Sender: TObject);
begin
  if Layer3DOptions.Width = 20 then begin
    Layer3DOptions.Width := 120;
    btnShowOptions.StyleLookup := 'arrowrighttoolbutton';
  end else begin
    Layer3DOptions.Width := 20;
    btnShowOptions.StyleLookup := 'arrowlefttoolbutton';
  end;
end;
```

We'll be back later to put some controls into this area to work with our 3D shapes. Speaking of which, let's get to the fun part by adding some simple shapes.

Adding basic and extruded shapes

The first 3D object I added to this application was a `TCube` object. When initially placed, it's in the center, or (0, 0, 0) in the 3D coordinate space. By changing its x and y position coordinates to -5 each, it moves the cube to the upper-left portion of the screen, which I've called **Quadrant 1**. By using the handles of the cube in Delphi's 3D Form Designer, I increased the size and rotated it slightly. Adjusting rotation and size with the mouse is a little clumsy at first—the handles rotate the shape in the direction of the long side of the handle. Sometimes, it's easier to adjust the `Height`, `Width`, and `Depth` properties manually to get the size to your liking; I set `RotationAngle.X` to 5 and `RotationAngle.Y` to 45 in the demo app.

Next, add a `TCone` object and set its `X` and `Y` position properties to (5, -5) to place it in the upper-right quadrant of the viewport. Again, adjust the size properties to your liking.

Another common shape is a `TSphere` shape. Add this with *x* and *y* coordinates of -5 and 5, respectively, to place it in the bottom-left quadrant.

Showing lines for the axes

For this app, I wanted to highlight the four quadrants of the viewport. To emphasize the *x* and *y* axes, I wanted to show horizontal and vertical lines. There are a few different shapes you could use to do that. One of the simplest is a `TPlane` object, which I implemented for the *x* axis. This is a 2D plane, and since I didn't need any depth for a simple line, it works perfectly. I set its `Height` property to 0.1 and its `Width` property to the width of the viewport.

> **Note**
>
> Since the `TPlane` object is a descendant of `TShape3D`, it has a `Depth` property but, by definition, does not have any depth; therefore, the `Depth` property of this component is set at 0.001 and cannot be changed.

The *y* axis is similar, but I wanted to emphasize the fact that its values decrease in the viewport by showing a down arrow. To do this, I built an arrow out of a thin cylinder with a small cone at the end. First, add a `TCylinder` with a `Width` property of `0.1` and a `Height` property just short of the viewport (to allow for the arrow tip at the bottom). Then, add a small `TCone` with its `RotationAngle.X` property set to `180` to turn it upside down; adjust its `Position.Y` property so that it is at the end of the cylinder just placed.

To further emphasize what the lines are for, I'd like to add some text to the *y* axis. This calls for a `TText3D` object, which is in a class of extruded shapes.

Extruded shapes

2D shapes can be useful in the 3D world, but all objects on `TViewPort3D` must be 3D objects. Therefore, in order to display a 2D shape, it must be extruded onto a 3D shape. What this does is create a 3D shape with the 2D shape on the front, a mirror image of the 2D shape on the back, and a section of the 3D shape connecting the two, called the *shaft*. The resulting object can be placed on a 3D form or viewport just like with any other 3D object—and, in fact, has all the same placement, sizing, and rotating properties found in other 3D objects.

The standard FireMonkey objects declared as descendants of TExtrudedShape3D are listed as follows:

- TRectangle3D (very similar to TCube)
- TEllipse3D (very similar to TCylinder)
- TText3D—allows text to be added to a 3D viewport
- TPath3D—extrudes a 2D user-defined polynomial onto a 3D shape

We want to add some text to the form to identify the *y* axis, so add a TText3D and set its Text property to Y-Axis and its Position.Y property to -2 to move up just above the *x* axis.

Your design form should now look something like this:

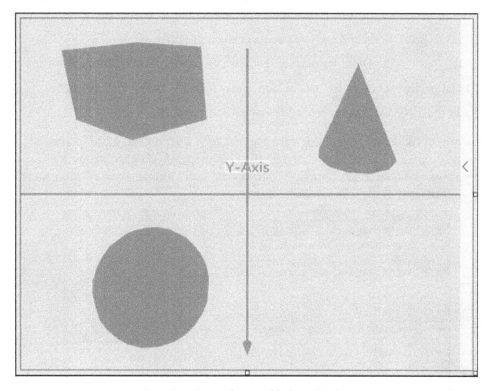

Figure 8.2 – Basic shapes added to a 3D form

We want to add one more object before moving on—a star. This isn't a predefined shape but one we must build ourselves using TPath3D.

User-defined shapes

TPath3D is a very versatile object because it's free from the predefined rectangular or spherical shapes; however, it does mean every line and curve must be explicitly specified with an array of TPathPoint coordinate instructions in its Path property. These graphical instructions are given in the common **Scalable Vector Graphics** (**SVG**) format supported by all modern web browsers and many graphical applications. Fortunately, you don't have to learn how to write SVG commands; you can download a free tool, such as Inkscape, draw an image, and simply export them.

For example, to add a star to our application, I followed these instructions:

1. Download and install Inkscape.

2. Using the **Star** tool, draw a star in a new document.

3. Select **Edit | XML Editor...** from Inkscape's menu.

4. Find the path node and copy the value of the d parameter.

5. Switch back to the Delphi project and add a TPath3D component to the form.

6. Click the ellipsis button for its Path property to bring up the **Path Designer**.

7. Paste the SVG code copied from Inkscape, then click **OK**.

The component suddenly looks like a star! And since it's a 3D object, it can be resized, rotated, and colored, just like any other extruded 3D shape. If you open up the Path property again, you'll see it has reformatted the coordinates into an array of TPathPoint strings, as illustrated in the following screenshot:

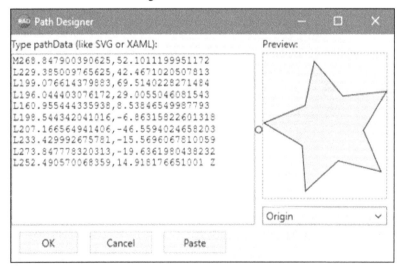

Figure 8.3 – SVG data for a star

Now that you can create almost any kind of 3D object (we'll save one more for later), it's time to get rid of all this red. Let's add some color and texture!

Adding color, lighting, and movement

Adding color is pretty simple. Let's start by coloring the axes black. Add a `TColorMaterialSource` component to the form. This won't show up as a clickable object, but you can see it in the **Structure** pane. Set its `Color` property to `Black`.

Now, select the `TPlane` object you added for the x axis, find its `MaterialSource` property, and drop down the **Property Editor** to select the black material source component you just added. The *x-axis* line should now be black. You can do the same for the *y-axis* cylinder and its cone arrow if you'd like; if you do, notice that the red *y-axis* text looks as though it goes through the *y-axis* line. To fix that, you can pull that `TText3D` object forward by setting its `Position.Z` value to `-0.5`.

How would you like to transform the sphere in the lower-left quadrant into a globe? Applying a material source with the right texture bitmap makes it simple, but finding the right texture source isn't always so simple. Fortunately, `https://visibleearth.nasa.gov` has some great freely downloadable images. Once you find one you like, follow these steps:

1. Add a `TTextureMaterialSource` component to the form.
2. Click the ellipsis button for the `Texture` property, then select **Edit...**.
3. When the **Bitmap Editor** comes up, click **Load** to load the earth image you downloaded, and then click **OK**.
4. Select the `TSphere` object and set its `MaterialSource` property to the `TTextureMaterialSource` component, with the earth image loaded.

The sphere in the lower-left quadrant should now look like this:

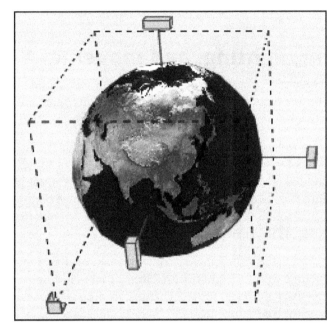

Figure 8.4 – A TSphere object textured with a world map

We haven't forgotten about the objects in the upper-left and upper-right quadrants. We'll add more colors and textures but with an added touch: light!

Adding a light source to colors and textures

The colors and textures we've applied so far are constants applied with the same intensity over the object to which they are applied. To get closer to photorealism, a light source can be added that affects the brightness of a texture or color depending on the angle at which the light source is hitting the object. This calls for a material that knows about light: a `TLightMaterialSource` component.

To add a light-affected blue color to the cube, follow these steps:

1. Add a `TLightMaterialSource` component to the form.

2. Set the `Ambient` property to `MidnightBlue`.

3. Set the `Diffuse` property to `SkyBlue`.

4. Set the `Emissive` property to `Blue`.

5. Set the `MaterialSource` property of the `TCube` object to the blue material source component we just added.

The three different color properties affect the color that will be applied to the object, but since we don't yet have a light source, the Emissive property is the only one affecting the cube.

Add a TLight to the form, and instantly, the color of the cube changes. Now, the Ambient and Diffuse properties come into play, the former affecting the shadowed areas and the latter setting the color when a direct white light is hitting the surface. Notice in the following screenshot that the top of the cube is dark, and the front of the cube, where the light is hitting it directly, is close to the skyBlue color:

Figure 8.5 – Our cube with a blue color applied that is affected by a light source

Light can also affect texture. Add another TLightMaterialSource component and this time, instead of setting the color properties, click the ellipsis button on the Texture property and load a green texture. I found one that has a rough look to it and from a distance could look like trees or bushes. Apply this material source to the TCone object in the upper-right quadrant. Notice the texture is brighter in the center and darker on the sides because the light source is hitting it from the front. You can add a brown cylinder at the bottom then duplicate these objects to simulate a forest. I added several small red spheres to make it look like a Christmas tree.

But a Christmas tree isn't complete without a star on top, right? So, let's attach the star we created out of TPath3D earlier to the top by making it a child of the Cone object and adjusting its Position properties. We'll leave it at the default color of red for now.

Now, it's starting to look how we want, as we can see here:

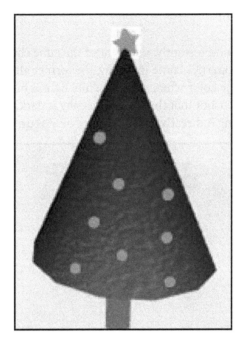

Figure 8.6 – A Christmas tree made from a textured cone, red spheres, a star, and a brown cylinder

Suppose you want to see what these objects look like with the light off. You can simply select the TLight object and set its Enable property to False. You can also do this at runtime. Follow these steps to set that up:

1. Expand the TLayer3D component on the right side of the form.

2. Add TGroupBox and change its caption to Light.

3. Add TSwitch inside the group box and TLabel below it with the Text property set to On/Off.

4. Right-click on the switch and select **Bind Visually....**

5. Hook up the switch's IsChecked property to the light's Enabled property.

6. Reset the Width property of TLayer3D to 20.

Now, when you run the program, you can expand the side panel and turn the light on and off to see how this affects the coloring of objects at runtime. This will become more interesting once we add animation later in the chapter.

Back over to the cube—it's a nice shade of blue but could use some colored text, which deserves a special mention.

Coloring extruded objects

Add another `TText3D` object to the form and set its `Text` property to **Quadrant 1**, then make it a child object of the cube. Adjust the `Depth`, `Height`, and `Width` properties so that it looks as though it sticks out from the side of the cube.

`TText3D` and other extruded objects have three material source properties, listed as follows:

- `MaterialSource`—The material applied to the 2D object on the front

- `MaterialBackSource`—The material applied to the mirror image on the back

- `MaterialShaftSource`—The material applied to the 3D shaft between the front and back of the 3D object

Since this text is partially embedded in the side of the cube, the back will never be seen, so we can leave the `MaterialBackSource` property blank. Set the other two material source properties to either ones already added to the form or new ones you create to uniquely color this text. I used the same material source with black applied to the *x* axis for the `MaterialSource` property and the brown material source used for tree trunks for the `MaterialShaftSource` property.

Now, my cube looks like this:

Figure 8.7 – My cube with colored text anchored to the side

While the text's position is relative to the center of the cube, the sizes of the cube and text are independent of each other; thus, you could increase or decrease the width or depth of the cube without changing the text's size properties to show or hide the child text object.

In fact, the text object could also be made to look completely outside of the cube, which might make it appear to be disconnected—until you rotate the cube and see the text rotate with it!

Speaking of rotating, it's time to add some movement to our objects.

Adding animation

We have a globe in the lower-left quadrant; let's spin it, as follows:

1. Add a TFloatAnimation component to the form.

2. Set the StartValue property to 0, the StopValue property to 360, and the Duration property to 20.

3. Using the **Structure** pane, drag the float animation component so that it's a child of the TSphere component with the world map texture.

4. Now, the PropertyName field can access one of the sphere's numerical properties; select RotationAngle.Y.

5. The default animation direction around the *y* axis is clockwise, so set the Inverse property to True to simulate the earth's actual rotation.

6. Set the Loop property to True so that it doesn't stop spinning once it's reached the StopValue property.

7. Set the Enabled property to True so that it starts when the application runs.

Now, run the application and see the globe spinning. What we've done should be obvious: the sphere's property being animated is RotationAngle.Y and it continuously loops from 0 to 360 degrees counter-clockwise around the vertical axis.

Let's add another TFloatAnimation, this time for the cube in **Quadrant 1**. Seeing the back of the cube isn't important, so set the StartValue and StopValue properties of the animation component to 10 and 150 and set both the Loop and AutoReverse properties to True. As before, drag the animation component in the **Structure** pane to be a child of the cube, select the RotationAngle.Y property for PropertyName, and enable and run it.

It's hard to capture animation in print, but here is a sequence of screenshots to indicate what's happening at runtime with the cube:

Figure 8.8 – Sequence of screenshots showing the cube's rotation at runtime

Notice the shades of blue change as the cube turns to face the light, which is coming from the center of the screen.

There are several more types of animation components. We'll look at one more that will make the star on top of the tree twinkle.

Animating color

Animation is simply changing a property through a sequence of values. If that property is a color, then instead of moving an object, it'll simply change its color.

Follow these steps to animate the color of the TPath3D component we used to form the star:

1. Add a TColorMaterialSource component.

2. You can leave its Color property at the default of Red or set it to Yellow or Null; this initial color will only be seen at design time.

3. Add a TColorAnimation component and make it a child property of the color material you just added.

4. Set PropertyName of the new color animation component to the Color field of the star object, its parent.

5. Set the StartValue property to Yellow and the StopValue property to Orange.

6. Set both AutoReverse and Loop to True.

7. Leave the Duration property at 0.2 and set Enabled to True.

8. Assign the TColorMaterialSource component to all three material properties of the star (MaterialSource, MaterialShaftSource, and MaterialBackSource).

9. Run the application.

Now, the Christmas tree has a blinking star on top. Just for fun, how would you like to also blink the lights on the tree?

In this case, let's just blink them on and off, so instead of changing their color, we'll just set `Opacity` to 0 and then back to 1, but there's a lot of them and we want them to blink at different times, so this will take a little coding. I named all the `TSphere` objects on the Christmas tree starting with a `bulb` prefix and appended these with an incrementing number, so this will be fairly simple. Call this procedure from the `OnCreate` event of the form, as follows:

```
procedure Tfrm3DQuadrants.StartTreeLightsBlinking;
var
  NewFloatAnimation: TFloatAnimation;
begin
  for var bulb := 1 to 8 do begin
    NewFloatAnimation := TFloatAnimation.Create(self);
    NewFloatAnimation.Parent := FindComponent('bulb' + bulb.
      ToString) as TSphere;
    if Assigned(NewFloatAnimation.Parent) then begin
      NewFloatAnimation.Duration := 0.5 + Random;
      NewFloatAnimation.Delay := 0.5 + Random;
      NewFloatAnimation.StartValue := 0.0;
      NewFloatAnimation.StopValue  := 1.0;
      NewFloatAnimation.Loop := True;
      NewFloatAnimation.AutoReverse := True;
      NewFloatAnimation.PropertyName := 'Opacity';
      NewFloatAnimation.Enabled := True;
    end;
  end;
end;
```

That's pretty cool! Now, the Christmas tree is complete.

There's one conspicuously empty space in our application: the lower-right quadrant. Let's import a complex model and display it there.

Importing 3D models

So far, we've shown how to add fairly simple objects in our 3D world—objects that can be defined with a few lines or curves. Even the `TPath3D` component with its array of path points is only a 2D object at its root.

To create a 3D object not constrained to a handful of lines requires a TMesh. Its parent class, TCustomMesh, is actually the base class for all non-extruded 3D shapes, with properties hidden and methods overridden to make them easy to use. A mesh allows a set of connected points and—optionally—textures to define a 3D object. What we want is something even more complex.

Enter TModel3D. This component has a MeshCollection property that connects several mesh objects into one comprehensive object. What's more, MeshCollection can import standard 3D model files built into many kinds of popular 3D modeling software, such as Autodesk or Blender. Three formats are supported: ASE, DAE, and OBJ.

3D modeling software has a steep learning curve and if that's not where you want to spend your time, there are websites that offer subscriptions to libraries of 3D models and people that specialize in creating them. There are also some free ones to be found. One such place is https://www.turbosquid.com/Search/3D-Models. They have a filter for price, and setting that to **FREE** will let you try this out without spending a dime—however, it will require you to create a free account. I found a model of an elk and downloaded that.

Once a file in one of the supported formats has been downloaded, place a TModel3D on the form, position it so it's in the bottom-right quadrant, and click the MeshCollection property's ellipsis button to bring up the **Mesh Collection Editor**. Click the **Load** button, select the 3D file, and then click **OK**. The model displays in its default color of red.

Remember that TModel3D is a collection of several TMesh objects, each of which can have a unique color or material. So, TModel3D does not provide a way to assign one at design time. The code to do it at runtime is fairly simple, though; I implemented it in the form's OnCreate event and assigned it to the brown material I created earlier, as illustrated in the following code snippet:

```
procedure Tfrm3DQuadrants.Form3DCreate(Sender: TObject);
begin
  for var AMesh in Model3DElk.MeshCollection do
    AMesh.MaterialSource := ColorMaterialSourceBrown;
end;
```

Now, when you run it, you should see a brown elk. Since I didn't use `TLightMaterialSource`, it's not very realistic as there are no shadows, just a solid brown elk. I tried setting its `Opacity` property to `0.8` and it looks a little better, as we can see here:

Figure 8.9 – A 3D model of an elk

To add a little interactivity in order to take a closer look at the model we imported, let's add some 2D controls to rotate and zoom the model, as follows:

1. Expand the `TLayer3D` side panel.

2. Add a `TGroupBox` and set its `Text` property to `Elk`.

3. Add a `TTrackBar` and corresponding `TLabel` with a `Text` value of `Rotate` in the group box.

4. Add another `TTrackBar` and corresponding `TLabel` with a `Text` value of `Scale`, also in the group box.

5. Select one of the `TrackBar` components, right-click, and select **Bind Visually....**

6. Add the `TModel3D` component to the **LiveBindings Designer** and add the following bindable members: `RotationAngle.Y`, `Scale.X`, `Scale.Y`, and `Scale.Z`.

7. Bind the `Value` property of the `TrackBar` used for rotating to the `Model3D` object's `RotationAngle.Y` property.

8. Bind the `Value` property of the `TrackBar` used for changing the scale to the `Model3D` object's `Scale.X`, `Scale.Y`, and `Scale.Z` properties.

9. Resize `TLayer3D` back to its original `Width` property of `20`.

Your **LiveBindings Designer** window should now contain these bindings, as illustrated in the following screenshot:

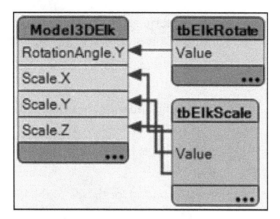

Figure 8.10 – Rotation and scaling TrackBar components bound to the elk model

Running the program, you will now be able to scale and rotate the elk.

What if you'd like to zoom in and show only the elk or a different object? This calls for a different camera.

Changing the camera

Every 3D view (`TForm3D` or `TViewPort3D`) has the concept of a camera to capture objects in the 3D world. The default camera used at design time points at the very center of the area but is located back from the screen, toward the viewer (-20 on the z axis). This gives a nice, viewable area of your 3D world.

But this is just the start—there can be several cameras.

Multiple cameras

You can add one or more cameras and switch between them and the default one. Their position and rotation can be customized to the specific views you're after, and if you want one of them to focus on one object, you can set the `Target` property to that object. Any change to a camera view at runtime requires a call to the camera's `Repaint` method to tell the output about the updated camera view.

Here are the steps to add a camera view that focuses on the elk—and how to switch it at runtime:

1. Add a TCamera component to the form and set its Position properties to (5, 4, -7) and its RotationAngle properties to (355, 30, 0).

2. Set the Target property of the new camera object to the Model3D object used for the elk.

3. Expand the TLayer3D side panel and add a TGroupBox component with a label of Camera.

4. Add a couple of TRadioButton objects inside the group box—one labeled Default, and the other labeled Elk.

5. Set the OnClick event of the radio button for the elk to the following code:

```
procedure Tfrm3DQuadrants.radCameraElkChange(Sender:
TObject);
begin
  UsingDesignCamera := False;
  Camera := CameraElk;
  Camera.Repaint;
end;
```

6. Set the OnClick event of the radio button for the default camera to the following code:

```
procedure Tfrm3DQuadrants.radCameraDefaultClick(Sender:
TObject);
begin
  UsingDesignCamera := True;
  Camera.Repaint;
end;
```

7. Reset the width of the TLayer3D side panel back to 20.

When you run the program and expand the side panel, you can now switch between a close view that focuses only on the elk and the default view of the whole screen.

You can apply these same concepts to add a camera that focuses on the cube. Set this camera's Position properties to (-3, -4.5, -10) and the RotationAngle properties to (350, 350, 0). Add another radio button to switch it at runtime with this code:

```
procedure Tfrm3DQuadrants.radCameraCubeClick(Sender: TObject);
begin
  UsingDesignCamera := False;
  Camera := CameraCube;
  Camera.Repaint;
end;
```

Think of your computer screen as a TV where you're watching a movie with different camera angles. If you think about it, the views are changing constantly to show close-ups of people or objects. Using multiple cameras can greatly enhance your 3D application if used properly.

In addition to cameras placed at stationary points of the scene, you can also place cameras that are relative to a specific object.

Satellite camera

If you add a camera as a child of another object instead of directly on the form as we've done here, its position and rotation properties are relative to that object. If there's any animation applied to the object, the child camera will (as any other child object would) move right along with it.

Try adding a camera as a child to the globe; it's `Position` and `RotationAngle` properties can all be left at 0 except for `Position.Z`, which you will want to bring away from the globe by setting it to `-15`. This camera will act like a satellite fixed on one spot on the globe, making it look as though the rest of the 3D space is spinning around it, as illustrated in the following screenshot:

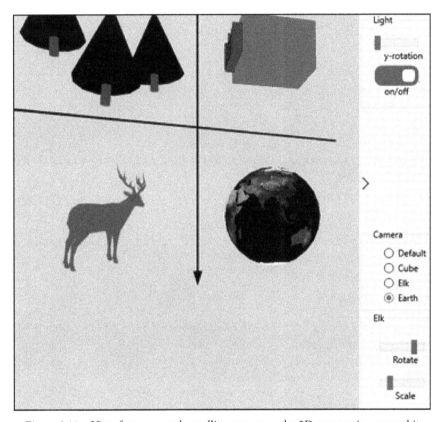

Figure 8.11 – View from an earth satellite camera as the 3D space spins around it

There are many aspects to manipulating objects, views, cameras, and lighting. In some ways, it's amazing this works so well and so similarly across platforms. Still, it's always a good idea to test on as many different devices as is practical.

Testing on phones

I ran all my initial tests of this application on a Windows machine—it's quick and easy. As I was nearing the end of development, I decided to make sure it worked on other devices and, as you might expect, found some things that needed to be addressed. The first thing I noticed was that when the app ran on a small phone in portrait mode, the sides of the viewport were cut off.

So, the first improvement was to go into **Project | Options** and force it to only use landscape mode. That setting is found under the **Application | Orientation** section, as illustrated in the following screenshot:

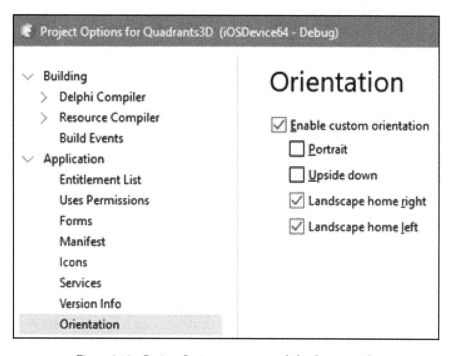

Figure 8.12 – Project Options set to use only landscape mode

Next, I noticed the 2D panel on the side was really crowded. On Windows and Mac desktops, the mouse is ubiquitous in selecting objects and can pinpoint the spot to click in radio buttons with relative ease. But fingers on a mobile device take up a much larger percentage of the area, requiring a lot more space between the options. The FireMonkey views show this, but I had failed to check them.

To free up space in the side panel, I changed how the camera views are selected from a group of radio buttons by clicking directly on the objects; a second click would reset the camera view back to the default view.

Here's how the OnClick event for the globe, and the procedure it calls, are coded:

```
procedure Tfrm3DQuadrants.SphereGlobeClick(Sender: TObject);
begin
  if Camera = CameraEarthSat then
    SwitchToCameraDefault
  else begin
```

```
    UsingDesignCamera := False;
    Camera := CameraEarthSat;
    Camera.Repaint;
  end;
end;
```

And the procedure it calls to reset the camera to default is also quite simple, as illustrated in the following code snippet:

```
procedure Tfrm3DQuadrants.SwitchToCameraDefault;
begin
  UsingDesignCamera := True;
  Camera.Repaint;
end;
```

For the cube, you need to hook into the `OnClick` event handlers for both the cube and its embedded text object in case the click occurs over the text instead. The elk didn't respond to clicks until I set the `HitTest` property to `True`; most 3D objects enable that property but `TModel3D` passes mouse events to objects (or the form) behind it by default.

Finally, I found the group boxes didn't show up very well on Android, so I loaded a style just for Android that made them stand out better, as illustrated in the following screenshot:

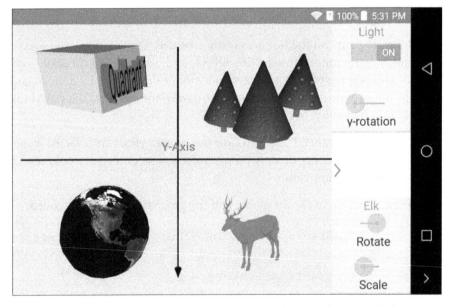

Figure 8.13 – Quadrant's demo app running on an Android phone showing styled group boxes

This has been a great introduction to several aspects of using 3D objects with Delphi, but there are several more things we should cover before leaving this topic. One way to explain them is to use them in more of a real-life application—for example, a game!

Let's write a game!

A vast majority of video games involve immersive graphics. We don't have the time or space to build a 3D world with **artificial intelligence** (**AI**)-controlled moving objects and lots of interactivity, but we can build a small "escape" game. The idea behind games of this genre is that you're stuck in a room that is pictured on the screen and you click on various objects to find clues to get out of the room.

As I looked through some old stock images that might be fun to implement, I came across one of a small room with a couple of computer racks, an old phone on the wall, and an engineer at a computer desk with his back to us. There's also a clipboard on the table, a perfect place to hold clues about how to set controls on one of the server racks. One of my favorite science-fiction movies is *The Matrix*, where the characters in the movie use old land phones to transfer their digital conscious between real and virtual worlds, so this will be the escape point, rather than a door as you'd find in most escape-type games, with the idea that we're stuck in 1985 and want to escape the time period back to the present day rather than just walk out of the room.

So, there are three clickable objects in my escape game: a notepad, a control panel, and a phone; the main screen is just a simple image of the room, as illustrated in the following screenshot:

Figure 8.14 – Escape game main screen

I started, of course, by creating a new multi-device application and choosing the **3D Application** template. Before I started placing objects on the form, I placed a container object for 3D controls called a TDummy object. This is a somewhat strange name for a component but represents the idea that it doesn't have separate functionality or controls like other 3D objects—it's just a dumb container, if you will. Its whole purpose is to group other 3D objects and allow you to perform manipulations on them as a whole by manipulating the dummy component instead of each of your objects individually.

On top of the dummy container, I added a TImage3D and loaded the image of the room. I stretched the image out and fixed the form size by setting its BorderStyle property to Single.

A couple of other setup things to mention before we explore the depths of the game include setting up a custom TCamera and adding a TLight. The camera is placed at (0, 0, -15) to simulate a person at the back wall looking at the room. The light is at (0, -5, -5) to represent a light on the ceiling. This will give a realistic look for a few objects we will add.

The interesting part of this game is implementing hidden clickable objects. We want the user to click around and find the clues without them being too obvious. Let's see how to do this.

Implementing hidden clickable areas

All 3D objects have an Opacity property, set by default at 1, which means the object is completely visible and hides objects behind it. Set it to 0 and the object is transparent. This means you can place an object on top of another and make it completely invisible, which is perfect for our needs because it can still react to a mouse click (or finger touch).

Now, let's pause our application building and think about the mechanics of the game. When you click on an object, you want the object to get big enough to see and interact with. For example, clicking on the notepad on the desk should look and act as though you've picked it up and are looking at it closely; when you're done, you can put it back down and move on to something else. The control panel should zoom forward (as if we walked up to it) so that we can see the individual controls and interact with it. Finally, the phone doesn't really need much interaction but we'll bring it forward as well for consistency in the **user interface** (**UI**) and as a visual confirmation to the game player that it was actually clicked.

The phone is our simplest clickable object, so let's start with that. We will use the TDummy object again to group each of the clickable objects, so create a dummy container for the phone and place it over the room image in the upper-left corner. Add a TImage3D on top of the phone container and load an image of the phone. As we're working with the objects, it's easier to work in the "zoomed-in" state and then hide them when the program starts, so build with the dummy container an image in the size you'd want to see once it's been clicked while playing the game, as illustrated in the following screenshot:

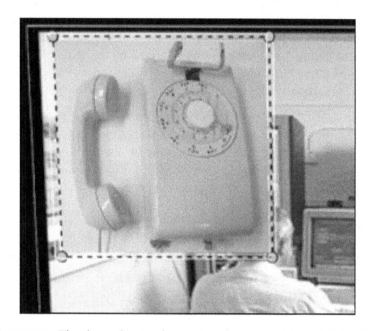

Figure 8.15 – The phone object and image in a dummy container, at design time

In addition to the full size, we also need to know the small size and assign it when the application starts so that the clickable area just barely covers the actual object in the room image. But instead of modifying the *x* and *y* properties, we can simply scale the phone's dummy container up and down, and all the components inside it (only one, in this case) will scale down with it. With the full size of the phone, as shown in *Figure 8.15*, the scale would be 1. Playing around with the scale a little bit, we find that setting the X and Y sub-properties of Scale to 0.4 will be the appropriate size for our small, hidden, clickable object. We won't need to adjust Scale.Z.

To provide a nice visual effect of zooming, instead of simply changing the scale from 0.4 to 1 and showing the image instantly zoomed, it's nice to show a quick animation of the object coming forward. This is fairly simple to accomplish with the animation components in FireMonkey we learned about earlier in the chapter.

Add a TFloatAnimation component for each of the two Scale properties we need to adjust as child components of the phone container, set StartValue to 0.4 and StopValue to 1.0 in each, and set one of them to affect the Scale.X property and the other to affect the Scale.Y property. We want the zoom to be quick and smooth, so set the Duration property of each to 0.4.

Lastly, remember that before the user clicks on the phone, it is invisible; the Opacity property is 0. We also need to animate that property to make it smoothly appear while it's zooming forward, so add another TFloatAnimation component, this one as a child of the phone image, as it's the image's transparency that needs to change. We set the duration of the zooming at 0.4 seconds, so the duration of the opacity change should be the same. I think it should start zooming just before we actually see it, so I set a Delay property of 0.1 and a Duration property of 0.3. You can set these values to your own liking.

We're almost ready—we just need to initialize the settings when the application starts. We'll work with these objects in their "full zoom" state at design time to make it easier to adjust properties, but we want the objects set up for actual use in their "unzoomed" state at runtime. So, I wrote a procedure called from the form's OnCreate event to set up the phone object, as illustrated in the following code snippet:

```
procedure TfrmEscape1985.SetupPhone;
begin
   DummyPhone.Scale.X      := FloatAnimPhoneScaleX.StartValue;
   DummyPhone.Scale.Y      := FloatAnimPhoneScaleY.StartValue;
   DummyPhone.Position.Z   := CLICKABLE_OBJECTS_Z;

   Image3DPhone.Opacity := 0;
end;
```

At design time, our scale values are 1 but we need to start at 0.4. The animation component's StartValue property is 0.4, so we can just assign that value—then, if we ever decide to change the scale, we can simply change the StartValue animation in the **Object Inspector** and won't have to worry about this code.

The Z position of these objects should always be just in front of the room image but behind any zoomed image because if any hidden object is at the same distance as the visible object, the click event might be attributed to the wrong image. I set the CLICKABLE_OBJECTS_Z value to -0.8.

So, then, what happens when the object is actually clicked?

Activating and deactivating an object

The OnClick event of an object should make the object appear only if it's not already showing. Clicking the object again should do something different—or nothing at all. Clicking on the room image should make the object disappear so that you can click on something else.

To allow for that, a variable—private to the form—is created for each of the clickable objects to indicate the state of the selected object; they are initialized in the form's OnCreate event to False. For the phone object, clicking on it when it's showing will try to escape the room—we'll get to that later. If it's not showing (and none of the other objects are showing either), then it should show the phone. The code is illustrated in the following snippet:

```
procedure TfrmEscape1985.Image3DPhoneClick(Sender: TObject);
begin
  if FPhoneShowing then
    TryEscape
  else if (not FNotepadShowing) and (not FPhoneShowing) and
          (not FControlsShowing) then
    ShowPhone;
end;
```

The interesting part is the ShowPhone procedure where we activate the animations, as illustrated in the following code snippet:

```
procedure TfrmEscape1985.ShowPhone;
begin
  if FPhoneShowing then begin
    FPhoneShowing := False;
    FloatAnimPhoneScaleX.Inverse := True;
    FloatAnimPhoneScaleX.Enabled := True;
    FloatAnimPhoneScaleY.Inverse := True;
    FloatAnimPhoneScaleY.Enabled := True;
    FloatAnimPhoneOpacity.Inverse := True;
    FloatAnimPhoneOpacity.Enabled := True;

    DummyPhone.Position.Z := CLICKABLE_OBJECTS_Z;
  end else begin
    FPhoneShowing := True;
```

```
        DummyPhone.Position.Z := ZOOMED_OBJECTS_Z;

        FloatAnimPhoneScaleX.Inverse := False;
        FloatAnimPhoneScaleX.Enabled := True;
        FloatAnimPhoneScaleY.Inverse := False;
        FloatAnimPhoneScaleY.Enabled := True;
        FloatAnimPhoneOpacity.Inverse := False;
        FloatAnimPhoneOpacity.Enabled := True;
    end;
  end;
```

This procedure is used to make the object both appear and disappear. When appearing, it will zoom forward and remove its transparency; that happens in the second half of the procedure, where it sets the FPhoneShowing variable to True and moves the object forward slightly by setting the Position.Z property to the ZOOMED_OBJECTS_Z (-1) constant.

Notice also that it sets the Inverse property to False in the second half of the procedure. Why is that? Because the first half of the procedure sets it to True in order to reverse the process when the object is disappearing. The same animation components are used, but with the Inverse property set to True, the values go in the opposite direction, from StopValue to StartValue. This allows us to use the same animation component for both zooming forward and backward at different times.

One important thing to note about the animation components is that they are short-lived animations, as opposed to the constant blinking lights or rotating globe we saw earlier in the chapter. Sure—they stop when the affected Property value reaches StopValue, but the animation component itself is still enabled. What this means is that the animation won't reset its values and won't start again until the Enabled property is set to False first. The animation components have an OnFinish event that activates when StopValue (or StartValue, in the case where Inverse is True) is reached, so it's pretty simple to hook into that and change the Enabled property. Furthermore, we can use the Sender property of the event handler to make this routine generic for all animation components that need to disable themselves, as illustrated in the following code snippet:

```
procedure TfrmEscape1985.FloatAnimationFinish(Sender: TObject);
begin
   (Sender as TFloatAnimation).Enabled := False;
end;
```

The phone object taught us a lot but the notepad is a little more complicated, mostly because we want the interface to show that it's being picked up from the desk at an angle and rotated into view. With this one, we not only scale it up and down but also animate the `Height`, `Width`, `Position.X`, and `Position.Y` properties, and even the `RotationAngle.Z` properties. The `SetupNotepad` and `ShowNotepad` procedures are similar in concept to the `SetupPhone` and `ShowPhone` procedures but are obviously much longer—we won't show them here.

The background of the notepad is a `TRectangle3D` object, a simple 3D object that can be rotated in 3D space; it has a white material applied and doesn't need any other capabilities. On the notepad, I placed a `TLayer3D` object holding some `TLabel` components to show important clues for the game—I had to set the `HitTest` property of these components to `False` to prevent them from hijacking the `OnClick` event the notepad rectangle uses to show itself. Finally, the `Position.Z` property of `Layer3D` had to be set to `-0.1` to raise it slightly from the notepad itself; otherwise, the label text would not show.

The most complex clickable area in our escape game is the control panel. This mixes 2D and 3D controls in a unique way to provide a nice interface that works across all devices.

Mixing 2D and 3D controls for best use of each

My idea for the game is that to escape, you have to use the phone while the power level is at exactly 115%, but there are protections for raising the power level that require a specific code to be entered—which changes every 60 seconds.

The control panel has `TRectangle3D` as the background and `TLayer3D` on top of it that holds all the other controls. Entering security codes is done by clicking on a row of square buttons whose caption changes each time you click it to the subsequent digit, wrapping around to 0 when it advances from 9. I based this design off of other escape games I've played on various devices; the simple interface works well on touch devices such as a phone. I originally tried to use `TCube3D`, but implementing a click where the depth is quickly reduced then increased and setting `TText3D` on top of it was slow and ugly—and took a lot of work. Using `TButton` was quite straightforward.

The power meter was initially TTrackBar, which would've been simple to implement, but on Mac and iOS devices, the track bar's width is very narrow. I wanted a wide meter and because of its size, it had to look 3D, so I implemented it with a dark outer cylinder and a colored inner cylinder. As the power level rises past certain points, its color changes from green to yellow to red. The challenging part of building that is keeping the bottom of the inner cylinder even with the bottom of the outer cylinder as the Position.Y property moves when the Height property is adjusted. Here's the procedure that took care of all of this:

```
procedure TfrmEscape1985.SetPowerLevel(const NewPwrPcnt:
Integer);
const
   PWR_100_PCNT_HT = 3.4;
   PWR_MAX_PCNT_HT = 3.8;
var
   CalcHt, CalcY: Single;
begin
   if NewPwrPcnt >= 0 then begin
     FCurrPwrPcnt := NewPwrPcnt;
     Text3DPowerPercent.Text := FCurrPwrPcnt.ToString + '%';

     CalcHt := Min((NewPwrPcnt / 100.0) * PWR_100_PCNT_HT, PWR_
        MAX_PCNT_HT);    CalcY := (PWR_MAX_PCNT_HT / 2.0) -
        (CalcHt / 2.0);

     CylPwrLvlInner.Height := CalcHt;
     CylPwrLvlInner.Position.Y := CalcY;

     if (FCurrPwrPcnt >= 0) and
        (FCurrPwrPcnt <= PWR_LVL_GREEN_MAX) then
       CylPwrLvlInner.MaterialSource := LitMatGreen
     else if FCurrPowerPercent <= PWR_LVL_YELLOW_MAX then
       CylPwrLvlInner.MaterialSource := LitMatYellow
     else
       CylPwrLvlInner.MaterialSource := LitMatRed;
   end;
end;
```

There are some variables and constants that are defined at the form level, but it should be obvious from the context and identifier names what they represent. When the power is over 90%, the meter turns yellow, and when it's over 100%, it turns orange, as illustrated in the following screenshot:

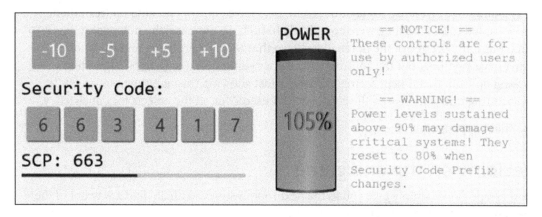

Figure 8.16 – The control panel in the escape game with the power level set dangerously high

Notice the power meter looks round because of the TLightMaterialSource applied. The TMemo on the right is partially transparent (Opacity is set to 0.7) to make it look like a faded label on an old machine.

The two rows of buttons look a little different—the top row has white text while the bottom row has dark text. How did I do that? Remember the discussion in *Chapter 7, FireMonkey Styles*? I created a custom style for TButton and applied it to the bottom row of buttons.

Once the right security code is entered and the power raised to the correct level, the user can click out of here and go to the phone object to "escape" and win the game.

You may be wondering how best to manage the editing of the various objects at design time. Let me share with you a trick I used.

Working with layered objects at design time

While working on various objects that will at different times be visible or invisible, it's handy to remember that you're working in a 3D space. In other words, in addition to height and width, you also have layers of objects closer and further back. In this game, we have the main image showing a room at the Z position of 0. So, any object with a *z* value greater than that (that isn't rotated on the *y* axis and has some point sticking out at us) is behind the room, rendering it invisible, and it follows that any *z* value less than that (for example, -1) is in front of the room image and thereby visible (if its opacity is not 0).

So, what I did is kept the dummy containers that I was not working with at a Z value of 2 to keep them behind the room. If I wanted to work on the notepad, for example, I would set the `Position.Z` property of the dummy container for the notepad to `-1` to bring it to the front. When I was done, I would set it back to 2.

Why 2 and not 1? Because some of the containers have controls sticking up from them and I wanted to be sure they were completely hidden. In our case, the notepad's `Layer3D` value was only -0.1 above the notepad itself, so that wasn't a problem, and the phone didn't have anything, but the control panel has a cylinder on it with depth, then text sticking up from that. I kept seeing part of the text showing through the notepad while I was testing until I increased the z value to 2. I used 2 for all the objects for consistency.

So, what happens when the game is over?

Deciding on the end game

Most escape games simply take you to the next room when you finally find a way out of the one you're in. Our game only has a single room, and there's no menu or score or anything after you finish—just a message saying **You escaped**, but I also wanted to indicate what happens if you try to leave the room with the power set too high. Both cases darken the room to indicate it's over and leave you with a message. In the case where the power is too high, there is large 3D text telling you such. Again, I used animation to pull the text forward; additionally, I animated the shaft of the 3D text, alternating between red and black, which, combined with the bright yellow face, gave it a pulsating, emergency-warning look.

Finally, I wanted to close the application down after the message was acknowledged, which works fine on Windows, macOS, and Android but is not allowed on iOS, so I used a compiler directive that specifically excludes that platform, as illustrated in the following code snippet:

```
{$IFNDEF IOS}
Close;
{$ENDIF}
```

This leaves iOS versions of the app just sitting there with a dark room after the game is over, probably not something I'd do for a game out on the app market.

Video games often include sounds, and this game would certainly be enhanced with some sounds at key points, but we'll save that discussion for a later chapter. You can download the code for what's been discussed here at the GitHub repository, found at `https://github.com/PacktPublishing/Fearless-Cross-Platform-Development-with-Delphi/tree/master/Chapter08/02_Escape1985`.

Play around with the game, tweak the settings, and change some things to see how the 3D controls work, and have fun continuing to explore the world of 3D.

Summary

The small demo at the start of this chapter packed a lot into it: a variety of 3D objects with techniques used to color, texture, and animate them; an area for 2D controls, a little interactivity, and some considerations when using it on various platforms.

These are great to get started, but to really learn the nuances of 3D programming and how humans interact with virtual objects, putting together a game reveals things you may not have considered. The escape game we wrote showcased many of those aspects, such as when and how to set the `Position.Z` property to work with controls at design time, how rotating an object on one axis affects the position and rotation along other axes, how to tastefully mix 2D and 3D objects for the best interactivity, how lights and colors can greatly enhance visual appeal, or when to disable `HitTest` to allow a click event to pass through to a parent control.

When working with 3D objects on mobile devices, it's important to remember there's often no keyboard or mouse and that gestures (not just simple taps) are increasingly expected to manipulate the objects with swipes or two-finger drags. We didn't get into that in this chapter, but the techniques you will learn in later chapters will be useful if you want to come back and try them out with the apps introduced here.

There are many more aspects to working with 3D controls that could be explored, several animation types we didn't cover, some lighting and camera properties left unmentioned—and we barely scratched the surface of 3D models.

This chapter was a fun digression into a niche area of programming but there are many more topics to explore in the cross-platform arena. The next chapter switches gears to focus on data storage and explores two popular, but very different, database libraries and what you need to know to work with them on various platforms.

Questions

1. Which are the three GPU engines used by the FireMonkey 3D objects?

2. Which property of a 3D object would you change to make the object appear closer or further away, and in which cases would the values be positive or negative?

3. Which component can be placed on a `TForm3D` to allow 2D FireMonkey controls? Conversely, which component can be added to a 2D FireMonkey form to allow 3D controls?

4. How do you turn a 3D object upside down?

5. If no light source is available, which of the `TLightMaterialSource` properties will be used to color the object to which it is applied?

6. How do you set a 3D object to be transparent?

7. How do you set the color of `TModel3D`?

8. How do you make a camera follow a 3D object when its movement is changing through animation?

9. How does an animation component's value get reset when it reaches the end value?

10. Which 3D component is used solely for grouping other 3D components?

11. What's a technique you can use to work on objects at design time that will be placed behind other objects at runtime?

12. On which platform are you not allowed to programmatically close an application?

Further reading

- *FireMonkey 3D:* `http://docwiki.embarcadero.com/RADStudio/Sydney/en/FireMonkey_3D`

- *FireMonkey Quick Start Guide - Creating a 3D Application:* `http://docwiki.embarcadero.com/RADStudio/Sydney/en/FireMonkey_Quick_Start_Guide_-_Creating_a_3D_Application`

- *Boost Mac performance with Metal and Delphi 10.4:* `https://blog.grijjy.com/2020/05/25/boost-mac-performance-with-metal-and-delphi-10-4`

- *Import Path from Inkscape:* `https://i-logic.com/delphi/path.htm`

- *Creating 3D scenes dynamically in FireMonkey:* `http://www.adug.org.au/technical/fmx/creating-3d-scenes-dynamically-in-firemonkey`

- Eight 3D demos in Delphi FireMonkey: `https://blogs.embarcadero.com/eight-3d-demos-featuring-volume-rendering-textures-shaders-materials-polygons-and-models-in-delphi-firemonkey`

- 3D models by TurboSquid: `https://www.turbosquid.com`

Section 3: Mobile Power

In this section, we concentrate on Android and iOS phones and tablets, showing where to store data, and how to access smartphone features such as the camera, location services, and mapping APIs. It also discusses Bluetooth technology, both classic and low-energy—with BLE being the basis for beacons and IoT devices, which opens up a world of device interactivity for the Delphi programmer. Finally, we tell you how to get a running Delphi app on a Raspberry Pi 3.

This section comprises the following chapters:

- *Chapter 9, Mobile Data Storage*
- *Chapter 10, Cameras, the GPS, and More*
- *Chapter 11, Extending Delphi with Bluetooth, IoT, and Raspberry Pi!*

9
Mobile Data Storage

As we will be focusing on mobile devices in the next few chapters, we'll need a way to save settings, pictures, audio clips, and a plethora of other pieces of information. Mobile applications often communicate with a server to send data back and forth, as we'll see in the next section of this book, but it's also useful to be able to store data locally. Sometimes, this is for looking up information, such as downloaded maps, while other times, you just want to store data temporarily to conserve cell phone data and then upload when you're in proximity to Wi-Fi. And, of course, there are times when no network is available at all –mobile apps are expected to be flexible and robust, no matter the situation.

If your data needs are simple, one or more files may be all you need to deal with. But sooner or later, the benefits of a relational database will encourage you to look at the powerful storage and retrieval mechanisms of using a database engine in your app.

Delphi has always had good database support. Its FireDAC library is a mature framework that supports both large, server-based enterprise databases and, as we'll discuss in this chapter, embedded ones that you can distribute on a handheld phone. We used FireDAC briefly in *Chapter 6*, *All About LiveBindings*, and we'll take things a step further here as we look at deploying to mobile devices.

In this chapter, we will show you how to use two database options available to Delphi developers: Embarcadero's embedded InterBase editions, IBLite or IBToGo, and the open source SQLite library. We'll explore free management tools, show you how to deploy your database and then load and display it, and share tips for providing a usable interface for editing data on a small device.

We will be covering the following topics:

- Comparing different approaches
- Managing databases
- Setting up access to tables and queries
- Deploying your database
- Updating data on a mobile device

Data storage is a very important step toward becoming an expert in cross-platform development.

Technical requirements

Delphi 10.4 on the Windows platform will be the starting point for building our apps and prototyping, but talking about the key points of deploying to mobile devices will be our main focus. Additionally, some discussion and code examples will involve **SQL**, the **Structured Query Language** that's used by modern databases, and familiarity with it will be assumed.

This chapter discusses two different databases: InterBase and SQLite. SQLite is free and open source and is supported on every platform. InterBase's embedded editions (IBLite and IBToGo) are geared for mobile platforms, and either can be used depending on the license you have. When developing and testing from Delphi in Windows, an InterBase server needs to be running – the Developer Edition that comes with Delphi serves this purpose. The Mac desktop will not be used in this chapter.

The code for the two projects we will cover in this chapter can be found on GitHub at the following URL: `https://github.com/PacktPublishing/Fearless-Cross-Platform-Development-with-Delphi/tree/master/Chapter09`.

> **Note**
> This book's GitHub repository does not include the two databases that were used as the data sources for the two applications. The one for InterBase, `EMPLOYEE.GDB`, comes with Delphi; the one for SQLite, `chinook.db`, can be downloaded from `https://www.sqlitetutorial.net/sqlite-sample-database`.

Let's get started.

Comparing different approaches

There are several database options available for Delphi programmers when you have the power of a desktop environment, but this isn't the case in the cross-platform arena. The two we'll look at in this chapter use very different approaches to provide database capabilities, and each has their pros and cons. First, we'll look at InterBase and its several editions.

Learning about InterBase's editions

InterBase is Embarcadero's database offering and has a long history of being a pioneer of small footprint databases requiring little management, while still providing solid performance with multiple connected users. It's renowned for being able to quickly recover from backups and provides 256-bit encryption, role-based security, scalable replication with change views, robust transactional support, and strict compliance with SQL standards. Its rich language supports building complex stored procedures and triggers with generators, custom exceptions, and event alerters.

There are several editions available. The first three listed here are considered "server" editions in that they utilize a server process that's running, and they differ mostly in their licensing:

- **Server Edition**: In the enterprise, the Server Edition, which runs on Windows or Linux, competes head-to-head with well-known database products. It provides strong encryption at both the database and column level, journal archiving, write-ahead logging, remote SSL support, hundreds or thousands of simultaneously connected clients, and SQL-based connection monitoring. It is sold on a per-connection basis.

- **Desktop Edition**: The Desktop Edition is basically the same database engine as the Server Edition but only runs on Windows. More importantly, it only supports one local connection. This is a great way to deploy an application that starts with a powerful database at a low cost, which can later be scaled up to the Server Edition if you wish to support multiple users in a client-server environment.

- **Developer Edition**: The Developer Edition, a free download and which comes with Delphi Professional and up, has many of the higher-end features found in the Server Edition – it even supports up to 20 simultaneous users. However, it is the only edition that cannot be deployed to end users – it is strictly for internal development and testing and requires a restart every other day.

- **IBToGo**: This is an "embedded" edition of InterBase, which means that it can be deployed without a separate server engine running. As with the other editions described here, it supports Windows and Linux, but is specifically targeted at iOS and Android, making it an excellent choice for mobile database storage. IBToGo provides database file encryption, can securely connect to remote databases, allows up to eight simultaneous connections, and supports change views. A license for deploying IBToGo on mobile devices comes with Delphi Enterprise and Architect or can be purchased separately on a per-user basis.

- **IBLite**: This embedded edition of InterBase also works on mobile devices. It is free and a royalty-free distributable license comes with all versions of Delphi. It is limited to one connection and one transaction at a time, does not provide any built-in encryption, and limits the database's file size to 100 MB.

This chapter will concentrate on mobile data storage, so when we discuss data access using InterBase, the embedded editions will be implied. Which one you should use depends on the license you have.

As you consider putting data on a mobile device, the security of the file itself is more of a concern than if the file is on a locked-down server behind a firewall. If your mobile device is lost or stolen and your database is not encrypted, the potential for exposing the data is much greater. This, and the 100 MB file size limit, restricts IBLite's general use to educational purposes or situations where data security is not a concern.

InterBase is a great solution for deploying professional applications with relational database needs as you can scale from mobile devices up to enterprise servers with a single code base, have a consistent SQL language, and have a full feature set.

Before we start showing you how to use it, let's talk about an alternative, known as SQLite, and why you might choose that instead.

Introducing SQLite

SQLite is a free library of database routines and exists on all mobile smart phones; it's also available for download for any platform. Unlike many other database offerings, SQLite does not have a server component for any platform. Instead, all data creation, access, and manipulation is done through library calls, requiring no configuration to get it running.

Similar to InterBase, the entire database, including all tables, indices, and data, resides in one cross-platform file; its maximum file size is virtually unlimited at 281 TB. You can access the data through blobs, thus improving performance often above raw disk file access. SQLite is open source, fully tested, and internationally supported. There are no "editions" of SQLite; it has the same functionality and limitations on every platform.

There are several considerations for using SQLite that need to be weighed up in light of your application needs. Because SQLite is purely serverless, it's well-suited for embedded devices or situations where no data administration or setup is possible. On the other hand, there is no security outside the statements that are called by your application as there is no connection to be made and authenticated. Since all SQLite's data is in one file, accessing a large one could be a bottleneck on some systems without any proper disk caching or spanning. Furthermore, while multiple reads can be made simultaneously, only one write can take place at a time.

SQLite has several other differences from "mainstream" databases that should be noted as well. They are listed here:

- Data fields are considered "*flexible*" in that string values can be stored in INTEGER fields; if a string value being saved to a numeric field contains digits and it can be converted into a number; otherwise, it's simply stored verbatim as if the field was a TEXT field. Additionally, TEXT values longer than the defined field will not be truncated when they're stored, so no error will be raised and retrieving it later gives you the full, original value. SQLite is designed to work this way, which can be considered a feature if all you work with is SQLite. However, this can cause problems if you're porting data to another system with the expectation of strict data types.

- There is no support for stored procedures.

- There are no GRANT or REVOKE commands as there is no user security.

- Table constraints, including primary and foreign keys, can only be established when a table is created with the CREATE TABLE statement; they cannot be added later.

- Columns can be renamed or added to the ALTER TABLE statement, but they cannot be dropped or modified in any other way.

- In SELECT statements, the RIGHT OUTER JOIN and FULL OUTER JOIN expressions are not supported, although LEFT OUTER JOIN expressions are.

- Row triggers are supported but statement triggers are not.

There are other quirks to using SQLite that, if you have experience with other database products, could be surprising and trip you up. Take a look at the links in the *Further reading* section to learn more.

Despite these limitations, as some may view them, SQLite is deployed on more devices than any other database and is used by many well-known products. Its simplicity, ubiquitous presence, and price makes it a natural choice for many scenarios.

Now, let's work with these two database products and see how they compare in an application.

Managing databases

To get started working with databases, we need a tool to manage them. Sure, you can write SQL to create tables, indices, views, and so on, but database management tools do that tedious work for us. There are many tools available that work with a variety of database products. First, we'll cover the one that comes with Delphi for managing InterBase databases.

Using the InterBase Server Manager and IBConsole

All versions of Delphi come with a tool for working with InterBase, **IBConsole**, and some sample databases (if you included InterBase when you installed Delphi). When working with InterBase on your Windows desktop, you must start the local InterBase server with the **InterBase Server Manager** (if it's not already running). This can be found in the InterBase2020\bin folder of your Delphi installation and is called IBMgr.exe; it may also be found in the Windows Start menu as "Embarcadero InterBase 2020." Its interface is very simple:

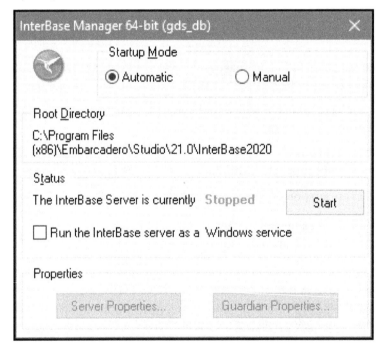

Figure 9.1 – InterBase Server Manager, ready to be started

The **Startup Mode** option allows you to automatically start the server when Windows starts, and its **Status** shows whether the server is currently running. Notice that the **Run the InterBase server as a Windows service** checkbox is currently disabled in the preceding screenshot. That's because this screenshot is of the Developer Edition of InterBase running on my development machine. The Server Edition has a few additional options, including running as a background Windows service. We'll cover these in greater detail in a later chapter.

> **Note**
>
> The InterBase tools mentioned here only run on Windows. Like Delphi, it is intended that you develop your application and database on Windows and then deploy it to other platforms.

Click **Start** to get your InterBase server running. Once it's running, start IBConsole, which can be found in the same folder as IBMgr:

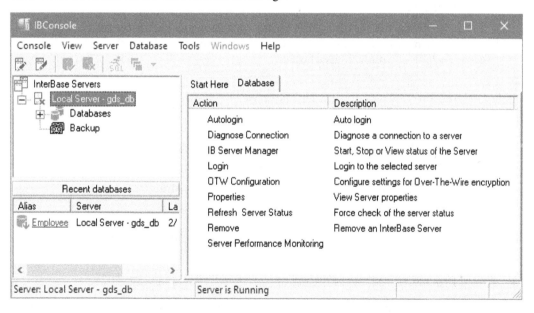

Figure 9.2 – IBConsole, a tool for managing InterBase databases

Now, you can create or add databases. Let's look at a sample one that comes with Delphi, most likely found in your public documents folder. Follow these steps to get it loaded into IBConsole:

1. Right-click on **Databases** and select **Add...**.

2. Click the ellipsis button next to the prompt for **File**, navigate to the `Embarcadero\Studio\21.0\Samples\Data` folder in your public documents area, and select `EMPLOYEE.GDB`.

3. Optionally, set an alias name that you can use to refer to the database later.

4. Use the default **User Name** and **Password** settings of `SYSDBA` and `masterkey`.

5. Check the **Save Password** checkbox for convenience and click **OK**.

You have now configured a database that you can work with.

Historical note

The `.GDS` default file extension for InterBase databases goes back to its routes in 1984, when it was started as Groton Database Systems, named after the Boston suburb of Groton, Massachusetts.

Double-click the `Employee` entry under **Databases** to connect to it. Here's a quick explanation of the actions that are available:

*   **Domains**: Allows you to create data definitions, package the data type, and use several attributes in field and parameter declarations

*   **Tables**: Creates and manages tables, keys, and check constraints

*   **Indices**: Defines the indices for your tables

*   **Views**: Creates pre-packaged SELECT statements that can be referenced like tables in your code

*   **Stored Procedures**: Manages the list of stored procedures in your database

*   **Triggers**: Establishes before/after triggers for your tables when records are inserted, updated, or deleted

*   **External Functions**: Loads DLLs containing additional functionality for use within your InterBase database

*   **Generators**: Defines autoincrementing values

*   **Exceptions**: Assigns global error message identifiers

- **Blob Filters**: Sets up special external functions that work with blob data

- **Roles**: Groups user permissions

- **User Permissions**: Controls fine-grained access to various database objects

If you click on these various actions in IBConsole, the right-hand side pane will show the list of that type of object in the selected database. Let's take a brief look at the CUSTOMER table; double-click on its entry under the **Tables** action to see its properties:

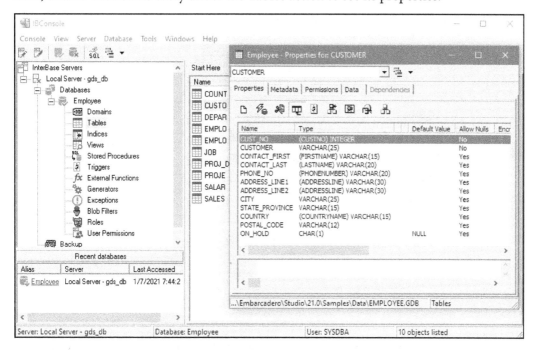

Figure 9.3 – The CUSTOMER table's properties in IBConsole

The first tab shows the fields that have been defined in this table; the **Metadata** tab shows all the SQL statements that were used to create this table and the surrounding infrastructure; **Permissions** shows who can access it and in what ways; the **Data** tab contains a grid you can use to browse the rows of data, allowing you to add, delete, and edit all the data in that table; and finally, the **Dependencies** tab shows objects that depend on the selected table.

Switch to the **Metadata** tab to view the SQL definition for the CUSTOMER table. At the top of the code shown here, there are several domain definitions:

```
CREATE DOMAIN ADDRESSLINE AS VARCHAR(30);
CREATE DOMAIN COUNTRYNAME AS VARCHAR(15);
CREATE DOMAIN CUSTNO AS INTEGER CHECK (VALUE > 1000);
```

```
CREATE DOMAIN FIRSTNAME AS VARCHAR(15);
CREATE DOMAIN LASTNAME AS VARCHAR(20);
CREATE DOMAIN PHONENUMBER AS VARCHAR(20);
```

Domains establish an alias to a predefined list of data type attributes that can be used in CREATE TABLE statements; for example, the ADDRESSLINE domain is defined as a VARCHAR(30) and is used in two fields of the table definition:

```
CREATE TABLE CUSTOMER
(
        CUST_NO              CUSTNO NOT NULL,
        CUSTOMER             VARCHAR(25) NOT NULL,
        CONTACT_FIRST        FIRSTNAME,
        CONTACT_LAST         LASTNAME,
        PHONE_NO             PHONENUMBER,
        ADDRESS_LINE1        ADDRESSLINE,
        ADDRESS_LINE2        ADDRESSLINE,
        CITY                 VARCHAR(25),
        STATE_PROVINCE       VARCHAR(15),
        COUNTRY              COUNTRYNAME,
        POSTAL_CODE          VARCHAR(12),
        ON_HOLD              CHAR(1) DEFAULT NULL,
    PRIMARY KEY (CUST_NO)
);
```

This table defines a primary key, CUST_NO, and then adds to the definition with a foreign key and a check constraint:

```
ALTER TABLE CUSTOMER ADD FOREIGN KEY (COUNTRY) REFERENCES
COUNTRY (COUNTRY);
ALTER TABLE CUSTOMER ADD CHECK (on_hold IS NULL OR on_hold =
'*');
```

There is no built-in autoincrement data type in InterBase, so a *before-insert* trigger is set up for the primary key field, CUST_NO, which then gets the next value of the CUST_NO_GEN generator:

```
CREATE TRIGGER SET_CUST_NO FOR CUSTOMER ACTIVE BEFORE INSERT
POSITION 0
AS
BEGIN
    new.cust_no = gen_id(cust_no_gen, 1);
END;
```

After this, the table definition is complete, so the transaction is committed to the database:

```
COMMIT WORK;
```

Permissions are assigned outside the definition of the table and are always immediate (no transaction is created or needs to be committed):

```
GRANT DELETE, INSERT, SELECT, UPDATE, REFERENCES, DECRYPT ON
CUSTOMER TO PUBLIC WITH GRANT OPTION;
```

Without IBConsole, or some other visual database management tool, the only way to create tables would be by executing these SQL statements from the supplied command-line tool, isql, or from your code. You can do that, and there are several scenarios where that is necessary, but you can also use IBConsole in Windows and then ship the database file with your application. Before we get to that, though, let's look at a tool for SQLite.

Trying out SQLite Studio

There is no "official" database management tool for Windows recommended by the developers of the SQLite project. It is a platform-agnostic library and all that is supplied is a command-line shell for running SQL statements (but it has some nifty extensions you may find useful). If you look around for a Windows-based tool, you'll find a great many of the general database management programs that are available that support SQLite, along with many other databases. You're free to use any of them to create and manage your SQLite databases. In this chapter, we'll use an open source one that is geared specifically to SQLite called **SQLite Studio**. You can download it from https://sqlitestudio.pl.

The quickest way to get started is by looking at a sample database. I found one at
`https://www.sqlitetutorial.net` called `Chinook`. After downloading and
extracting the database file, run SQLite Studio and select **Database** | **Add a database**
from the menu, click the folder icon, navigate to the extracted file, and click **OK** to add
`chinook.db` to the list of databases managed by SQLite Studio. Double-click on the
`chinook` database to see a list of tables and views defined in the database. Double-click
on the `customers` table to see the structure:

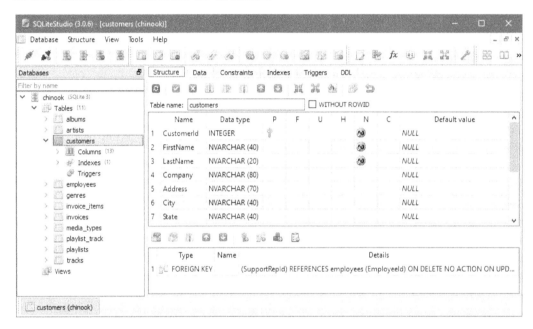

Figure 9.4 – SQLite Studio showing the customers table of the sample chinook database

Similar to what we saw when viewing table properties in IBConsole, there are multiple
tabs in the right-hand pane. The **Structure** tab lists the fields and various attributes; the
Data tab contains a grid to browse the rows of data, allowing you to add, delete, and edit
all the data in that table; the **Constraints** tab lists primary and secondary keys and check
constraints; the **Indexes** tab allows you to view, create, modify, and delete indices for the
table; the **Triggers** tab manages before/after triggers for your tables when records are
inserted, updated, or deleted; and the **DDL** tab shows the SQL in the data description
language for that table.

There are no domains in SQLite. It supports autoincrement integers; any constraints for a table must be established when it's created, and there are no SQL-level permissions supported in SQLite, so the CREATE TABLE statement for the **Customers** table is much simpler than for the similar one we saw in the sample InterBase database – it is contained in just one statement:

```
CREATE TABLE employees (
    EmployeeId INTEGER PRIMARY KEY AUTOINCREMENT NOT NULL,
    LastName NVARCHAR (20) NOT NULL,
    FirstName NVARCHAR (20) NOT NULL,
    Title NVARCHAR (30),
    ReportsTo INTEGER,
    BirthDate DATETIME,
    HireDate DATETIME,
    Address NVARCHAR (70),
    City NVARCHAR (40),
    State NVARCHAR (40),
    Country NVARCHAR (40),
    PostalCode NVARCHAR (10),
    Phone NVARCHAR (24),
    Fax NVARCHAR (24),
    Email NVARCHAR (60),
 FOREIGN KEY (ReportsTo) REFERENCES employees (EmployeeId) ON
 DELETE NO ACTION ON UPDATE NO ACTION
 );
```

Now that we have a sample database for both InterBase and SQLite and a tool to manage each, let's set up grids and data controls on the screen to hook them up.

Setting up access to tables and queries

Connecting to databases from Delphi is very straightforward. We mentioned FireDAC in *Chapter 6, All About LiveBindings*, when we showed you how to hook up controls to live data, but now, let's give it a proper introduction.

Utilizing FireDAC, Delphi's cross-platform Data Access Component

FireDAC provides a lightweight, flexible, yet highly optimized data access layer for a multitude of database systems, with many of the differences between them abstracted to greatly simplify working with them in Delphi. The initial connection often provides the most variance as user credentials, server ports, and encryption settings often define many aspects of the underlying engine. After that, the components provide a consistent and seamless interface to the data your application needs while still allowing vendor-specific options to pass through or optimized SQL to be executed.

> **Note**
> The exception to the idea of consistency when using various database products is, of course, when you're writing SQL. The differences in SELECT statements, JOIN clauses, and a host of expressions between the database vendors are many, and there's no getting around learning those nuances for complex or optimized queries.

To get started using FireDAC with both InterBase and SQLite, let's create two applications, one for each, and see how this all works. Follow these steps to set up the base of each application:

1. Create a multi-device application and select the **Tabbed with Navigation** template.
2. Select a place to save your application and call it MobileSalesIB.
3. Add a new data module to the project and give it a good name.
4. In the Data Module, add a TFDConnection component and a TFDPhysIBDriverLink.
5. In the main form, add the new data module to the uses clause in the implementation section.

6. Repeat *steps 1* through *5,* but in *step 2,* give this application the name
    `MobileSalesSQLite`. Then, in *step 4,* add a `TFDPhysSQLiteDriverLink`
    instead of the one for IB.

Before you go any further, make sure that both applications compile and run on whatever
various mobile devices you have available without any databases – it's always a good idea
to establish a baseline for testing before introducing new concepts. Here's what it looks
like on an Android phone:

Figure 9.5 – Standard "Tabbed with Navigation" template app running on an Android phone

We'll only use two of the main tabs, so delete the third and fourth ones if you so wish. Set the `Text` values of the first two to `Sales` and `Customers`. Note that the first tab item of the main tab control contains another tab control with two tabs in it, which makes that a total of four tabs we will be using.

> **Tip**
>
> With multiple tabs and with a tab control inside one of the tabs, it can be difficult selecting the right tab to work with at design time. One technique I often use is to make the tabs visible at design time by setting `TabPosition` to `Top` or `Bottom`, and then programmatically setting it to the desired visual appearance (`None`, `Dots`, or `PlatformDefault`) at runtime from the form's `OnCreate` event. Another way that doesn't involve adding code is to use the **Structure Pane** window.

To prevent confusion as we discuss placing components, we need to rename the tab controls and tabs. Since the template project came with code that uses them, I suggest modifying the code that references them as you make each change shown here:

- Outer (main) tab control: `tabCtrlMobileSalesIB`
- First outer tab of **tabCtrlMobileSalesIB**: `tabSales`
- Second outer tab of **tabCtrlMobileSalesIB**: `tabCustomers`
- Inner (second-level) tab control, residing on **tabSales**: `tabCtrlSales`
- First tab of **tabCtrlSales**: `tabSaleList`
- Second tab of **tabCtrlSales**: `tabSaleDetails`

Now, let's fill them in.

Getting table and query records from InterBase

First, let's focus on getting InterBase data inside the `MobileSalesIB` project:

1. Double-click on the `TFDConnection` component in the data module to pull up **FireDAC Connection Editor**.
2. Set **Driver ID** to `IB` (or `IBLite` if that's the version you have).

3. In the **Parameter** list, select EMPLOYEE.GDB for **Database** from Embarcadero's sample data.

4. Next, in the **Parameter** list, enter sysdba and masterkey for the **User_Name** and **Password** parameters, respectively.

5. Test the connection and click **OK**.

6. In **Object Inspector**, uncheck the **LoginPrompt** property.

7. Add a TFDTable component to the data module and ensure that its connection gets automatically assigned to the TFDConnection component.

8. Set the Table property of TFDTable to SALES by selecting it from the drop-down list of tables found in the InterBase database, and set its Active property to True.

9. On the main form, add TStringGrid to tabSalesList and set its Align property to Client.

10. Right-click on the grid and select **Bind Visually...** to bring up the LiveBindings Designer.

11. You should see both the grid and the table with fields from the SALES table in the LiveBindings Designer. Connect the asterisks of both to bind the grid to the sales data. The grid should instantly populate with data.

The default grid view, when connected to a data source, is to show all columns, so right-click on the grid, select **Columns Editor...**, and then click the **Add All Fields** button and remove the columns you don't want, adjusting the rest to suit your preference. I also set the TabPosition property of our main TTabControl to Bottom as I think it looks better and set the StyleLookup property of TSpeedButton in the top-right corner of the toolbar to nexttoolbutton to change it to an arrow. As a final touch, I added a TStyleBook and loaded the WedgewoodLight style for all four platforms.

Run this in your Windows development environment to confirm it looks similar to this:

Figure 9.6 – Sales from the InterBase sample database running on Windows

It may seem like it takes a lot to get to this point, but once you do it a few times, the steps almost become automatic.

Next, we want to see the details of a sale with some customer fields, so set up a query by dropping a TFDQuery component on the data module, right-click on the component and select **Query Editor ...,** and then enter the following SQL in the **SQL Command** area:

```
SELECT * FROM CUSTOMER WHERE CUST_NO = :CUST_NO
```

In the **Parameters** tab, set a couple of options for the CUST_NO parameter:

- **Param Type**: ptInput

- **Data Type**: ftInteger

You can optionally set **Value** to one of the CUST_NO values in the database and then click **Execute** to see your data right in **FireDAC Query Editor**. I entered 1004 and up came "*Elizabeth Brocket of Central Bank*" from my sample data:

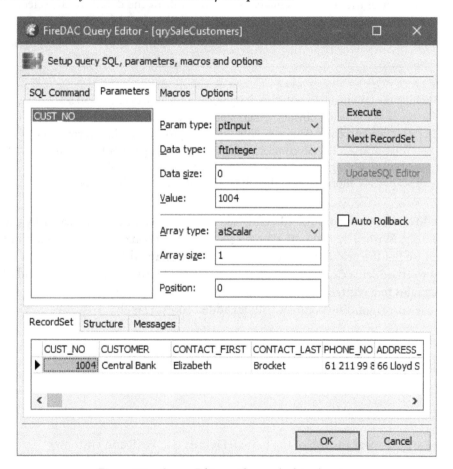

Figure 9.7 – Query Editor with sample data showing

Click **OK** to save the changes and close the Query Editor.

To set this up in a master-detail relationship with the sales table, create a `TDataSource` and set its `DataSet` property so that it points to the sales table component we established earlier. Then, set the `MasterSource` property of our new query component to that new data source and select `CUST_NO` for the `MasterFields` property. Since the master field matches the `CUST_NO` parameter of the query, we don't have to write any code to get the master-detail link to work – it happens automatically.

When a row on the sales grid is selected and double-clicked, or the right-arrow button in the top-right corner of the screen is clicked, we want to show the details of the selected sale, along with its associated customer information. We'll switch to **tabSaleDetails** programmatically at runtime to do that, so let's add the fields we want to that tab.

> **Tip**
>
> Switching the tabs of a tab control at design time with `TabPosition` set to `Dots` or `None` is not obvious – there's no visible tab to click on. You must select the tab control and change its `ActiveTab` property to the tab you want to see. You can select the tab control in the `Structure` pane or by clicking on the grid and hitting **Escape** to select its owner, and then the tab item, and then **Escape** once more to select the tab's owner; that is, the tab control.

First, let's add a few fields from the currently selected sale. We can do that easily by using the LiveBindings Wizard by right-clicking on the `BindingSourceDB` component we added when we put the grid in place. (I renamed my autonamed `BindSourceDB1` component to `BindSourceSales` to make it easier to identify.) Select **Link a field of BindSourceSales to a control** in the LiveBindings Wizard for each of the following fields and create the corresponding controls with an added control label:

- PO_Number: `TLabel`
- SHIP_DATE: `TDateEdit`
- ORDER_STATUS: `TEdit`
- PAID: `TCheckBox`

Arrange these fields near the top of the **tabSaleDetails** tab. We'll add customer details next.

> **Note**
>
> The LiveBindings Wizard will add controls to the form, not the currently selected tab. So, after adding these controls, you'll need to move them by simply dragging them into the **Structure** pane. If you cut and paste the controls, it'll break the LiveBinding link, which you won't discover until you save the form or run the app

To link up customer fields, we'll need another `TBindSourceDB` component. The easiest way to do that is to simply make a copy of the first one that was used for linking the sales data to the grid and then change its `DataSet` property so that it points to the data module's FDQuery component we created for getting a customer's fields. I named my second copy of the BindSourceDB `BindSourceSaleCustomer`.

> **Note**
>
> If the dataset being pointed to by the BindSourceDB component that's used for LiveBindings does not have any fields defined, the LiveBindings Wizard will not be able to create and link any controls. Simply set the table or query to active at least once – then all the fields will be available for linking.

Link the following fields from `BindSourceSaleCustomer` to new `TEdit` controls with control labels. Then, put them on the second tab under the fields containing the sales fields we just added:

- CUSTOMER
- CONTACT_FIRST
- CONTACT_LAST
- PHONE_NO
- ADDRESS_LINE1
- ADDRESS_LINE2
- CITY
- STATE_PROVINCE
- COUNTRY
- POSTAL_CODE

After arranging them nicely, the Sales Detail tab should look similar to this at runtime:

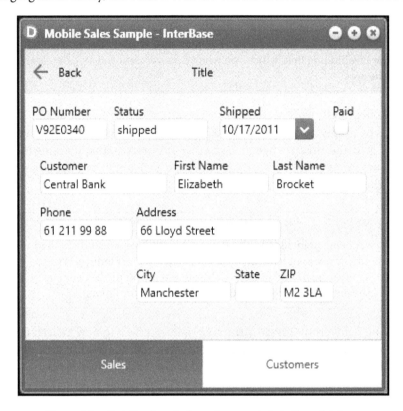

Figure 9.8 – Details of a sale from the sample InterBase database

We're now showing data from both tables and queries using FireDAC from InterBase in a Windows app – and we've not added any Delphi code yet, just one line of SQL. Before we show you how to deploy this to a mobile device, let's get our SQLite project up to the same point.

Getting table and query records from SQLite

This process is similar for preparing the *MobileSalesSQLite* project, but the initial set of instructions will be repeated here for completeness and to cover a few unique aspects of SQLite:

1. Double-click the `TFDConnection` component in the project's data module.
2. Select **SQLite** for `Driver ID`.
3. Select the `chinook.db` file you downloaded earlier for the `Database` parameter.
4. Click the **Test** button (you'll be prompted for user credentials, just like you were for the InterBase connection, but you can leave the fields blank as there is no user security for SQLite).
5. If all is well, clicking **OK** will confirm that the database can be accessed.
6. Uncheck the `LoginPrompt` property in the **Object Inspector** window.
7. Add a `TFDTable` component to the data module and ensure its connection gets automatically assigned to the `TFDConnection` component.
8. Set the `Table` property to **invoices** by selecting it from the dropdown list of tables found in the SQLite database, and set its `Active` property to **True**.

    We'll use a `TStringGrid` again as it plays better with LiveBindings. We can use the LiveBindings Wizard to set this up for us, but since the tab control is an aligned client, we can't right-click on the form to get to the LiveBindings Wizard in the right context; launching the wizard from the LiveBindings Designer gives us the options we need, so we'll go there first.

9. Pull up the LiveBindings Designer and click the **LiveBindings Wizard** button.
10. Select **Link a grid with a data source**, select `TStringGrid` from the **New grid** tab, and select the table component from the data module that points to the **invoices** table. Then, click **Finish**.
11. Align the new grid to the client of the tab and modify its columns.

As with the InterBase version of this app, I set the `TabPosition` property of our main `TTabControl` to `Bottom` and added a `TStyleBook`, this time loading the *Emerald Dark* style.

A quick preview of the app running in Windows looks like this:

Figure 9.9 – Sales from our SQLite sample database running on Windows

Again, we want to see the details of a sale with some customer fields. Add a TFDQuery component to the data module and enter the following SQL inside the Query Editor:

```
SELECT * FROM customers WHERE CustomerId = :CustomerId
```

Set the **Param Type** section of the `CustomerId` parameter to `ptInput` and **Data Type** to `ftInteger`. Then, click OK to save the changes and close the Query Editor.

Create a `TDataSource` and set its `DataSet` property so that it points to the sales table component. Then, set the `MasterSource` property of our new query component to the new data source and select **CustomerID** for the `MasterFields` property to establish a master-detail relationship. Then, right-click the BindingSourceDB component for the invoices table and use the LiveBindings Wizard to link the following two fields and controls from the invoice:

- InvoiceID: `TText`
- InvoiceDate: `TDateEdit`

Arrange these fields near the top of the **Sale Details** tab.

To link up customer fields, add another `TBindSourceDB` component by making a copy of the first one that was used for linking the sales data to the grid, and then changing its `DataSet` property to point to the data module's FDQuery component we created for getting a customer's fields. I named my second copy of the BindSourceDB `BindSourceInvoiceCustomer`.

Link the following fields from `BindSourceSaleCustomer` to the new `TEdit` controls with control labels. Then, put them on the second tab under the fields containing the sale fields we just added:

- Company
- FirstName
- LastName
- Phone
- Email
- Address
- City
- State
- PostalCode
- Country

After arranging them nicely, the **Sale Details** tab should look similar to this at runtime:

Figure 9.10 – Sale details from our SQLite sample database running on Windows

As we did with our InterBase data, we're now showing data from both tables and queries using FireDAC by utilizing SQLite in a Windows app. Now, it's time to actually deploy these to a mobile device!

Deploying your database

Database deployment could be a deciding factor for which database you select for your mobile app. If you own the Enterprise or Architect versions of Delphi, you have a license to deploy IBToGo on mobile devices (IBToGo for desktop platforms is sold separately); otherwise, you either need to purchase distributable licenses from a reseller, use IBLite, or use a different database such as SQLite.

Regardless of which one you choose, the database file itself will need to be deployed – unless you're creating it from scratch once the application has been installed on the mobile device. The project deployment feature of Delphi automatically places support files where they need to go, so all we need to do is add the file to the project and then use a platform-aware function in the IOUtils unit, TPath.GetDocumentsPath, which embeds the Application ID as part of the folder name on mobile devices. That function should be called in the BeforeConnect event of the FDConnection component in our data module, like this:

```
procedure TdmSQLiteSales.FDConnSQLiteBeforeConnect(Sender:
TObject);
var
  DataPath: string;
begin
  {$IF DEFINED(iOS) or DEFINED(ANDROID)}
  DataPath := TPath.GetDocumentsPath;
  {$ELSEIF DEFINED(MSWINDOWS)}
  DataPath := TPath.Combine(TPath.GetPublicPath,
    'MobileSalesSQLiteData');
  {$ENDIF}
  FDConnSQLite.Params.Values['Database'] := TPath.
    Combine(DataPath, 'chinook.db');
end;
```

This code was written for SQLite, but the same concept applies to IBLite or IBToGo database deployments.

> **Note**
>
> For SQLite, if you do not deploy the database file or it can't be found, it will simply be created automatically. So, either check for its existence before connecting or handle the exception when it tries to open a table that doesn't exist – the tables are not automatically created.

Adding the database file to the Delphi project triggers a deployment aid called **Featured Files**, which is for detecting database file types. Delphi realizes it will also need to deploy some library packages to support it and presents a list of what you might need, prompting you to select the appropriate options. This list is different based on the type of file you're adding. For SQLite, the libraries are already present on both iOS and Android devices, so you only need to add support files if you're using a ClientDataSet (which uses MIDAS. DLL) or dbExpress.

There are more files you need to distribute for InterBase applications, but the process is quite similar.

Deploying IBLite and IBToGo

SQLite is an open source product and was designed to be small and self-contained. InterBase started as a large enterprise server product and has recently added a small portable version. Due to its background, and to provide as much compatibility as possible, there are more files to be deployed with an IBLite or IBToGo distribution.

Here's what the **Featured Files** prompt includes when you're including an InterBase database file:

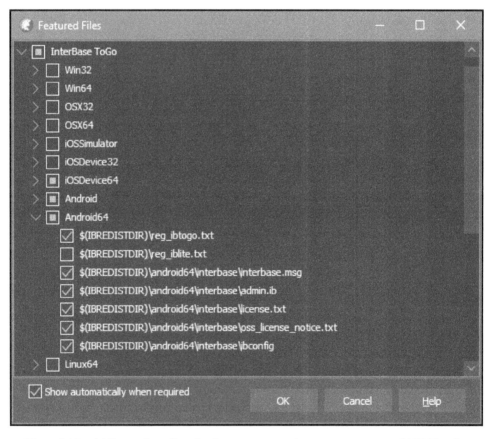

Figure 9.11 – Adding an InterBase database to your project prompts you to add IB libraries

The first two files listed under each platform show the license type you have. Either IBToGo or IBLite should be selected but not both.

If you need to modify this list later, you can select **Project | Deployment** to bring up the Deployment list for your project and click on the **Add Featured Files** button. The list of featured files will not be context-sensitive to the file you're adding, so it will list all the database drivers that are available.

After selecting the files and license type, Delphi takes care of compiling the appropriate libraries, bundling the application and database file, and putting them into your mobile device pretty seamlessly. Our deployed InterBase application on an Android phone looks like this:

Status	Date	Qty	Total
shipped	3/22/12	1	47.50
shipped	8/9/12	40	399,960.50
shipped	10/10/12	1	490.69
open	12/12/12	15	450,000.49
waiting	12/18/12	1	999.98
shipped	8/27/12	10	10,000.00
open	12/12/12	20	5,980.00
open	12/12/12	3	16,000.00
shipped	8/1/12	3	9,000.00
open	10/27/12	5	2,693.00
shipped	10/30/12	3	210.00
shipped	8/20/12	16	18,000.40
shipped	12/18/12	2	1,500.00
shipped	10/27/12	4	120,000.00
open	1/17/13	100	3,399.15
shipped	2/7/13	17	422,210.97
open	1/4/13	1	3,999.99
open	2/13/13	10	9,000.00

Figure 9.12 – The MobileSalesIB app running on an Android phone

We've done it! A database-enabled application is actually running on a mobile device. And to think this database technology was once relegated to large, back-office servers not too long ago.

But we're not quite done yet. We have shown you how to deploy a database to a phone, but there are some big differences to editing data on a small device.

Updating data on a mobile device

Now that you have a database on your mobile device, we will mention some of the techniques you can use to edit and save data as there are a few differences from desktop-based applications.

First, you're probably accustomed to presenting a screen filled with edit fields and allowing the customer to click **OK** or **Cancel** to decide whether to keep the changes they've made or discard them all. Mobile device users expect everything to always be saved automatically. In fact, the Apple guidelines for app design encourage fluidity to help users interact with their devices in a smooth and seamless manner; stopping to ask for confirmation should only be done when essential to prevent the loss of data or to alert the user to an important notice or decision to be made. For example, a point-of-sale application should allow a receipt-in-progress to be tendered or canceled, depending on the customer's choice to proceed with the purchase, but a change in font size should take effect immediately.

At this point in the discussion, we'd typically talk about transactions, but in a single-user mobile application, that's seldom a worry. Both InterBase and SQLite have good transaction support, and FireDAC sets up auto-start and auto-commits by default so that editing and saving your data is as seamless as possible.

Second, and of much greater concern, is the user interface – the actual layout and controls that are used for editing data on a handheld computer.

Understanding touch-oriented interfaces

Arguably the biggest paradigm shift when moving from writing desktop applications to mobile devices is getting used to programming for touch-centric user interfaces, where the fingertip is often the smallest pointing device in use and there are no "mouse-over" events or scroll wheel actions to capture. Instead, you need to be cognizant of the fact that a virtual keyboard will likely cover the bottom half of your screen, drop-down lists and radio buttons need to be big enough that the user won't get frustrated by constantly "fat-fingering" a different control, and everything fits on a variety of form factors.

As a final addition to our two mobile apps, let's add a list of customers to the second tab and make some edits. I won't list all the steps here as they should be very familiar by now. Put a `TFDTable` on the data module for each app. In the InterBase app, select the *CUSTOMER* table; in the SQLite app, select the *customers* table. Use LiveBindings to connect these customers to a `TStringGrid` and open the table at runtime, along with the other ones in the `DataModuleCreate` procedure.

At runtime, try to navigate the customers on your phone using your finger. Gestures work really well, but if you have too many columns or they're too wide, not only will you scroll up and down but also side to side. And take note of the keyboard that comes up when you click in a cell to edit a value.

Back in Delphi, move the grid to the bottom half of the tab and put a couple of `TEdit` boxes with labels in the top half, hooked up to the first and last names of the customer. Also, mark the grid as `ReadOnly` – we want all the editing to happen in the top half of the screen in bigger edit controls, and we wish to prevent the keyboard from inadvertently covering the grid while we're scrolling records.

Looks pretty decent, right? Try switching the design-time view to *iPhone 4* and take a look. You will likely have to adjust things. Try other views of the form, adjusting the grid height and the placement and width of the other controls so that it looks appropriate for each device type. When I switched to an Android 10" tablet, the grid was not even showing on the screen – I had to scroll down the view to find it and adjust the height up. Don't forget landscape mode – the screenshots in this chapter have all been in portrait mode, but your users may want the option to change this.

Once you're satisfied, run it on your mobile device. Notice that after making an edit in one of the name fields, the grid is updated immediately after the focus leaves the control. And if you shut down the app and restart it, those changes are persisted. That's an indication of LiveBindings working and both FireDAC's and the underlying database's automatic transaction support working well to save your data.

Here's what MobileSalesSQLite looks like while we're editing the `First Name` field on an iPhone:

Figure 9.13 – Editing a field in the MobileSalesIB app on an Android phone

Making your app usable and presentable while looking natural on each type of device is an acquired skill that can be tedious, but it pays off when your apps are well-accepted.

Summary

In this chapter, we've examined ways to store and retrieve relational data on mobile devices and showed how FireDAC can work with a variety of products. We learned about several editions of the enterprise-grade InterBase database that have been made available for mobile devices, thus providing advanced security and scalability within an application that can be held in your hand. SQLite, a robust and open source library that is available everywhere, was shown to be quite capable, despite its simplicity and price.

Free database management tools were then introduced for each of these products. After that, we loaded those databases and showed you how to access tables and queries from your Delphi environment, as well as how to set up the connections to edit the same data when it's deployed to a mobile device. Finally, considerations for working with edit controls were presented, giving you a great start toward managing mobile data in a user-friendly way.

Now that you're comfortable with storing and displaying data on a mobile device, we can use this in various ways as we move forward in our study of mobile devices.

Questions

1. Which edition(s) of InterBase come with Delphi Professional?

2. What happens if you try to store a character string in an integer field in SQLite?

3. How are autoincrement values supported in InterBase?

4. At what point can table constraints be established for tables in SQLite?

5. What function does Delphi provide to give convenient access to an application's data folder?

6. Do you need to distribute a license file with IBLite?

7. Where should edit fields be placed in a mobile app to prevent them from being overlaid by a virtual keyboard?

Further reading

*   InterBase 2020: `https://www.embarcadero.com/products/interbase`

*   History of InterBase: `https://en.wikipedia.org/wiki/InterBase#History`

*   Command-line isql tool: `http://docwiki.embarcadero.com/InterBase/2020/en/Command-line_isql_Tool`

*   InterBase for RAD Studio Developers: `https://youtu.be/hFAyAqoZLrE`

*   Getting started with InterBase for RAD users: `https://blogs.embarcadero.com/getting-started-with-interbase-for-rad-users/`

*   InterBase Editions: `http://docwiki.embarcadero.com/InterBase/2020/en/InterBase_Editions`

- InterBase ToGo Quick Start Guide: `http://docwiki.embarcadero.com/InterBase/2020/en/ToGo_Quick_Start`

- IBLite and IBToGo Deployment Licensing: `http://docwiki.embarcadero.com/RADStudio/Sydney/en/IBLite_and_IBToGo_Deployment_Licensing`

- What is SQLite? `https://sqlite.org`

- Quirks, Caveats, and Gotchas in SQLite: `https://sqlite.org/quirks.html`

- SQLite Tutorial: `https://www.sqlitetutorial.net`

- SQLite Studio: `https://sqlitestudio.pl`

- Using SQLite with FireDAC: `http://docwiki.embarcadero.com/RADStudio/Sydney/en/Using_SQLite_with_FireDAC`

- Quickly Learn How to Connect and Manage a SQLite Database for Delphi: `https://blogs.embarcadero.com/quickly-learn-how-to-connect-and-manage-a-sqlite-database-for-delphi-c-builder-with-sqlite-sample-app/`

- Embarcadero's FireDAC – Universal Enterprise Data Connectivity: `https://www.embarcadero.com/products/rad-studio/firedac`

- Using InterBase ToGo to Secure Mobile Data: `https://blogs.embarcadero.com/using-interbase-togo-to-secure-mobile-data`

- Deployment Manager – Add Featured Files: `http://docwiki.embarcadero.com/RADStudio/Sydney/en/Deployment_Manager_-_Add_Featured_Files`

- Apple's Human Interface Guidelines for iOS: `https://developer.apple.com/design/human-interface-guidelines/ios/overview`

10
Cameras, the GPS, and More

In this chapter, we'll look specifically at smartphone features, such as accessing the built-in camera, utilizing location and mapping services, and sharing data with other applications.

There are several sample projects available from Embarcadero that come with Delphi or can be downloaded via the GetIt Package Manager. We will not cover these specifically but will build our own app throughout this chapter that takes code from these projects and puts them together in a fun, park-visiting app.

The idea for this app will be that we want to build a list of all the parks in our neighborhood, take pictures, and save their locations to a small database. You can easily tweak this app to apply to museums or clients or whatever you want—or even a combination of these by adding a category. This will be a thorough introduction to accessing mobile app services.

We'll keep it simple, building it through the following sections:

- Establishing a base
- Getting permission
- Capturing your neighborhood
- Marking your spot

- Mapping your way
- Sharing your pictures

Let's learn how to access your phone's features programmatically.

Technical requirements

The Windows platform will play a minor role in this chapter, being used only as the development base. Android and iOS devices will be the focus of our discussion and the target for the deployment and testing of our applications, showcasing various features of these constantly evolving platforms. For the most part, any Android- or iOS-based phone or tablet supported by Delphi 10.4 will work—we will mention a few things to keep in mind about device capabilities along the way.

You can find the code present in the chapter on GitHub at `https://github.com/PacktPublishing/Fearless-Cross-Platform-Development-with-Delphi/tree/master/Chapter10`.

Setting up

Before we get into the meat of this chapter, let's quickly set up the app we'll work with. Create your app for this chapter by starting, as usual, with a multi-device application; select the **Tabbed with Navigation** template, then follow these steps:

1. Remove the fourth tab and name the three remaining tabs of the main `TTabControl` property `tabParkList`, `tabParkEdit`, and `tabParkMap`; rename the tab control `tabCtrlMain`.

2. Remove the tab control residing on the `tabParkList` property and add a `TToolBar` aligned to `bottom`, and to that, add a left-aligned button with a `StyleLookup` property of `additembutton`.

3. Also, on that tab, drop a `TListView` in the center of the tab area and align it to `Client`.

4. We'll be calling `NextParkTabAction` from an event on the list view, so remove the **Next** button in the top toolbar and change the title's `Text` property to `My Parks`.

5. On `tabParkEdit`, add a `TToolBar` aligned to `bottom`, just as you did on `tabParkList`, and then add a left-aligned button with a `StyleLookup` property of `pagecurltoolbutton` and a right-aligned button with a `StyleLookup` property of `trashtoolbutton`.

6. We don't want to see the tabs at runtime but we don't want to hide them at design time, as it's much more convenient to move around when they're visible. So, put this code in the form's OnCreate event:

```
procedure TfrmMyParks.FormCreate(Sender: TObject);
begin
   tabCtrlMain.TabPosition := TTabPosition.None;
   tabCtrlMain.ActiveTab := tabParkList;
end;
```

7. Add a data module and drop both a TFDConnection component and a TFDPhysSQLiteDriverLink component on it. We'll use SQLite to store our data for the app, building on the skills we learned in the previous chapter, so set up the connection for a SQLite database in a cross-platform manner; we'll name this database MyParks.db.

8. Bring up your favorite SQLite database management tool and create two tables. The first, Parks, will hold the park name, an *x* and *y* **Global Positioning System** (**GPS**) location, a picture of the park, whether the park has a playground, and some notes. Here's the **Structured Query Language** (**SQL**) code to create it:

```
CREATE TABLE Parks (
    ID               INTEGER PRIMARY KEY AUTOINCREMENT
UNIQUE NOT NULL,
    ParkName         STRING (50) UNIQUE NOT NULL,
    LocX             REAL,
    LocY             REAL,
    MainPic          BLOB,
    HasPlayground BOOLEAN
);
```

9. Add the data module to the implementation uses clause of the main form, and then use the LiveBindings wizard to hook the TListView up to the ParkName field of the Parks table.

10. Optionally, add a TStyleBook and choose a style for both Android and iOS platforms. (I chose EmeraldCrystal, which you will see in the screenshots in this chapter.)

11. Use the SQLite database management tool to add one or two records in the `Parks` table, setting the `ParkName` field value so that we'll have something to see when we start the app, and then disconnect.

12. Add the database file to the project so that it will get deployed to your mobile device.

13. Run it and make sure you see something like this:

Figure 10.1 – The start of the "My Parks" app on an Android phone, with two parks in the database

To edit and save entries to the park database from the mobile app, we'll also need to set up the Edit tab. Switch to the `tabParkEdit` tab and add a `TEdit` with a label for the `ParkName` field, and a checkbox for the `HasPlayground` Boolean field (I used panels and alignment to make it look good on various device sizes). Use LiveBindings to hook these two new controls to the `tblParks.ParkName` and `tblParks.HasPlayground` fields, respectively. In the toolbar at the bottom of the page, add a right-aligned button to delete the current park. Create an action to delete a park after prompting to make sure the user really wants to do it, and assign it to the new **Delete** button.

Here's how I implemented mine:

```
procedure TfrmMyParksMain.actDeleteParkExecute(Sender:
TObject);
begin
  TdialogServiceAsync.MessageDialog(
    'Are you sure you want to delete this park?',
    TMsgDlgType.mtWarning, [TMsgDlgBtn.mbYes, TMsgDlgBtn.mbNo],
    TMsgDlgBtn.mbNo, 0,
    procedure (const AResult: TModalResult)
    begin
      if AResult = mrYes then begin
        dmParkData.tblParks.Delete;
        ParkEditDoneTabAction.Execute;
      end;
    end);
end;
```

Your edit screen should now look similar to this:

Figure 10.2 – The park edit screen on an Android phone

The final piece is to create an action to add a park; hook it up to the button at the bottom left of `tabParkList`, as follows:

```
procedure TfrmMyParksMain.actAddParkExecute(Sender: TObject);
begin
  dmParkData.tblParks.Insert;
  NextParkTabAction.Execute;
end;
```

Run the application and make sure you can add, edit, and delete park entries.

To keep the edits in the mobile database, you need to make one deployment modification. In Delphi's **Project | Deployment** screen, modify the **Overwrite** setting for the MyParks.db file you added earlier and change it from its default of Always to Never. Otherwise, every time you compile and redeploy the project, the database file on the mobile device will get overwritten with the initial database you set up on your Windows machine. Setting it to Never will still deploy it if it doesn't already exist.

Having built this app, you've successfully mastered several prior chapters of this book—congratulations on expanding beyond Windows! I recommend writing a few different apps such as this yourself, to get used to working with FireMonkey, LiveBindings, FireDAC, and mobile device deployment. To compare what you're building, the finished source for this app can be found on GitHub at https://github.com/PacktPublishing/Fearless-Cross-Platform-Development-with-Delphi/tree/master/Chapter10/01_MyParks.

Now that we have a basic database-enabled app on our mobile device, let's put its services to work for us.

Establishing a base

Not every mobile device supports every service and some people don't update their devices, meaning they could have an old version of the **application programming interface** (**API**) that doesn't support a new service you'd like to use.

You need to be aware of the various versions of devices your users will have and pick a minimum-supported API. You may recall reading in *Chapter 4, Multiple Platforms, One Code Base,* the minimum versions supported by Delphi for each platform. There are a lot of smartphones out there and while some people are constantly upgrading to the latest and greatest, a larger percentage hang on to their devices for several years. Additionally, not all devices support all features.

Instead of building a long list of devices and versions and which ones have which capabilities, there's a far simpler way to figure out what you can do on a device: just ask it! The FireMonkey platform has done all the hard work for you and provides a class in the FMX.Platform unit called TPlatformServices, which contains a SupportsPlatformService Boolean function, returning whether or not the given service is available. Here's how you would check whether your device supports modal dialogs (hint: iOS devices do, while Android ones do not):

```
if TPlatformServices.Current.SupportsPlatformService(
    FMX.Platform.IFMXDialogServiceSync) then ...
```

Not all the services available are in the `FMX.Platform` unit. What may be more interesting is to see whether the device is capable of making a phone call. That particular service is listed in the `FMX.PhoneDialer` unit, which means you'd have to include both units to make this check, as illustrated in the following code snippet:

```
if TPlatformServices.Current.SupportsPlatformService(
        FMX.PhoneDialer.IFMXPhoneDialerService) then ...
```

This will return `True` for Android phones and iPhones, but `False` for iPads. Here are a couple of units we'll need for this chapter:

- `FMX.MediaLibrary.IFMXCameraService`
- `FMX.Maps.IFMXMapService`

To see a complete list of services available, follow the *FireMonkey Platform Services* link in the *Further reading* section at the end of this chapter.

Now that you have established what you can do, make sure you have all the permissions needed to access those services.

Getting permission

One of the first things you'll encounter when trying to use mobile services is that some require express permissions to be granted by the user before allowing your app to access them. That's a good thing—a person's tablet or phone is often a huge collection of personal information. Things such as photos, addresses, emails, passwords, and your current location are not the kind of data that should just be blindly accessed by anyone or any app. Users should be in charge of who gets access to which information.

It used to be that when a new app was being installed, apps would ask the user to confirm all the possible permissions the app would need in order to be installed. These days, the apps are installed, and if and when they need to do something sensitive, that permission is requested and either denied, preventing access for that feature, or approved, allowing the app to proceed. This second method gives the user a more fine-grained control over what apps can and cannot do—for example, they may allow an app to take pictures but deny location services. We'll need to take this into account and make sure our app works for the user in the capacity the user has granted.

Setting up permissions in Android and iOS is done quite differently. Let's look at what it takes to allow an Android app to do what you need.

Setting up permissions for Android apps

A FireMonkey's project options has a section under **Application** called **Uses Permissions** that, for Android targets, presents a long list of permissions for you to check off according to what your app needs to do, as illustrated in the following screenshot:

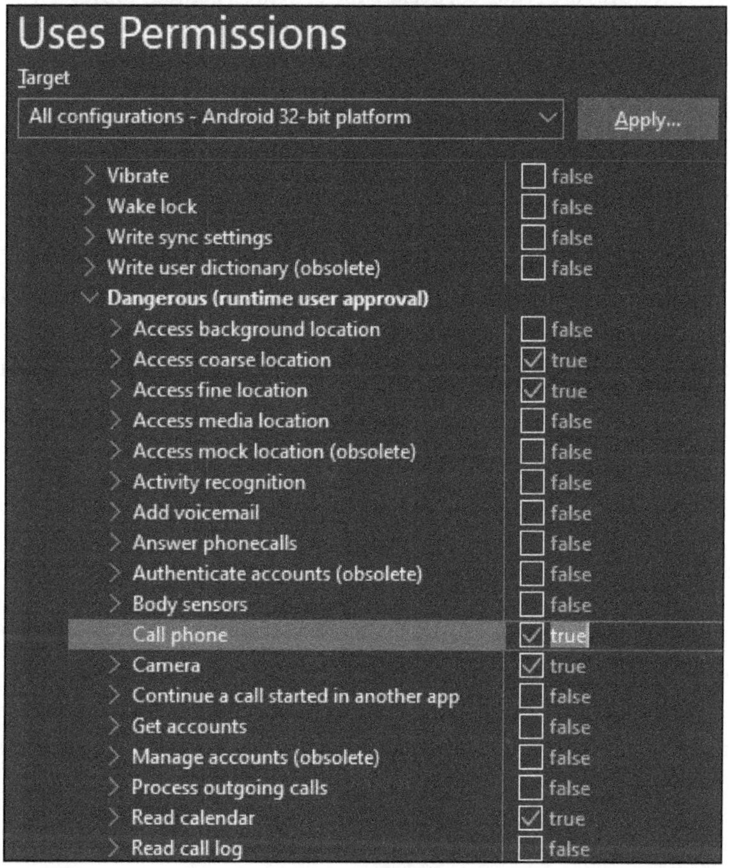

Figure 10.3 – Some of the Android permissions available in Project Options

As you can see, there are many, very specific, permissions. The early days of permissions saw much more generalized levels such as accessing external storage, which is now split to reading and writing external storage. The permissions section under **Dangerous (runtime user approval)** requires user approval before the app is allowed to perform the named function; the other permissions are allowed by default if selected, and cannot be denied. You can read about what each permission does by following the link titled *Android Developers documentation – Manifest.permission* in the *Further reading* section at the end of this chapter. These permissions are put into a **manifest file** associated with Android apps and automatically built and packaged for you by the Delphi **integrated development environment** (IDE).

Note

Manifest files are **Extensible Markup Language (XML)** files that accompany every Android project and describe several aspects of how the app interacts with the Android **operating system (OS)**. They specify which hardware and software features are required, declare the activities, content providers, broadcast receivers, and other components of the app, and establish the permissions needed to access these features and components. You will likely never need to worry about the details of an Android manifest file as Delphi builds it for you based on several parts of the project options. However, if you're interested, there is a default manifest file template in your `%APPDATA%\Embarcadero\BDS\21.0` folder that gets copied to new project folders.

Once the permissions required are defined in the project, you can check to see whether one has been allowed by the user. If you do not check one of the permissions considered dangerous, it doesn't give the user the chance to approve it and it is therefore automatically denied, resulting in an ugly error, as illustrated in the following screenshot:

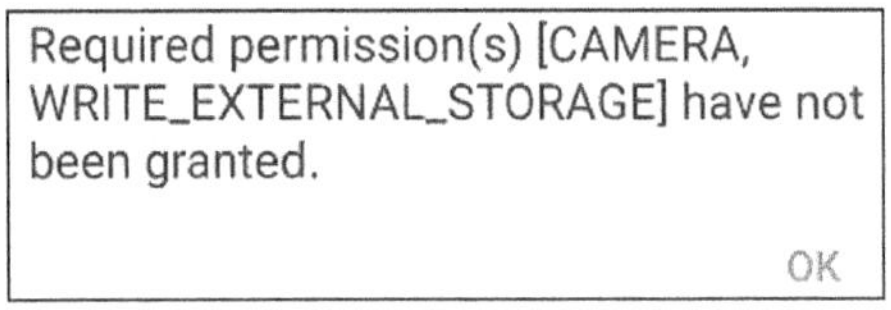

Required permission(s) [CAMERA, WRITE_EXTERNAL_STORAGE] have not been granted.

OK

Figure 10.4 – Error message if permission not requested to take and store pictures on Android

The FireMonkey framework provides a way to check permissions and take action if granted, all in one statement, by calling `PermissionsService.RequestPermissions` from the `System.Permissions` unit. This procedure has three parameters, listed as follows:

- `Permissions`: A dynamic array of strings, each of which must match a permission constant

- `Request result event handler`: A non-blocking procedure that should check to see whether the permissions were granted and take the appropriate action

- `Rationale event handler`: A non-blocking procedure to display the reason a permission is being requested if the first request was denied

The permissions constants can be found in the *Manifest.permission* link mentioned earlier. If you click on a permission link, look down until you find the constant value. For example, to request access to the camera, you would pass `['android.permission.CAMERA']` in for the first parameter.

Many demo programs show an alternate way that you may prefer; these use some helper units that come with Delphi, as illustrated in the following code snippet:

```
uses
{$IFDEF ANDROID}
  Androidapi.Helpers,
  Androidapi.JNI.JavaTypes,
  Androidapi.JNI.Os,
{$ENDIF}
```

With these units available, you can assign the string constant pulled from the manifest file to a local variable, as illustrated in the following code snippet:

```
var
  FPermCamera: string;
...
{$IFDEF ANDROID}
  FPermCamera := JStringToString(TJManifest_permission.
    JavaClass.CAMERA);
{$ENDIF}
```

Then, pass that string variable in for the permissions parameter.

The TRequestPermissionsResultEvent parameter is a special event handler procedure that, in addition to the traditional Sender: TObject parameter used by many event handlers, has two more parameters, the first containing an array of permission constants and the second containing an array of results for each of the permission constants. This lets you go through each permission and discover whether it was granted or not; you can act independently on these results if you'd like.

Once the permission constants and event handlers are set up, the rest is done in one statement, as follows:

```
PermissionsService.RequestPermissions(['android.permission.
CAMERA'],
      TakePicturePermissionRequestResult);
```

We will discuss how to actually take a picture from the request result event handler procedure later in this chapter, in the *Capturing your neighborhood* section.

The `PermissionsService` class is very useful for checking permissions on the Android platform. So, what do you have to do for iOS?

Nothing.

Using sensitive services on iOS

On iOS devices, any call to services that require user permissions is automatically handled by the OS. If the user should be queried, the OS pauses the app and waits for the user to answer a message—without any extra coding to handle it. In other words, you can simply attempt to take a picture, make a phone call, read from the address book, write to external storage, or carry out another of the other functions that might require specific user permissions, and not do any checking as to whether the user wants you to do it or not.

But if your app is truly cross-platform, you can use the same `PermissionsService` class used by Android apps to keep your code simple—the iOS implementation of the class does nothing more than call the result event handler.

> **Note**
>
> Be aware that your app may not act in the way you expect if a requested permission is denied by a user—for example, if you attempt to access the camera and the user does not allow you to, the app will still appear to use the camera but the picture will be black. When attempting to get contacts from `TAddressBook`, a disallowed iOS device will inform the user the operation could not be performed. Different services will act differently when denied.

While there's a lot of setup and coding for Android devices, there's more control in how your app responds to user choices. Most of the time, users just give apps whichever permissions they request; this puts a responsibility on you, the developer, to follow the Apple guidelines on requesting permissions, which advise to keep requests for sensitive information to the minimal amount necessary.

The rest of this chapter will check for permissions, as described previously, but assumes they will be given for the functionality we'll be implementing. So, without further ado, let's take some pictures!

Capturing your neighborhood

Delphi has made it very simple to take pictures regardless of which platform you're on. This capability is provided in one of the standard actions. Double-click on a `TActionList` and click on the little arrow to the right of the **New Action** list to drop down the menu of new action types, select **New Standard Action...** and scroll to the **Media Library** section, where you'll find `TTakePhotoFromCameraAction`. Click to add this new action to your action list, set its name to something appropriate, and leave all the properties at their default. The one thing you want to do with this is create an event handler for the `OnDidFinishTaking` event. In here, you'll add the code to do something with the image returned by the camera-taking action.

Before you hook up this new action to a button, remember that the camera can only be accessed if it's been granted permissions. What I like to do is create a custom action that will be linked to a button or menu item, and that action's `OnExecute` event will ask for permissions and then call the camera action. You caught a glimpse of this technique in the *Setting up permissions for Android apps* section, where we looked at `PermissionsService.RequestPermissions`. The second parameter of that call is a procedure parameter, and instead of creating a separate procedure, as was indicated previously, we can simply create an anonymous procedure in its place. This way, we can ask for the permissions, check the results, and call `TakePhotoFromCameraAction` all in one step. Here's the `OnExecute` event for my custom action that I have hooked up to a button on `tabParkEdit`:

```
procedure TfrmMyParksMain.actTakeParkPicExecute(Sender:
TObject);
begin
  PermissionsService.RequestPermissions(
    ['android.permission.CAMERA',
     'android.permission.WRITE_EXTERNAL_STORAGE'],
    procedure (const APermissions: TArray<string>;
               const AGrantResults: TArray<TPermissionStatus>)
    begin
      if (Length(AGrantResults) = 2) and
         (AGrantResults[0] = TPermissionStatus.Granted) and
         (AGrantResults[1] = TPermissionStatus.Granted) then
        TakePhotoFromCameraAction.Execute
      else
        TDialogServiceAsync.ShowMessage('Camera denied
          access');
```

```
    end);
  end;
```

Here are a few things to note about the preceding block of code:

- We requested two permissions—access to the camera and writing to external storage. Using the camera also requires reading from external storage, and many demos will show all three permissions requested. The Android documentation says that if permission is granted to write to external storage, then reading from it is implicitly granted; so, I left it out.

- Many sample apps will show more detailed messages, with the code looping through all permission requests and checking each corresponding grant result, showing a message for each; I skipped that and combined them all together in one pass-or-fail case.

- The call to `TakePhotoFromCameraAction.Execute` is how we call the standard `TAction` that does the photo-taking for us and calls the `OnDidFinishTaking` event when the picture has been confirmed by the user for use.

- Android event handlers must be non-blocking, even when displaying messages. The simplest way to show a non-blocking message is to include the `FMX.DialogService.Async` unit and call `TDialogServiceAsync.ShowMessage`.

Using this technique and some of the other standard actions, you can see how simple it is to expand your app to include many other features. Before we explore those techniques, let's look at the event handler that saves the camera's image.

Saving an image to the database

The standard action, `TTakePhotoFromCameraAction`, does all the work of utilizing the camera on various platforms. When a picture has been taken, it provides the image in a simple event handler, aptly named `DidFinishTaking`. The single parameter is a `TBitmap`, allowing you to simply assign it to an image component on your form.

I created one called `imgParkPic` on `tabParkEdit`, so my event handler looks like this:

```
procedure TfrmMyParksMain.
TakePhotoFromCameraActionDidFinishTaking
          (Image: TBitmap);
begin
  imgParkPic.Bitmap.Assign(Image);
```

```
    SaveImageToDatabase;
  end;
```

As you can see, assigning it to the image is straightforward; however, saving it to the database—not so much. LiveBindings works to pull the image out and display it on the form, but when you hook it up, the `Direction` property of `TLinkControlToField` is automatically set to `linkBidirectional`; you need to change that to `linkDataToControl` so that LiveBindings only reads from the database and doesn't try to update it.

Saving the image to the database requires a bit of code involving `TMemoryStream`, as illustrated in the following snippet:

```
procedure TfrmMyParksMain.SaveImageToDatabase;
var
  PicStream: TMemoryStream;
begin
  PicStream := TMemoryStream.Create;
  try
    imgParkPic.Bitmap.SaveToStream(PicStream);
    PicStream.Position := 0;

    dmParkData.tblParks.Edit;
    dmParkData.tblParksMainPic.LoadFromStream(PicStream);
    dmParkData.tblParks.Post;
  finally
    PicStream.Free;
  end;
end;
```

If you're not familiar with using streams in Delphi, the bitmap's `SaveToStream` method fills the memory stream and leaves the pointer to the memory stream at the end of the stream; so, it needs to be reset to the beginning, or 0. Then, we put the table into **Edit** mode, use the bitmap field's `LoadFromStream` method to read from the memory stream we just wrote out, and save the modified record in the database with the `Post` method.

And that's it—you now have a mobile app working on both iOS and Android that takes pictures and saves them in a database.

> **Note**
>
> If you encounter an error containing `java.lang.NullPointerException` on Android with advice to modify your manifest file, go to your project's options and under **Application | Entitlement List**, make sure **Secure File Sharing** is checked.

What if you had already taken a picture and just wanted to load it from your photo library?

Loading previously taken images

While you were scrolling down through the standard actions, you may have noticed another one in the **Media Library** section: `TakePhotoFromLibraryAction`. This one is very similar to the camera-taking one we just looked at but only needs one permission: reading from external storage. Here's how I implemented my custom action to load a picture:

```
procedure TfrmMyParksMain.actLoadParkPicExecute(Sender:
TObject);
begin
  PermissionsService.RequestPermissions(
      ['android.permission.READ_EXTERNAL_STORAGE'],
      procedure (const APermissions: TArray<string>;
                 const AGrantResults:
                   TArray<TPermissionStatus>)
      begin
        if (Length(AGrantResults) = 1) and
           (AGrantResults[0] = TPermissionStatus.Granted) then
          TakePhotoFromLibraryAction.Execute
        else
          TDialogServiceAsync.ShowMessage('Storage access was
            denied.');
      end);
end;
```

As you can see, this code is nearly identical to the code that takes a photo with the camera, which we explained earlier. The OnDidFinishTaking event handler of TakePhotoFromLibraryAction has the exact same signature as TakePhotoFromCameraAction, so we can simply reuse it.

Assign your two new custom actions to a couple of buttons at the bottom of tabParkEdit and then go take a picture of a park. Here's a screenshot of a picture of a park, taken with the app:

Figure 10.5 – A park picture taken in the app with an Android device

This is the simplest way to take a picture but doesn't allow very much control. It's also limited to still pictures using the default camera on the phone.

Expanding your use of the camera

You can take videos, switch to the front-facing camera (if the device has one) and adjust other settings of your mobile device's built-in camera so that you can use a `TCameraComponent`. There aren't many properties surfaced in the **Object Inspector** but there are plenty you can adjust manually in code, and one important event handler: `OnSampleBuffer`.

Instead of calling a method to snap a picture and getting that completed picture back in a callback event, the camera component must be enabled with the `Active` property. In that state, you control several aspects of the image being captured, such as whether the flash is on, the resolution of the image, and which camera to use if there's both a front and back camera. While it's active, you capture images in the `OnSampleBuffer` event. This allows you the ability to save one frame, several frames, or a whole series in a video, or simply display it on screen as a preview.

Here are some of the public properties you can manipulate at runtime:

- `Kind`: This determines which camera is in use and can be set to `Default`, `FrontCamera`, or `BackCamera`.

- `Quality`: The quality can be set to `PhotoQuality`, `HighQuality`, `MediumQuality`, `LowQuality`, or `CaptureSettings` (custom).

- `CaptureSetting`: This encapsulates `Height`, `Width`, `FrameRate`, and other properties that determine overall resolution.

- `TorchMode`: The torch, also referred to as the flash, can be set to automatically turn on when needed, turned on and left on to work as a flashlight, or manually disabled. The property values are `ModeOff`, `ModeOn`, or `ModeAuto`.

With this increased control also comes the responsibility of knowing which settings work with each other. For example, on my Android phone, the torch is only available for the rear-facing camera. So, if in the following code the `Camera` variable is of type `TCameraComponent`, an exception would be raised on the second line if run on such a phone:

```
Camera.TorchMode := ModeOn;
Camera.Kind := FrontCamera;
```

We won't incorporate the use of video or advanced features of camera manipulation in our park app, but there's a great sample `CameraComponent` project that comes with Delphi in the `Object Pascal\Mobile Snippets` folder that showcases these capabilities.

So, now that we have a picture of the park, wouldn't it be nice to automatically record the location?

Marking your spot

There are three sensors on Delphi's **Tool Palette** and they deal with location, motion, and direction. Three sample apps that get installed with Delphi in the `Object Pascal\ Mobile Snippets` folder demonstrate their use: `Location`, `Accelerometer`, and `OrientationSensor`, respectively. The one we're interested in for this chapter is the first one.

If you look over this simple app, `TLocationSensor` is the key to recording our geographical coordinates. Run it to get a sense of how it works, and you'll notice that no location information is gathered until the component is made active. Furthermore, after it's turned on, there's a bit of delay while it gathers location details to calculate its exact position.

Drop a `TLocationSensor` component onto the main form of our park app and on `tabParkEdit`, add a button that will be used to record our current location.

When viewing the list of parks on the `tabParkList` tab, we don't need location services to be active. It's only when we're editing data that we might want to record the location of the park, so a good location in our app for activating the component would be in the `OnUpdate` event for the `TAction` that goes to the `tabParkEditMain` tab. But we also need to remember to ask for permission.

Getting permission for location services

In the project's options, check the **Uses Permissions** section for both Android platforms and select **Access fine location**.

Note

Some users are cautious about granting fine-grained location data to apps, so you may want to select **Access course location** instead, which gives an approximate location and is more acceptable for those concerned about privacy because it's less accurate. For the purposes of the code in this demo, I'll assume **Access fine location** has been selected—adjust your code accordingly.

In addition to setting the Android user permissions, when iOS devices ask the user to allow access to location services, they also include an explanation as to why the user should do so. This explanation is pulled from special **Version Info** keys in Delphi. There are two types of access granted, each with a special key name and outlined as follows:

- NSLocationWhenInUseUsageDescription: This explanation is for when the app asks whether you want to use location services.

- NSLocationAlwaysAndWhenInUseUsageDescription: This explanation shows up in iOS settings when you select **Privacy** | **Location Services** on your app; it shows just under the **Always** option.

You should set these descriptions in your project's settings, overriding the default ones initialized by Delphi.

Once permission has been set at the project level, we need to ask for it at runtime for Android devices and then turn the service on for all devices. We could ask for this permission right when the application starts—and that would be simplest—but I prefer waiting until a service is actually needed before attempting to use it. The problem (discovered while building this demo) is that the code executed in the event handler happens simultaneously with the slide transition, which stops the transition effect halfway through, leaving the park list view still partially on the screen and the park edit tab not fully slid into place.

This will never do, but is easily resolved by putting our permission-requesting code in a separate execution thread so that both the slide transition and the permission request happen without stepping on each other. By simply including the System.Threading unit, we can wrap our permission request in an anonymous method and pass it off to the TTask class. Here's how this is done in the OnUpdate event for NextParkTabAction:

```
procedure TfrmMyParksMain.NextParkTabActionUpdate(Sender:
TObject);
begin
  tabctrlParkEdit.ActiveTab := tabParkEditMain;


  TTask.Run(procedure
    begin
      PermissionsService.RequestPermissions([
          'android.permission.ACCESS_FINE_LOCATION'],
        procedure(const APermissions: TArray<string>;
             const AGrantResults:
               TArray<TPermissionStatus>)
```

```
      begin
        if (Length(AGrantResults) = 1) and
          (AGrantResults[0] = TPermissionStatus.Granted)
            then
          LocationSensor.Active := True
        else
          TDialogServiceAsync.ShowMessage(
            'Park location data will not be available.');
      end);
    end);
end;
```

Running it on an iOS device for the first time will now look something like this:

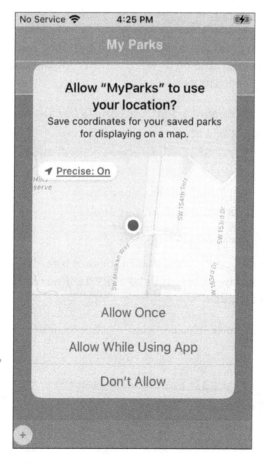

Figure 10.6 – iPhone asking permission to use location services with a custom explanation

Now that location services are available, we can capture the current longitude and latitude and store that data with our park.

Saving coordinates in the database

The location sensor component does not store coordinates for us to access any time we need them. Instead, an event provides that data whenever it changes—which could be frequently if the device is in motion, or not at all if it's stationary. What we'll do then is capture the data every time the event fires and, when a button to save the park's location is clicked, use the last-saved coordinates as the location for that park.

In the `private` section of the main form of our app, create a couple of variables of type `Double`, as follows:

```
FParkLongitude: Double;
FParkLatitude : Double;
```

The event to use is `OnLocationChanged`, and we'll grab the `NewLocation` parameter passed in (we won't use `OldLocation` as we're not charting our course over a map), as illustrated in the following code snippet:

```
procedure TfrmMyParksMain.LocationSensorLocationChanged(
          Sender: TObject;
          const OldLocation, NewLocation: TLocationCoord2D);
begin
  FParkLongitude := NewLocation.Longitude;
  FParkLatitude  := NewLocation.Latitude;
end;
```

We'll now be able to save those coordinates when the user is ready for it in the database. Add a `TButton` to the bottom toolbar on `tabParkEditMain` and attach a `TAction` to it with the following event handler:

```
procedure TfrmMyParksMain.actSaveParkLocExecute(Sender:
TObject);
begin
  dmParkData.tblParks.Edit;
  dmParkData.tblParksLocX.AsFloat := FParkLatitude;
  dmParkData.tblParksLocY.AsFloat := FParkLongitude;
  dmParkData.tblParks.Post;
```

```
TDialogServiceAsync.ShowMessage(Format(
        'The park''s location (%0.3f, %0.3f) has been saved.',
        [FParkLatitude, FParkLongitude]));
end;
```

The short message The park's location (x,y) has been saved. at the end of this procedure is nice to show that something happened and to verify the coordinates saved. We'd also like a way to know which parks have saved location data later when we come back to the app. Depending on your preferences and device size, you could do that in the park edit tab, but I'd like to show the coordinates in the list view on the main park list tab.

Showing the location in the list view

Back on the tabParkList tab, let's modify TListView to indicate whether location data has been saved. Right-click on the list view component and select **Toggle DesignMode**. Expand the ItemAppearance property and change the ItemAppearance sub-property from ImageListItem to ImageListItemBottomDetail. This adds a detail line for each item that we can use to display the coordinates, if set. It also exposes this newline to LiveBindings so that we can easily hook it up to our table. The problem is there's one newline but two coordinate values. We could modify the list view and add another field, or add a field to our table that produces what we want and just use that. Let's do the latter.

In the data module, open up the field list of the Parks table component, add a new string field called Coordinates, and set its **Field type** setting to Calculated. In the table's OnCalcFields event, add the following code:

```
procedure TdmParkData.tblParksCalcFields(DataSet: TDataSet);
begin
  if tblParksLocX.IsNull or tblParksLocY.IsNull then
    tblParksCoordinates.AsString := EmptyStr
  else
    tblParksCoordinates.AsString := Format('(%0.3f, %0.3f)',
              [tblParksLocX.AsFloat, tblParksLocY.AsFloat]);
end;
```

It will show nothing when the park location is not set and a small parenthetical set of coordinates for the park when its location is known. Now, hook it up with LiveBindings to the detail line of the list view. Here's what my **LiveBindings Designer** looks like:

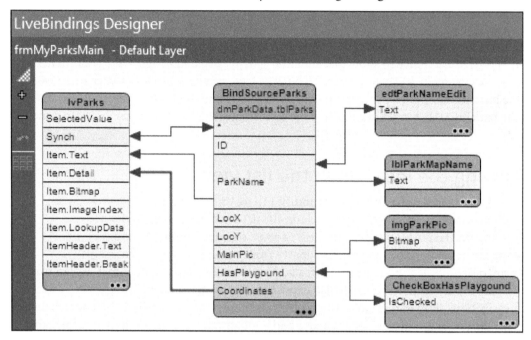

Figure 10.7 – LiveBindings Designer showing coordinates hooked up to the list view's Item.Detail property

So, with location services granted, captured, saved, and bound to the list view, we should be able to go to some parks, grab the coordinates, and see them on the main form of our app, as follows:

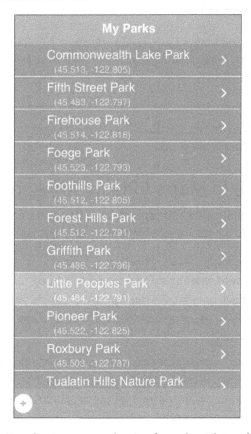

Figure 10.8 – Coordinates are now showing for parks with saved location data

There are other ways to indicate location. You could add a glyph and enable it or disable it based on this data instead of showing the actual coordinates. In any case, now that location data is stored with the parks, let's put it to use.

Mapping your way

Plotting coordinates on a map is where the usefulness of saved coordinate data really comes into play. This service also requires the permission of location services, but we covered that in the previous section so we don't need to address it again. However, viewing coordinates on a map introduces another layer of APIs, detailed as follows:

- For iOS, the **MapKit framework** is ready to use, with no additional setup.

- For Android, the **Google Maps Android API** requires a special API key you have to request through a Google account.

These are the only two platforms supported by Delphi's TMapView component.

> **Note**
>
> Not all devices based on Android support Google Maps. For example, Amazon Kindle Fire tablets use a variant of Android but provide their own Amazon Maps API; apps that try to access the Google Maps Android API on such devices will raise an error.

Since there's nothing extra to do for iOS devices, let's go straight to getting set up to view maps on Android phones before we proceed.

Setting up a Google Maps API key for Android

You'll need to register an application and generate an API key for each app you build that uses Google Maps. These API keys are free for mobile apps but take a few steps to set them up. Proceed as follows:

1. Create a **Google account** at `https://console.developers.google.com` if you don't already have one. This is free but since Google provides many services, some paid, it will ask you to set up a billing account. The mapping service we'll use is free for mobile devices.

2. Once signed in to your Google account, create a new **Google project**. You'll need to give it a unique project name (make it a good one, as it cannot be changed later). Assign it an organization name and/or a billing account (depending on how you configured your Google account).

3. After your Google project is established, you'll need to create credentials with a credential type of **API key**. There can be multiple API keys per project but we'll only need one; there are also other types of credentials, **OAuth client** and **Service account**, which we won't use here.

    Once your API key is created, you can optionally restrict it to specific APIs and application types. This is highly recommended to prevent unauthorized use of your API key, which could result in usage quota overages and unwanted charges. Additionally, it prevents you from inadvertently calling non-free APIs or calling APIs that are free on mobile platforms but not elsewhere. The next two steps tell you how to do this.

4. To restrict your API key usage to the Maps API we'll use in this project, select the API key to edit its settings, then select **Restrict key** under **API restrictions** and select **Maps SDK for Android** from the list.

5. Your API key is not yet secured—it needs one more restriction. Select **Android apps** under **Application restrictions** and add an item. Android app restriction items require two pieces of information that uniquely identify your app: a certificate fingerprint and a package name. The package name is simply the `package` key from the version information section of your Delphi project. You should prepend a reverse domain name that identifies you or your business to the default of `$(ModuleName)` for all apps you produce. For samples in this chapter, I'm using `book.FearlessCrossplatformDelphiDev.$(ModuleName)`, where `ModuleName` gets assigned the name of the Delphi project when it's built. Therefore, an API key restricted to our sample parks app would be assigned the package name of `book.FearlessCrossplatformDelphiDev.MyParks` (spaces and dashes are not allowed but underscores are, if you prefer; it is case-sensitive). The certificate fingerprint is far more complicated.

There are two different fingerprints you'll need to generate, one for debug mode and one for release versions of your app. We'll explain the lengthy process of generating release-mode fingerprints in *Chapter 15, Deploying an Application Suite*; for now, we'll concentrate on the debug fingerprint, which can be used for all the apps you build on your machine.

The **Java Development Kit (JDK)** that is installed with the Android platform development tools contains a command-line utility called `keytool.exe` that is used to manage key and certificate information for your Android projects. Using a Command Prompt shell, go to Embarcadero's `APPDATA` folder for your version of Delphi; for Delphi 10.4, you can get there by typing this from Command Prompt:

```
cd "%APPDATA%\Embarcadero\BDS\21.0"
```

If you've built at least one Android app in debug mode, you should find a special file in this folder that contains the certificate information we need: `debug.keystore`. Once that file is located, run this command (assuming the JDK is in your `PATH` variable):

```
keytool.exe -list -v -keystore debug.keystore -alias
androiddebugkey -storepass android -keypass android
```

This lists a whole bunch of information on the screen, including three lines of certificate fingerprints, as illustrated in the following screenshot:

```
Certificate fingerprints:
     MD5:  35:95:78:7A:07:B7:F4:A7:B8:87:98:D1:C1:AE:FF:03
     SHA1: 3C:B0:DF:02:25:4D:77:8D:B5:B5:B1:8D:48:BF:11:A2:DA:D4:F2:04
     SHA256: 4C:9F:8B:43:87:85:97:EF:79:F6:8B:30:0C:4E:7B:BE:EC:53:91:6C:A3:AF:4E:5B:9B:01:C0:3D:0A:B2:96:96
```

Figure 10.9 – Certificate fingerprints extracted from the debug.keystore file

The one we're interested in is SHA1. Copy the **hexadecimal (hex)** digits for that entry, and we can finish *Step 5* by filling in the **SHA-1 certificate fingerprint** box for the new restricted Android app item.

The Google API key is now set up and ready to be used. Let's switch back to Delphi.

Setting up your Delphi project to use Google Maps

The final step to use the Google Maps API involves three adjustments to settings in your Delphi project for Android targets, all made in **Project | Options**. Proceed as follows:

1. Under **Application | Entitlement List**, check the **Maps Service** checkbox.

2. Under **Application | Uses Permissions**, check **Access network state**, **Access course location**, and **Access fine location**.

3. Under **Application | Version Info**, set the `apiKey` value to the new API key generated in your Google account for the new Google project.

> **Note**
>
> It is important to protect your applications' API keys. If your project file is published on a public site, as this project is to GitHub, then Google will send you an email warning that your private key has been made publicly accessible and gives you options to remedy the situation. In my case, I simply regenerated the API key on my Google account after sending the updates to GitHub, updated the project with the new key, and continued developing, knowing the previously generated key would be useless within 24 hours.

You're now ready to show maps on both Android and iOS.

Plotting park points

Let's set up the tab that will display our park map, as follows:

1. On `tabParkMap`, add a small `TPanel` aligned to `Top`.

2. Inside the panel, add a `Left`-aligned button with a `StyleLookup` of `arrowlefttoolbutton`; create a `TPreviousTabAction`, assigning its `TabControl` to `tabCtrlParks`, and assign an action to the new button.

3. On the right side of the panel, add a `TComboBox` with three string items: `Map`, `Satellite`, and `Hybrid`. We'll add code to an event handler later.

4. Between the two controls in the panel, add a `TLabel` and use LiveBindings to hook it up to the park name.

5. Below the panel, fill the rest of the tab area with a `TMapView`, aligned to `Client`.

The map view component, once all the permissions and APIs are set up, is really pretty simple to use. We simply give a location and it takes care of showing itself, even providing finger gestures for zooming, rotating, and moving, without any extra coding. (You can disable this functionality if you'd like to provide more of a static map interface; just uncheck the **Zoom**, **Tilt**, **Scroll**, and **Rotate** properties of `GestureOptions`.)

To show a park's location on the map, create a `TAction`, `actMapPark`, and assign it to a new button in the toolbar on `tabParkEdit`. Then, fill in the `Execute` event for `actMapPark`, as follows:

```
procedure TfrmMyParksMain.actMapParkExecute(Sender: TObject);
begin
  if dmParkData.tblParksLocX.IsNull or
    dmParkData.tblParksLocY.IsNull then
    TDialogServiceAsync.ShowMessage(
      'There are no coordinates saved for this park.')
  else begin
    var SavedLocation: TMapCoordinate;
    SavedLocation.Latitude := dmParkData.tblParksLocX.AsFloat;
    SavedLocation.Longitude := dmParkData.tblParksLocY.AsFloat;
    MapViewParks.Location := SavedLocation;

    NextParkTabAction.Execute;
  end;
end;
```

In the preceding code snippet, we first check to make sure we actually have coordinates for the selected park. Next, we create a temporary `TMapCoordinate` to contain the saved latitude and longitude values of the park's location. Then, simply assigning that `TMapCoordinate` to `TMapView` brings it to life, with a display of the park map.

But one thing is missing. When viewing mobile maps that have been given specific coordinates, there's usually a location marker to pinpoint those coordinates. We can add that to our action event handler with just a few more lines of code, as follows:

```
var ParkMarker: TMapMarkerDescriptor;
ParkMarker := TMapMarkerDescriptor.Create(SavedLocation,
                 dmParkData.tblParksParkName.AsString);
ParkMarker.Draggable := True;
ParkMarker.Visible := True;
MapViewParks.AddMarker(ParkMarker);
```

This code should be inserted immediately after assigning SavedLocation to the map view. Creating a TMapMarkerDescriptor with the saved location and giving it a name associates a labeled pin with the exact coordinates, producing a view like this:

Figure 10.10 – Map with a labeled marker, showing the park's saved coordinates

The map view component has a LayeredOptions property that allows you to add points of interest, buildings, and traffic to the display, and even show a special marker highlighting your current location.

So, what is the combo box for?

Changing the map style

Both the Apple and Google map APIs support three map types, as follows:

- **Normal**, which is just a basic street map, excellent for following directions while driving.

- **Satellite**, which is a bird's-eye view of the location.

- **Hybrid**, which overlays a street map onto the **Satellite** view.

- Additionally, the Android map API provides a fourth: **Terrain**, which shows elevation on the map.

Let's add an `OnChange` event handler to the combo box on `tabParkMap` to change the map type, as follows:

```
procedure TfrmMyParksMain.cmbMapTypeChange(Sender: TObject);
begin
  case cmbMapType.ItemIndex of
    0: MapViewParks.MapType := TMapType.Normal;
    1: MapViewParks.MapType := TMapType.Satellite;
    2: MapViewParks.MapType := TMapType.Hybrid;
    3: MapViewParks.MapType := TMapType.Terrain;
  end;
end;
```

As you may recall, there were only three string items added when we set this page up. We don't want the fourth option to be there unless we're on an Android device, so let's take care of that when the application starts up. Add these three lines somewhere in the form's `OnCreate` event:

```
{$IFDEF ANDROID}
cmbMapType.Items.Add('Terrain');
{$ENDIF}
```

On an Android device we now have a fourth map type, and if I visit a park that has some hills, the **Terrain** map type looks quite interesting, as we can see here:

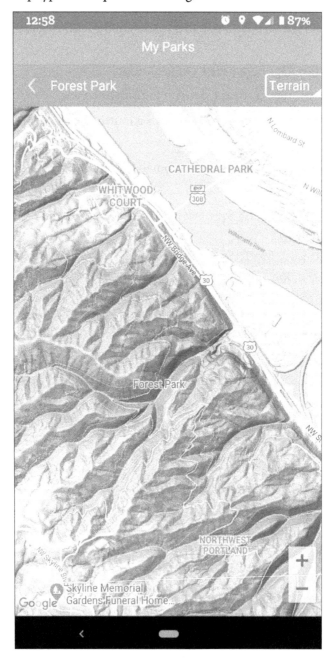

Figure 10.11 – Android's Terrain map type

This app is really starting to be useful! Now that you are building a collection of parks with pictures and map locations, wouldn't it be nice if you could easily share them with your friends?

Sharing your pictures

Almost every mobile app has some way of sharing pictures and text on social media services, through file-sharing sites and as attachments in text messages or emails. It's been made super-simple to do so in your Delphi app by simply using another one of the standard actions similar to taking a photo. From the action list in your app, add a new standard action and select TShowShareSheetAction in the **Media Library** section. Assign it to a new button in the toolbar at the bottom of tabParkEdit.

We'll hook into just one of the events, OnBeforeExecute, which lets us set up what we're going to share, as follows:

```
procedure TfrmMyParksMain.
ShowShareSheetActionBeforeExecute(Sender: TObject);
begin
  ShowShareSheetAction.Bitmap := imgParkPic.Bitmap;

  ShowShareSheetAction.TextMessage := Format('%s (%0.5f,
    %0.5f)',
                       [dmParkData.tblParksParkName.AsString,
                        dmParkData.tblParksLocX.AsFloat,
                        dmParkData.tblParksLocX.AsFloat]);
end;
```

The main goal of this sharing action is to share an image. We're sending text along with it, but not all sharing services support both images and text. In this event, we don't know which sharing service will be selected, so we just assign what we have—an image and some information to go along with it—and let the user's selected service determine what it can use.

Sending a park picture via text message from our app will look something like this:

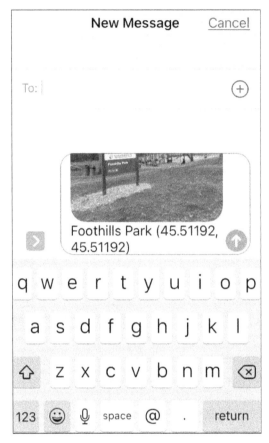

Figure 10.12 – Sharing a park picture with its name and coordinates in a text message on an iPhone

Sharing photos is one of the simpler mobile programming tasks, but it's limited to the capabilities of the services to which the data is shared. Later in this book, we'll explore other ways of sharing data that are a lot more involved. In fact, we'll come back to this app and show how the park information you've collected can be synchronized to a server, allowing you to share much more than just a picture and a line of text.

Summary

There are many ways you could change or add to this app. For example, instead of a collection of parks, perhaps you'd rather have a list of craft stores or ski resorts. A database table could be added to store an unlimited number of pictures instead of just one main image per entry. The possibilities are limitless with your newfound skills of utilizing mobile services and APIs.

There are other capabilities of mobile devices we did not explore, such as setting appointments and alarms, auto-answering certain phone calls, tracking your speed and elevation while moving, or vibrating the phone. The information you've learned in this chapter should be a springboard to learning more advanced features and helping you understand how to utilize other related services that you discover.

This chapter also leads nicely into the next, where we continue to work with unique device features, exploring how to put our code onto credit-card-sized computers, communicate with Bluetooth devices, use do-it-yourself electronic kits with our projects, and access devices through various technologies previously relegated to obtuse libraries. Delphi's mobile power continues to go deeper!

Questions

1. Which two things does an Android app need to do in order to use *dangerous* or sensitive services?

2. Which three permissions are needed to take pictures?

3. How do you get an image from a standard action that takes a picture or loads a file from your photos?

4. If you wanted to chart your course over a map, which component would you use?

5. Which two mapping APIs are supported by Delphi's `TMapView` component?

6. In addition to acquiring an API key for the Google Maps API, which three things do you have to do to your Delphi project to use a `TMapView` on Android devices?

Further reading

- *Android Version List: A Complete History and Features :*`https://www.temok.com/blog/android-version-list/`

- *iOS version history:* `https://en.wikipedia.org/wiki/IOS_version_history`

- *FireMonkey Platform Services:* `http://docwiki.embarcadero.com/RADStudio/Sydney/en/FireMonkey_Platform_Services`

- Android Developers documentation: `Manifest.permission`: `https://developer.android.com/reference/android/Manifest.permission`

- *Guide to Android App Permissions & How to Use Them Smartly:* `https://www.avg.com/en/signal/guide-to-android-app-permissions-how-to-use-them-smartly`

- *GT Explains: Understanding Android App Permissions:* `https://www.guidingtech.com/67272/android-app-permissions/`

- Apple's Human Interface Guidelines: *Requesting Permission:* `https://developer.apple.com/design/human-interface-guidelines/ios/app-architecture/requesting-permission/`

- *Taking Pictures Using FireMonkey Interfaces:* `http://docwiki.embarcadero.com/RADStudio/Sydney/en/Taking_Pictures_Using_FireMonkey_Interfaces#Taking_a_Picture_with_a_Device_Camera`

- *Mobile Tutorial: Using Location Sensors (iOS and Android):* `http://docwiki.embarcadero.com/RADStudio/Sydney/en/Mobile_Tutorial:_Using_Location_Sensors_(iOS_and_Android)`

- *Delphi: Taking and Sharing Pictures and Text (iOS and Android):* `https://youtu.be/eMSKyt5cmvw`

- *Configuring Android Applications to Use Google Maps:* `http://docwiki.embarcadero.com/RADStudio/Sydney/en/Configuring_Android_Applications_to_Use_Google_Maps`

11
Extending Delphi with Bluetooth, IoT, and Raspberry Pi

As electronic devices proliferate and communication between them becomes more ubiquitous, the ability to collect and share information among nearly anything offers boundless opportunities. **Delphi** embraces this new field of the **Internet of Things (IoT)** wholeheartedly. We'll start by examining **Bluetooth** technologies, the foundation for connecting small devices in many ways. It not only provides wireless connection capabilities to nearby paired devices but also allows ad hoc information polling from IoT devices, allowing many specialized applications in a wide variety of fields. We'll also demonstrate how to put your apps on a **Raspberry Pi**.

We'll explore these platforms in the following sections:

- Starting with Bluetooth Classic
- Learning about low-energy Bluetooth
- Utilizing beacons

- Doing more with the Internet of Things
- Using a Raspberry Pi

Pull out your discarded phones and ask a friend to donate theirs—we're going to fill our desk with electronic gadgets!

Technical requirements

Once again, the **Windows** platform will not be a major part of the discussion in this chapter. After using it as the starting point for compiling your project, you will be deploying apps to other devices running a variety of OSes. Each section will introduce what will be covered and expected. The list will include the following:

- **Bluetooth**: Your app will need to run on a platform that supports Bluetooth, and you will need to have two Bluetooth-capable devices that can connect and run your apps.
- **Raspberry Pi**: You will need a Raspberry Pi Model 3 or 4 or newer and the ability to burn an image onto a **microSD** card for booting the Pi.

You may want to have more than two Bluetooth devices, preferably ones on which you can install your own apps, in order to gain the most experience, as we talk about distance calculation and beacons.

The complete code for this chapter can be found online on GitHub:

```
https://github.com/PacktPublishing/Fearless-Cross-Platform-
Development-with-Delphi/tree/master/Chapter11
```

Starting with Bluetooth Classic

There are two broad categories of Bluetooth technology, **Classic Bluetooth** and **Bluetooth LE** (or low energy). The classic version only became known as "classic" once the LE version came into widespread use. Classic uses more energy than its newer LE version and is therefore not the best solution for mobile devices where battery size and weight are big considerations. However, it provides a much higher transfer rate. Classic Bluetooth is not accessible on **Apple iOS** devices and is not supported on Windows Servers but is available on all the other platforms we've been working with so far.

When you work with Bluetooth, think of it as short-range wireless networking. It doesn't know about your application and your application uses it just like any other method of sending data. Two applications that support Bluetooth but don't know anything about each other can share information if they can work with the same data structure.

Bluetooth applications use a client-server method of communication. One of the Bluetooth-enabled applications serves, or publishes, a service for other Bluetooth devices to connect to. A Bluetooth client will then be able to connect to and then send and receive data with the Bluetooth server application. Both devices can act as both a client and server simultaneously.

Classic Bluetooth requires a little setup before you can use it. Let's cover that before we go on.

Configuring Classic Bluetooth

In order to use Classic Bluetooth, the device to which you're connecting must be *discovered* and *paired* to establish a communication pipe between the two. Of the two devices, one must be discoverable, and the other discovers it (it doesn't matter which does which). This process of pairing can be done either with the standard Bluetooth configuration settings of your device or you can provide this capability as a convenience directly within your application.

Let's create a sample application and walk through the steps:

1. In Delphi, create a **Multi-Device Application** using the **Tabbed** template.

2. Name the four tabs tabBTDiscover, tabBTDevices, tabBTServices, and tabBTChat; modify the FormCreate procedure to set the ActiveTab to tabBTDevices.

3. On both the tabBTDiscover and tabBTDevices tabs, add a toolbar with a button using a StyleLookup of refreshtoolbutton.

4. On the tabBTDiscover tab, add a TListView named lvDiscoveredDevices and aligned to Client, and set its ItemAppearance.ItemAppearance property to ImageListItemBottomDetail. Do the same on the tabBTDevices tab, naming the ListView, lvBTPaired.

5. Add two instances of TFDMemTable to the form and call one tblFoundDevices and the other tblPairedDevices. Edit the list of fields of each and add DeviceName and Address as string fields.

6. Right-click on one of the memory tables and select **Edit DataSet...** from the pop-up menu. Add a record with a sample device name and address so that you'll be able to see something show up on the screen in the next step. Repeat for the other table component (keep these datasets active at design time to preserve the sample data).

7. On `tabBTDiscover`, right-click on **ListView** on `tabBTDiscover` and select **LiveBindings Wizard...**, linking the ListView's `Item.Text` property with the `DeviceName` field of `tblFoundDevices` and the ListView's `Item.Detail` property with the `Address` field. Repeat the linking with the ListView on `tabBTPairedDevices` and `tblPairedDevices`.

8. Finally, drop a `TBluetooth` component on the form.

Now you should have a ListView on each of the first two tabs showing a couple of records from a memory table. We will empty and refill those memory tables at runtime. In fact, we should make sure that when the application starts, the tables are cleared from the design-time data we have in there, so add these two lines to the `FormCreate` procedure:

```
tblFoundDevices.EmptyDataSet;
tblPairedDevices.EmptyDataSet;
```

Most of the sample Bluetooth applications I've seen do all the Bluetooth manipulation in code. It's not very complicated, but we'll use the component out of convenience. It automatically pulls in the right units for us and puts a wrapper around the global Bluetooth classes, which saves us a few lines of code. The only thing we need to do is enable it. You could do this at design time by simply checking the `Enabled` property of the component, but I prefer to use the `OnActivate` and `OnDeactivate` form events to manage it:

```
procedure TfrmBTC.FormActivate(Sender: TObject);
begin
  Bluetooth.Enabled := True;
end;
procedure TfrmBTC.FormDeactivate(Sender: TObject);
begin
  Bluetooth.Enabled := False;
end;
```

Next, we want to get the list of paired Bluetooth devices to display in the ListView. Create a form-level variable to hold the list:

```
private
  var
    FPairedDevices: TBluetoothDeviceList;
```

Now, add an `ActionList` and create an action, name it `actBTCDeviceRefresh`, and assign it to the **Refresh** button on `tabBTDevices` to refresh the paired Bluetooth device list, adding each one found to the memory table:

```
procedure TfrmBTC.actBTCDeviceRefreshExecute(Sender: TObject);
begin
  if Bluetooth.CurrentManager.Current.ConnectionState =
    TBluetoothConnectionState.Connected then begin
    tblPairedDevices.EmptyDataSet;
    FPairedDevices := Bluetooth.CurrentManager.Current.
      GetPairedDevices;

    for var LPairdDevice in FPairedDevices do
      tblPairdDevices.InsertRecord([LPairdDevice.DeviceName,
        LPairdDevice.Address]);
  end;
end;
```

Since the table is hooked up to the ListView with LiveBindings, that should be all you need to run the app on a Bluetooth-enabled device. Click the **Refresh** button, and see the Bluetooth devices already paired with it:

Figure 11.1 – Paired Bluetooth devices on an Android phone

Now that we can get the list of currently paired devices, let's discover and pair a new device.

Discovering and pairing devices

Bluetooth pairing is a two-way street—both devices must agree to be connected to each other. It doesn't matter which one looks for the other, but one of them needs to search for and initiate the connection. Once you enable the Bluetooth component, it is immediately discoverable by other Bluetooth devices. One nice thing to do in a Bluetooth-enabled app is to let others know the name of your device so that it can be found by other Bluetooth devices. Let's do that right after we enable the component.

First, add a label in the toolbar on the `tabBTDiscover` tab called `lblDiscoverableName`, so we can see what the name is, and then add this to the `OnActivate` event handler:

```
procedure TfrmBTC.FormActivate(Sender: TObject);
begin
  Bluetooth.Enabled := True;
  if Bluetooth.CurrentManager.Current.ConnectionState =
    TBluetoothConnectionState.Connected then begin
    lblDiscoverableName.Text := 'Device''s BT Name: ' +
      Bluetooth.CurrentManager.CurrentAdapter.AdapterName;
  end else
    lblDiscoverableName.Text := 'No Bluetooth device Found';
end;
```

Now that the name is clearly visible in the app, try running it and then using a different Bluetooth app to search for it.

> **Note**
>
> Some Bluetooth devices may not be discoverable in both directions. For example, my **iPad** cannot see my **Kindle Fire**, but the Kindle can discover the iPad and initiate the connection.

In addition to being discoverable, we also want to be the discoverer! Searching for Bluetooth devices can take several seconds, so add a `TAniIndicator`, named `AniIndicatorDiscover`, to provide a standard way to let the user know they should be patient during a long process. We'll make it visible when the discovery process starts and invisible when it ends.

The Bluetooth discovery mechanism is an asynchronous process, so we'll need to provide an event handler that is called when the discovery process is finished. The event handler will need a `TBluetoothDeviceList` to store its results in, similar to the paired device list we created earlier, so add that in at the form level first:

```
private
  var
    FDiscoverDevices: TBluetoothDeviceList;
```

Now, to create the event handler, simply double-click on the Bluetooth component's `OnDiscoveryEnd` event to create the method:

```
procedure TfrmBTC.BluetoothDiscoveryEnd(const Sender: TObject;
  const ADeviceList: TBluetoothDeviceList);
begin
  TThread.Synchronize(nil, procedure begin
    AniIndicatorDiscover.Visible := False;
    AniIndicatorDiscover.Enabled := False;
    actBTCDeviceDiscover.Enabled := True;

    tblFoundDevices.EmptyDataSet;
    FDiscoverDevices := ADeviceList;
    for var i := 0 to FDiscoverDevices.Count - 1 do
      tblFoundDevices.InsertRecord([FDiscoverDevices[i].
        DeviceName,
            FDiscoverDevices[i].Address]);
  end);
end;
```

Create a field in the form class declaration that will contain a handle to the current Bluetooth adapter:

```
private
  var
    FAdapter: TBluetoothAdapter;
```

The new action, `actBTCDeviceDiscover`, will be attached to the **Refresh** button on `tabBTDiscover` and should disable itself, enable the activity indicator, save a handle to the current adapter, and start the Bluetooth discovery:

```
procedure TfrmBTC.actBTCDeviceDiscoverExecute(Sender: TObject);
begin
  btnBTDiscoverDevices.StyleLookup :=
    'transparentcirclebuttonstyle';
  AniIndicatorDiscover.Visible := True;
  AniIndicatorDiscover.Enabled := True;
  actBTCDeviceDiscover.Enabled := False;
```

```
if Bluetooth.CurrentManager.Current.ConnectionState =
            TBluetoothConnectionState.Connected then begin
    FAdapter := Bluetooth.CurrentAdapter;
    Bluetooth.CurrentManager.Current.StartDiscovery(10000);
  end;
end;
```

The `StartDiscovery` method's single parameter is the maximum number of milliseconds to search.

> **Note**
>
> I've found that the Bluetooth discover functionality in Delphi does not always return the list of devices that its OS-level functionality does. If this functionality doesn't work in your app, use the device's native Bluetooth preferences to search for and pair a device.

The list returned is pretty simple—on a Mac with five devices found, it looks like this:

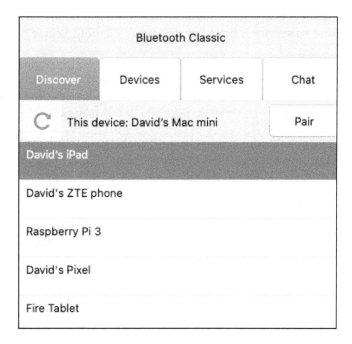

Figure 11.2 – Several discovered Bluetooth devices ready to be paired on a Mac

When the discover process ends, the list of available Bluetooth devices is shown in the ListView. Add an action and assign it to a new button on the toolbar to pair the newly discovered device. The code to do that is really simple:

```
procedure TfrmBTC.actBTCPairDeviceExecute(Sender: TObject);
begin
  if (lvDiscoveredDevices.ItemCount > 0) then
    FAdapter.Pair(FDiscoverDevices.Items[lvDiscoveredDevices.
      ItemIndex])
end;
```

Once a new device has been paired, you can go back to the **Devices** tab in the app, click the **Refresh** button, and see the updated list.

So, what can you do with a paired Bluetooth device?

Publishing Bluetooth services

Bluetooth applications publish services for other Bluetooth applications to use. When an app publishes a service, it acts as a **server** and the Bluetooth app that finds and connects to it is its **client**. Publishing a service is done by opening a Bluetooth **Server Socket** and then listening for incoming requests.

The sample we're building will be a simple chat program, allowing users on two devices to type messages and send them to each other. Either one of the devices can be the server or the client—our application will handle both. Let's add to our user interface to support this:

1. On the `tabBTChat` tab, add an edit box with a label prompting for text to send and a button called `btnSendChatMsg`, which will be used to actually send the message with an attached `TAction` later. Put these controls near the top of the tab area.

2. Below them, add another label with the `Text` property set to `"Conversation:"` and in the rest of the area, place a memo aligned to `Client`.

3. Back on `tabBTDevices`, add another button on the toolbar called `btnStartChat`. This will be assigned an action to connect as a client to another device's chat service.

When the app is acting as a server, it's best done as a multithreaded process so it can be listening for messages from the connected client in the background while allowing the user to continue using the app. Here's the thread class we'll declare in the `interface` section:

```
type
  TBTServerThread = class(TThread)
  strict private
    FServerSocket: TBluetoothServerSocket;
    FSrvrClientSocket: TBluetoothSocket;
  protected
    procedure Execute; override;
  public
    property ServerSocket: TBluetoothServerSocket read
      FServerSocket write FServerSocket;
    property ClientSocket: TBluetoothSocket read
      FSrvrClientSocket;
    constructor Create(ACreateSuspended: Boolean);
    destructor Destroy; override;
    procedure ConnectClientDevice(ADevice: TBluetoothDevice;
      ServiceGUID: TGUID);
  end;
```

The `constructor` simply initializes the local client socket field. The bulk of the work is done in the `Execute` procedure:

```
procedure TBTServerThread.Execute;
var
  Msg: string;
  LData: TBytes;
begin
  while not Terminated do begin
    while (not Terminated) and (FClientSocket = nil) do begin
      FClientSocket := FServerSocket.Accept(100);
      if Assigned(FClientSocket) then
        Synchronize(procedure
          begin
            frmBTC.btnSendChatMsg.Action := frmBTC.
              actSendServerChatMsg;
          end);
```

```
      end;
    while not Terminated do begin
      LData := FClientSocket.ReceiveData;
      if Length(LData) > 0 then
        Synchronize(procedure
          begin
            frmBTC.ChatConversationAdd(TEncoding.UTF8.
              GetString(LData));
          end);
      Sleep(100);
    end;
  end;
end;
```

A brief explanation of this procedure is in order. The first inner `while` loop waits for a client to connect—if it never does, the rest of the procedure is never encountered. A timeout of 100 milliseconds is passed to the `Accept` method of the server socket, which is attempted in a continuous loop in this background thread. If it ever returns a socket, that socket becomes the server's connection to the client Bluetooth device and is used in the next loop. We synchronize a call to the application's main thread to assign an action to a button that we'll describe shortly.

The second inner `while` loop takes effect after a client is connected and incoming messages are encountered and synchronized with the application's main thread to show the message on the screen in the memo control. The `ChatConversationAdd` method simply adds the given `string` parameter as a new line in the memo.

All of the rest of the code in this section will rely on Bluetooth being active and connected. It's always a good idea to double-check this is the case before attempting (and timing out) any communication, so let's create a convenience function we can call from many other places:

```
function TfrmBTC.BluetoothActive: Boolean;
begin
  Result := Bluetooth.CurrentManager.Current.ConnectionState =
              TBluetoothConnectionState.Connected;
end;
```

Publishing a Bluetooth service requires a service name and a **universally unique identifier**, or **UUID**. Create constants for these values in the `private` section of the form's class as they'll be needed in a few different places, as seen in the following code block:

```
const
    CHAT_SERVICE_NAME = 'Classic Chat';
    CHAT_SERVICE_GUID = '{61FD1F5E-4945-4A09-B7F5-
66E64F5BF69D}';
```

We're now ready to publish the service, which is done in the thread we set up. So, when does the thread get created? You could provide a button to start and stop listening for client connections, and that's what some sample programs do. We've got a lot going on in this app already, so let's just make it automatically start listening when the app starts. I added a call to `StartChatService` from the `OnActivate` event of the main form:

```
procedure TfrmBTC.StartChatService;
begin
  if (FChatServerThread = nil) and BluetoothActive then
    try
      FAdapter := Bluetooth.CurrentAdapter;
      FChatServerThread := TBTServerThread.Create(True);
      FChatServerThread.ServerSocket := FAdapter.
        CreateServerSocket(CHAT_SERVICE_NAME,
        StringToGUID(CHAT_SERVICE_GUID), False);
      FChatServerThread.Start;

      mmoConversation.Lines.Clear;
      ChatConversationAdd('Started chat service: ' + CHAT_
        SERVICE_NAME);
      ChatConversationAdd('   ' + CHAT_SERVICE_GUID);
    except
      on e:Exception do begin
        ChatConversationAdd(E.Message);
      end;
    end;
end;
```

This creates the server socket that publishes the Bluetooth service for this app. It's now finally time to actually use the service!

Connecting and communicating

As mentioned earlier, one of the Bluetooth devices publishes a service and becomes the server, while the other connects to it as a client. We just saw how to publish a service, so now let's see how to connect to it as a client. Before we do that, remember there could be several Bluetooth devices paired, each with multiple services published. We need to connect to a specific service on a specific device.

In the `tabBTDevices` tab of our app, we can get a list of paired devices with the **Refresh** button that lists the available devices in a ListView. When the app is running, the user will select a device and then click `btnStartChat`, the button we created in *Step 3* of the previous section.

Just before we show the action event handler for that button, we need to write a lookup function. The ListView shows the names of the devices, but the actual devices are stored in `FPairedDevices`. This function looks through that list for the device with the name passed in and returns the actual device:

```
function TfrmBTC.FindBTDevice(BTDeviceList:
TBluetoothDeviceList; const SearchDeviceName: string):
TBluetoothDevice;
begin
  Result := nil;
  for var LBTDevice in BTDeviceList do
    if SameText(LBTDevice.DeviceName, SearchDeviceName) then
begin
      Result := LBTDevice;
      Break;
    end;
end;
```

Now, create the `actStartChatClient` action and assign it to `btnStartChat` on the toolbar on `tabBTDevices`. Its `Execute` method should look like this:

```
procedure TfrmBTC.actStartChatClientExecute(Sender: TObject);
var
  LDevice: TBluetoothDevice;
begin
```

```
  if BluetoothActive and (lvBTPaired.ItemIndex > -1) then begin
    LDevice := FindBTDevice(FPairedDevices, lvBTPaired.
      Items[lvBTPaired.ItemIndex].Text);
    if Assigned(LDevice) then begin
      tabctrlBTC.ActiveTab := tabBTCChat;
      ChatConversationAdd('Connecting to "' + CHAT_SERVICE_NAME
        + '" on ' + LDevice.DeviceName);
      FClientSocket := LDevice.CreateClientSocket
        (StringToGUID(CHAT_SERVICE_GUID), False);
      if FClientSocket <> nil then begin
        FClientSocket.Connect;
        ChatConversationAdd('Connected.');

        btnSendChatMsg.Action := actSendClientChatMsg;
        actSendClientChatMsg.Enabled := True;
        tmrClientChatRcvr.Enabled := True;
      end else
        ChatConversationAdd('Could not connect to chat
          service.');
    end;
  end
end;
```

After checking to make sure Bluetooth is active and a paired device is selected, this method switches to the chat tab and attempts to connect to the chat service of the selected device. If that's successful, the action for btnSendChatMsg is set to actSendClientChatMsg as we've just connected as a client.

Another important thing this method does is start a timer to receive chat messages. If we're acting as the server, the thread we set up earlier is already listening for messages. We could've set up another thread to listen for messages as a client, but I chose this technique as it's simpler for demonstration purposes. Here's the OnTimer event handler for the chat message timer:

```
procedure TfrmBTC.tmrClientChatRcvrTimer(Sender: TObject);
var
  LData: TBytes;
begin
  tmrClientChatRcvr.Enabled := False;
```

```
  LData := FClientSocket.ReceiveData;
  if Length(LData) > 0 then
    ChatConversationAdd(TEncoding.UTF8.GetString(LData));
  tmrClientChatRcvr.Enabled := True;
end;
```

Now that everything should be hooked up and listening for messages, let's write the last two actions we'll need to send data between the two devices. When the app acts as a client to another device's service, a client Bluetooth socket in the main application's thread is used, but when it acts as a server, the background thread watching for client connections will create its own client Bluetooth socket. Therefore, two different actions will be used to send text to the other end of the connection. First, create actSendClientChatMsg to send a message from the app as a client to a connected server and write its Execute method like this:

```
procedure TfrmBTC.actSendClientChatMsgExecute(Sender: TObject);
begin
  if BluetoothActive then
    try
      var ToSend: TBytes;
      if FClientSocket <> nil then begin
        ToSend := TEncoding.UTF8.GetBytes(edtChatText.Text);
        FClientSocket.SendData(ToSend);
        ChatConversationAdd('Cli > ' + edtChatText.Text);
      end else
        ChatConversationAdd('[No server connected]');
    except
      on e:Exception do
        ChatConversationAdd(e.Message);
    end;
end;
```

Since the other client socket was created in the background server thread, the action will need to call a property of the thread class to send the data in the `Execute` method for `actSendServerChatMsg`:

```
procedure TfrmBTC.actSendServerChatMsgExecute(Sender: TObject);
begin
  if BluetoothActive and (FChatServerThread <> nil) then
    try
      if FChatServerThread.ClientSocket <> nil then begin
        var ToSend: TBytes;
        ToSend := TEncoding.UTF8.GetBytes(edtChatText.Text);
        FChatServerThread.ClientSocket.SendData(ToSend);
        ChatConversationAdd('Srv > ' + edtChatText.Text);
      end else
        ChatConversationAdd('[No client connected]');
    except
      on e:Exception do
        ChatConversationAdd(e.Message);
    end;
end;
```

With these two actions ready, we assign one or the other to `btnSendChatMsg`, depending on how we're connected to the other Bluetooth device. This was done in the code shown earlier—first, when a server detects a connection from a client in the `TBTServerThread.Execute` method, it assigns the `actSendServerChatMsg` action, and second, when connecting as a client to a published service in the `actStartChatClientExecute` method, it assigns the `actSendClientChatMsg` action.

Running it on an **Android** phone and communicating with an **Amazon Fire Tablet** looks like this:

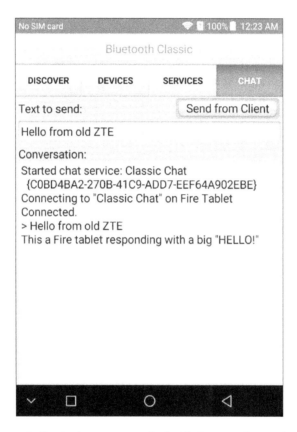

Figure 11.3 – Bluetooth Classic chat app on an Android phone sending and receiving messages

This has been quite a lot of work and it has given me a greater appreciation for what goes on behind the scenes of Bluetooth devices that seem to just work—and more patience for when they don't.

The source code for this app can be found on GitHub:

```
https://github.com/PacktPublishing/Fearless-Cross-
Platform-Development-with-Delphi/tree/master/Chapter11/01_
BluetoothClassic
```

Now, let's turn our attention to a newer variant of Bluetooth that is less work to use.

Learning about low-energy Bluetooth

With our understanding of what it takes to use Classic Bluetooth, we can build on that to understand the newer version, **Bluetooth Low Energy** (**BLE**), also known as **Smart Bluetooth**. Here are the differences:

- BLE works well with low-powered devices.

- BLE runs on Macs and Android devices like its classic predecessor, but also on iOS devices; for Windows, you need Windows 10 or newer.

- BLE publishes discoverable services, just like Classic Bluetooth, but there is no need for pairing—clients simply connect and start using a service.

- BLE's published services are defined with a profile, rather than just the raw stream of bytes that you get with Classic Bluetooth. A widely used profile is the **Generic Attribute Profile** (**GATT**).

- BLE is primarily used for information reporting from IoT devices such as heart rate or blood pressure monitors, or cycle speed and cadence measuring, but is not limited to one-way communication.

There are many published BLE profiles and manufacturers can create their own custom ones. The GATT profile has several dozen services called **sub-procedures**; you can download specifications for the ones you want to implement from the *Bluetooth Specifications List* found in the *Further reading* section at the end of this chapter. A pair of sample applications that comes with Delphi in the `Samples\Object Pascal\Multi-Device Samples\Device Sensors and Services\Bluetooth\ProximityClientServer` folder implements three of them:

- **Link Loss Service**: Defines what happens when a link is lost between two devices.

- **Immediate Alert**: Allows a connected device to cause an immediate alert by setting the server device's alert level.

- **TX Power Service**: Reports the current device's power level.

The server application uses the `BluetoothLE` component and the client application simply calls the `TBluetoothLEManager` class in the `System.Bluetooth` unit to use the BLE functionality. When the **Start Announce** button is clicked on the server app, the Bluetooth Manager is established, all three services are published, and it begins listening for clients to connect to. The services are assigned specific names and UUIDs and must conform to their specifications.

When the client app finds the server and connects to the services identified with the same UUID, it starts a timer, fired every 2 seconds, that checks the distance to the server, and sends an alert level to the server based on how far away it is—the further away, the higher the alert level. The server responds by logging messages and setting the color of an alert box. Here's what a mild alert level looks like:

Figure 11.4 – Bluetooth LE sample, ProximityServer, running on Android

We won't cover the details of the implementation but the code is readable and can be followed or even expanded for your own learning without much difficulty. The **Embarcadero** documentation (in the *Using Bluetooth Low Energy* link in the *Further reading* section at the end of this chapter) explains additional information you'll need to explore this topic further.

Building on BLE technology, devices that advertise their location are called **beacons**, which we'll discuss in the next section.

Utilizing beacons

The previous section introduced the GATT profile as a way for Bluetooth LE devices to communicate once connected. But BLE devices can also send data without being connected by using the **Generic Access Profile (GAP)** in "advertising" mode. In this mode, data is sent out in specially formatted data packets on a periodic basis, thus acting as a beacon for any device that is listening.

There are currently three general beacon formats:

- **iBeacon**: Apple was the first to create a beacon format; it is simple and robust.
- **AltBeacon**: This open format, created by **Radius Networks**, is very similar to iBeacon but has a little more data available and is not company-specific.
- **EddyStone**: Google created this open source, cross-platform format that defines different types of frames for a variety of applications; it is part of their **Physical Web** initiative.

To use beacons in your Delphi app, you may be thinking that you have to learn about the underlying protocols and sub-formats and then figure out how to implement them in the BluetoothLE component. But Delphi provides the TBeaconDevice component to create a beacon server app and TBeacon to create a beacon client app, both with very little effort. What's more, these two components support all three of the aforementioned formats.

You won't often build your own beacon server device—you're much more likely to build apps that use pre-made beacon devices. But it's nice to be able to set up your own test lab. Let's show how simple it is to do this by writing a beacon server app. We'll use the iBeacon format as it's the simplest to implement and does all we need right now.

Setting up a beacon server app

Follow these steps to build an app that will act as a beacon server:

1. Create a new **Multi-Device Application** using just the blank template.
2. Place a few labels aligned to the top. We'll use these to display the beacon server parameters.
3. Add a couple of buttons—one will be used to start the beacon and the other to stop it. I put them in a one-row grid-panel layout.
4. Add a TActionList and create a couple of actions assigned to the two buttons you just added; I called mine actBeaconStart and actBeaconStop.

5.   Underneath the buttons, add a memo that will be used to display log messages.

6.   Finally, add a `TBeaconDevice`. We don't need to set any of its properties at design time—we'll just use the defaults.

As explained earlier in this chapter, Bluetooth services are identified by a UUID. Since we're building both a client and a server, they'll need to reference the same UUID to prevent finding beacons that are not relevant to our app. To make this simple, we'll share a unit between the two apps with a global constant declaring the UUID we'll generate and assign it at runtime.

1.   Add a unit called `uBeaconConsts.pas` and add it to the `uses` clause in the `implementation` section.

2.   Create a constant in the `interface` section with a new UUID (a GUID will suffice for our purposes).

> **Tip**
>
> The default keyboard shortcut to create a new GUID in the Delphi editor is *Ctrl + Shift + G*. It puts square brackets around the string, which you need to remove.

3.   Create another constant in that unit for a manufacturer ID of your choosing; it must be four characters.

4.   Back in the main form, create an event handler for the form's `OnCreate` event and assign the BeaconDevice's `GUID` and `ManufacturerId` properties from the new constants.

There are three other properties of the BeaconDevice that need some explaining. The `Major` and `Minor` properties may seem like they specify a version, but they're used instead as an identifier for a particular beacon. The idea is that you might have several rooms full of beacons (like in a museum) and as you walk around the room and to other rooms, each beacon needs to be uniquely identified from all the others. Use these two properties to do that. One strategy is to use the `Major` property to specify a room and the `Minor` property for a sequential numbering of the devices in the room. For our simple demonstration, we'll simply hardcode a number, deploy it to a device, then change the hardcoded number and deploy it to a different device. However, a real-life scenario would have some way to define these two properties externally for properly mapping the devices in a logical manner.

The `TxPower` property is the power level that is output by the device the beacon server app is running on. This value must be carefully measured and calibrated as it is used to help calculate the distance to the beacon. It's beyond the scope of this book to explain how this is done—follow links at the end of this chapter for further study on this topic. For now, just assign a random number for testing.

1. Continue the `OnCreate` event handler by assigning the `Major`, `Minor`, and `TxPower` properties of the BeaconDevice as discussed.

2. Finish the `OnCreate` event handler by assigning appropriate values for the `Text` property of the labels on the form to display our beacon server's parameters.

One property that we left at its default, but which is very importantm is the `BeaconType` property. This specifies which of the formats is used and has the following options:

- **Standard**: This uses the iBeacon format from Apple.

- **Alternative**: This uses the open format from Radius Networks.

- **EddystoneURL** and **EddystoneUID**: The two formats from Google—the first broadcasting a compressed URL and the second broadcasting a Namespace and an Instance ID; both of these require additional properties to be set in the BeaconDevice component.

The default is **Standard**, which is what we want.

The last part of our simple beacon app is to start and stop the BeaconDevice. We'll do this in the two actions we created earlier. Here's the one that starts it:

```
procedure TfrmBeaconServer.actBeaconStartExecute(Sender:
TObject);
begin
  try
    BeaconDevice.Enabled := True;
    actBeaconStart.Enabled := False;
    actBeaconStop.Enabled := True;
    mmoBeaconMessages.Lines.Add('Beacon started at ' +
              FormatDateTime('yyyy-mm-dd hh:nn:ss', Now));
  except
    on E: Exception do begin
      mmoBeaconMessages.Lines.Add('Problem starting: ' +
        E.Message);
      BeaconDevice.Enabled := False;
```

```
      end;
    end;
  end;
```

The event handler to stop the beacon is very similar and not given here.

Before you run the app, you must check a couple of permissions in the project options for either or both of the Android platforms you want to run:

- Bluetooth
- Bluetooth admin

These two permissions flag the Android system that Bluetooth will be used by this app.

That's all there is to it. There are no timings or other properties to set as the frequency and data broadcast from beacons is established in the protocol selected, iBeacon in our case. Once enabled, it broadcasts the UUID, major and minor IDs, and the power level at which it was calibrated every 100 milliseconds, or 10 times per second, until disabled. When running on an **iPhone**, it looks like this:

Figure 11.5 – Beacon server broadcasting from an iPhone

You can download the source for this section's apps from GitHub:

```
https://github.com/PacktPublishing/Fearless-Cross-Platform-
Development-with-Delphi/tree/master/Chapter11/02_Beacons
```

With the server broadcasting its identity, it's time to create a client that can read it.

Finding and reacting to beacon messages

The user interface of the client beacon app is very similar. We won't need labels to identify a server but we will need two memos, one for the log as usual and the other to display a continuously updated list of the beacons we find. There is one more obvious exception to setting up the user interface and components of the client app—instead of a BeaconDevice, drop a TBeacon component on the form.

Before we start writing code, select the beacon component and look at the properties in the **Object Inspector**. You'll notice fewer properties and a new one called MonitorizedRegions. That's a strange name for holding a collection of beacon search parameters. Click the ellipses button on that property to edit the items in the list and add one. It adds a TBeaconRegionItem with some properties that resemble what you saw for the BeaconDevice component in the server app.

What happens when the beacon component starts scanning for beacons is that this list defines what types of beacons it will search for. There could be a plethora of beacons for many different applications, all broadcasting their device info, and if you don't have a meaningful way of sifting through that, you would be flooded with unnecessary noise. The first obvious filter we'll apply was alluded to when we built the server—the UUID. This will immediately limit the discovered beacon list down to just the ones running the server we wrote. A further filter we should already be familiar with is IDManufacturer (called ManufactureId in the server's BeaconDevice).

The Major and Minor properties can be left at -1, which tells the beacon to scan for all values of those properties. Your application could set these values as well to further limit the returned list of beacons found. EddyInstance and EddyNamespace are only used for the Google format.

The KindofBeacon property specifies, as the name implies, what kind of beacon to find. The options should be familiar except for a strange one, iBAltBeacons. Choosing that option allows this BeaconRegionItem to include both iBeacons and AltBeacons in its result list. I chose to limit the list to just iBeacons.

With the understanding of how the beacon component will search and filter the beacons we want to see, create the OnCreate event handler. However, setting the UUID isn't so straightforward as the beacon search filters are in a list. Just in case you ever want to add another item to the list, it's best to write the code in a flexible way now:

```
procedure TfrmBeaconClient.FormCreate(Sender: TObject);
begin
  for var i := 0 to Beacon.MonitorizedRegions.Count - 1 do
    Beacon.MonitorizedRegions.Items[i].UUID   := FEARLESS_
```

```
BEACON_UUID;
end;
```

So why would you want more than one item in the `MonitorizedRegions` list? Perhaps there is more than one type of beacon your app supports? When the client app gets the list of beacons, it can branch on the beacon information and provide custom functionality for each type, each UUID, each major or minor ID value, and so forth.

Before we discuss starting and stopping the beacon scan, let's look at the events we'll want to capture. The BeaconDevice had no events at all—broadcasting beacon information as a server is very simple and well defined. Listening for them has a lot more opportunity for custom functionality and event branching.

Each of the entries in the `MonitorizedRegions` list can be thought of as a group or region of beacons. Whenever a new beacon is detected that matches the filters in one of those regions, it fires the `OnBeaconsEnterRegion` event, and for the first beacon found in a region, it fires `OnEnterRegion`. This would allow you to, for example, detect when a person walks into a new room and encounters a new set of beacons. Similarly, the `OnBeaconExitRegion` and `OnExitRegion` events notify when a region (or group) of beacons is no longer reachable.

This sample beacon client app makes use of the `OnBeaconEnter` and `OnBeaconExit` events, which conveniently provide the list of current beacons. Here's one of them and the associated procedure that adds the beacon information to one of the memos:

```
procedure TfrmBeaconClient.BeaconBeaconEnter(const Sender:
TObject; const ABeacon: IBeacon; const CurrentBeaconList:
TBeaconList);
begin
  UpdateBeaconList(CurrentBeaconList);
end;
procedure TfrmBeaconClient.UpdateBeaconList(const ABeaconList:
TBeaconList);
const
  ProximityStr: array[TBeaconProximity] of string =
                ('Immediate', 'Near', 'Far', 'Away');
begin
  mmoBeacons.Lines.Clear;

  for var i := 0 to Length(ABeaconList) - 1 do
    mmoBeacons.Lines.Add(Format('Beacon ID %d-%d:   ' +
```

```
                    'RSSI = %d  Distance = %f  Proximity = %s',
               [ABeaconList[i].Major, ABeaconList[i].Minor,
                ABeaconList[i].Rssi, ABeaconList[i].Distance,
                ProximityStr[ABeaconList[i].Proximity]]));
end;
```

Now that we have something to catch broadcast information from a beacon, we can start scanning for available beacons. But remember, part of the reason we're accessing beacons is to determine proximity, an aspect of location, which is sensitive information for a personal mobile device. So, we have to ask permission from the user on an Android device (and you should recall from an earlier chapter that permissions are automatic on iOS devices—the code is transparent). Once permission is granted, starting the scan on a client is just as easy as starting the broadcast from the server:

```
procedure TfrmBeaconClient.actBeaconStartScanExecute(Sender:
TObject);
begin
  PermissionsService.DefaultService.RequestPermissions(
      ['android.permission.ACCESS_FINE_LOCATION'],
      procedure(const Permissions: TArray<string>; const
        GrantResults: TArray<TPermissionStatus>)
      begin
        if (Length(GrantResults) = 1) and (GrantResults[0] =
          TPermissionStatus.Granted) then
          try
            Beacon.Enabled := True;
            Beacon.StartScan;

            actBeaconStartScan.Enabled := False;
            actBeaconStopScan.Enabled := True;

            AddLog('Scanning for beacons...');
          except
            on e:Exception do
              AddLog('Problem starting scan: ' + e.Message);
          end;
      end,
```

```
    procedure(const Permissions: TArray<string>; const
      PostRationaleProc: TProc)
    begin
      ShowMessage('no permission to scan for beacons!');
    end);
end;
```

Just like the server app, make sure both **Bluetooth** and **Bluetooth admin** permissions are checked in the project options. In addition to that, the client app will also need location permissions, so check off **Access course location** and **Access fine location** as well.

Running the client app from a **Fire HD** tablet and finding two devices nearby running the server version of the beacon app looks like this:

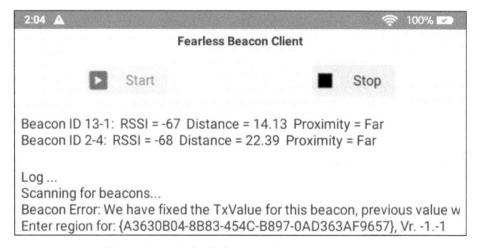

Figure 11.6 – Android client app showing two beacons

I have not calibrated the TX Power values for these apps so got quite a few errors—this screenshot only shows the top portion of the tablet's screen.

Both the client and server beacon sample apps can be found on GitHub:

```
https://github.com/PacktPublishing/Fearless-Cross-Platform-
Development-with-Delphi/tree/master/Chapter11/02_Beacons
```

Now you can detect a beacon and get some minimal data from it. Managing multiple beacons, calculating their distance, and triangulating your position is the next logical step, but sounds like a bunch of work. Fortunately, Embarcadero offers a couple of components to help us out.

Fencing your application

To make the most of beacons with their distance data, **BeaconFencing** components manage multiple beacons and provide accurate positional data. You can use either the `TBeaconZonesFencing` or the `TBeaconMapFencing` components, with the latter providing an interactive map for your application. These build on `TBeaconManager` to allow you to monitor beacons, trigger proximity events, place your position on a map, and calculate routes between points using optimized algorithms.

> **Note**
>
> Use of the BeaconFencing components is not free—and they're not even available for the **Professional** edition. **Delphi Enterprise** lets you download the BeaconFencing components and demo apps from the **GetIt Package Manager** and test your application with a limited number of devices, but to deploy and use these on a larger scale requires additional licensing from Embarcadero.

These components will save you a lot of work building professional applications that offer location-sensitive features and positional event handling.

The point at which we started discussing the concept of the "Internet of Things" necessarily started with the introduction of Bluetooth LE since, by definition, it's the basis of all low-energy device communication we'll discuss in this chapter. Beacons are one type of IoT device not always thought of as part of this category. The next section gets into the more popular concepts of IoT, wearable devices and gadgets of all sorts, and how they are simply an extension of the BLE concepts we've covered thus far.

Doing more with the Internet of Things

Bluetooth LE opens the door to a broad spectrum of applications and devices. The protocols we've explored are just the beginning of the many options for sharing data. Small devices lurking in corners of the building, throughout your car, on exercise equipment, in your appliances and entertainment system, all broadcasting and sharing little bits of information that can be picked up and read in various ways, serve specific functions and work without the need for manual interaction to enhance our lives. To some, this proliferation of information sharing is a little disconcerting, but by understanding the types of information being shared and how it's being used, we can see that there's far less magic and quite a bit of usefulness in these technologies.

The Bluetooth, Bluetooth LE, Beacon, and BeaconDevice components come with all versions of Delphi. As mentioned, Bluetooth LE, with its GATT and related profiles, is the basis of several IoT components that you can expand to work with any IoT device. To save you a lot of research and work, there are many that can be downloaded using the GetIt Package Manager. Each one comes with custom components and most come with a demo program. They are installed in your tool palette under the **Internet of Things** category:

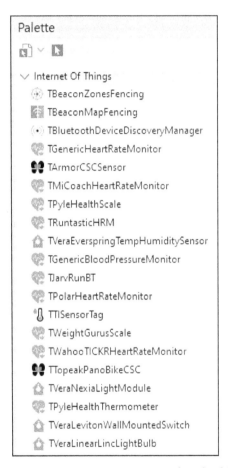

Figure 11.7 – Some IoT components you can download from GetIt

The first two, dealing with beacon fencing, are ones we covered in the previous section (the absence of the beacon components gives credit to the confusion over where, exactly, the idea of the IoT begins). The rest of the components in that list make up the bulk of what Embarcadero considers part of the "Internet of Things."

Discovering and managing your device

You can spend time studying specs and constructing BLE components and libraries to work with the myriad devices out there that speak Bluetooth, but why do that if that work has been done for you? Embarcadero has several dozen devices supported on the GetIt Package Manager and you can download these components along with a demo of how to use each one. These components are affectionately known as **ThingConnect** components and they each have one component in common, the `TbluetoothDeviceDiscoveryManager` component. Every time you download a ThingConnect component, it will come with one of these discover managers and install it if you don't already have one. Each of the device-specific components needs this component connected.

Even though these components encapsulate much of the work needed to deal with the various profiles, you'll still have to understand not only the peculiarities of the devices to which you connect but also the platform your app will be running on. There are differences between how Windows and other platforms scan and connect to BLE devices and some devices may respond differently, necessitating different ways of reacting to data events.

Speaking of reacting to data events, let's take a peek at one of these IoT device components and get a sense of how to use it.

Getting data from IoT devices

If you had to write your own methods for getting and parsing data from IoT devices, you'd have to dig deeply into the GATT profile, learn about BLE profile **characteristics**, and then study the specification for the device you're supporting and write the code to parse the bytes returned. With the high-level ThingConnect components, this work is done for you.

For example, the **Blood Pressure Service** specification is 31 pages long, but there are only three events in addition to the standard discovery, connection, and error events:

- `OnBloodPressureFeatureUpdate`
- `OnBloodPressureMeasurementUpdate`
- `OnIntermediateCutoffPressureUpdate`

These events, along with the supplied helper classes, make parsing the variety of data quite simple.

The topic of IoT and the many applications of BLE communication could take up a book by itself. The goal of this chapter was to give you a glimpse of information-sharing capabilities in the world of small devices and whet your appetite for expanding your programming skills beyond monolithic desktop applications.

You may suddenly be interested in acquiring more electronic gadgets on which to test ideas with your newfound skills, whether that involves capturing images of your front door when you're not around, building your own array of beacons around your yard, or monitoring temperature in different parts of your home.

This new hobby could get expensive if you had to use regular Windows or Mac computers. You may be able to find some used phones or tablets, but they're not well-suited for long-term, unattended, single-purpose use, such as broadcasting location and ID data, or capturing pictures on a regular basis.

However, there are small devices that are often used for such applications, and we should cover them before completing this book's section on mobile devices. And even though they're not always mobile, they are low-power and support BLE quite nicely. They're also very inexpensive.

Using a Raspberry Pi

Before the 1980s (and even for some time after that), computers were huge, taking up large air-conditioned rooms, and requiring massive amounts of energy and a highly trained staff. Today's hand-held smartphones are far more powerful than those behemoths and require almost no training at all. As the power of computers continues to increase and the size continues to decrease, the physical devices humans use to interact with a computer (such as keyboards and mice) can take more space than the computer itself. The term "fat finger" means more than just making a mistake on the keyboard; it now represents the limiting factor on user interfaces.

The Raspberry Pi is one of the more popular examples of this great reduction of computer size where the actual computer—CPU, memory, and interface ports—is the smallest component of the system, with the keyboard and monitor dwarfing the tiny case housing the electronics. Indeed, there are some keyboards that come with a complete computer built inside, and I have used Velcro to attach a Raspberry Pi case to the back of a monitor. The entire board upon which its **ARM**-based CPU and supporting chips reside is about the length and width of a credit card—and depending on what you add on top of the card, it starts at barely an inch high.

These small computers started out running a slim version of **Linux,** but as these devices have become popular, chips have advanced in power, and more RAM has been added, specialized versions of Android and even an ARM version of Windows can now be installed. So how can we utilize these with Delphi?

Deploying Delphi apps on Windows or Linux is currently relegated to the **Intel x86** architecture (there may be ways around this if you research and have the time to experiment), and you can't separate Apple from its hardware. So, the only directly supported method for deploying an app to a Raspberry Pi is to do it with Android.

Let's show how this is done.

Using Android to run your apps on a Raspberry Pi

We won't go into great detail on installing Android on a Raspberry Pi as that's beyond the scope of this book. But there's an excellent tutorial in the *Further reading* section at the end of this chapter called *How to FULLY Install Android 9 on Raspberry Pi 3* (you can search for other **YouTube** videos that may be more relevant to your needs). It really is as simple as it looks and only took me a few minutes on the first try.

You'll need a microSD card and a way to burn a **.ISO** image onto it. There are many places from which to download these .ISOs and links are constantly being upgraded, so do a quick internet search, follow the directions using one of several different free image-burning tools, and soon you'll be booting up your Raspberry Pi with a fresh copy of Android.

My Raspberry Pi 3 is shown in the following figure, next to my blue **Logitech** mouse:

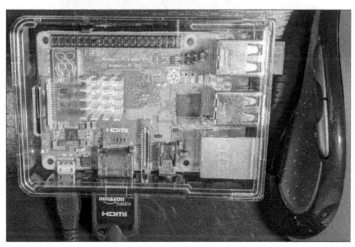

Figure 11.8 – Raspberry Pi 3 single-board computer with a heat sink

To debug Delphi apps on true Android phones and tablets, you connect a USB cable directly, turn on Android's debug mode in the device, and Delphi sees it. When using a Raspberry Pi, you can do the same thing but it might not always be as convenient, especially if you've mounted a Pi behind a TV or under the eaves of your house and you don't want to move it just to give your app an update. You can, however, deploy and debug Android apps over the network, either wired or wireless depending on which Pi you have and your configuration. To do this, use the freely available **Android Debug Bridge** (**adb**).

There are two parts to this. The first, **adb**, is the **daemon**, or background process, running on the Android devices. This provides the connection to external devices. You probably don't need to download adb because most Android installations come with it built-in. The version of Android I installed on my Raspberry Pi, **LineageOS 16.0**, already has it built-in, and I followed the same instructions laid out in *Chapter 4*, *Multiple Platforms, One Code Base* (under the section *Preparing your PC to deploy to an Android device*) to enable USB debugging. While in the **Developer Options**, also enable *ADB debugging over network* so that you can deploy your Android apps directly from Delphi.

The second part to getting this set up is to run adb.exe from your Windows machine to connect to the adb process on Android and provide visibility to it from within Delphi. This command-line utility can be found in the platform-tools folder where your Android **SDK** was installed. (To recall the location of your SDK, you can check in Delphi's options under **Deployment | SDK Manager**) You'll need to know the IP address of your Raspberry Pi. Then, bring up a **Command Prompt** and run adb.exe connect <IP_Address>, and put in your Pi's address:

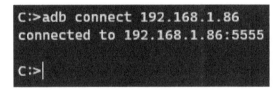

Figure 11.9 – Windows PC connected to Android debugging on a Raspberry Pi over the network

Once adb is connected to your Pi, it'll act just the same as if a phone or tablet were directly connected to your PC, so just select the target like you would any other Android device:

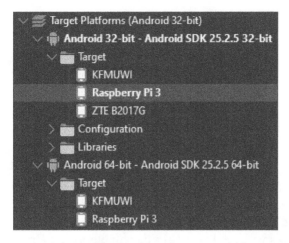

Figure 11.10 – Selecting the Raspberry Pi 3 target platform in Delphi

You can now run your own custom-written Android apps on a Raspberry Pi! To see this demonstrated, check out the YouTube video link in the *Further reading* section at the end of this chapter called *Delphi and C++Builder on Raspberry Pi and SBC*.

I took an open source project, *DelphiVersions*, and ran it, unmodified, on mine:

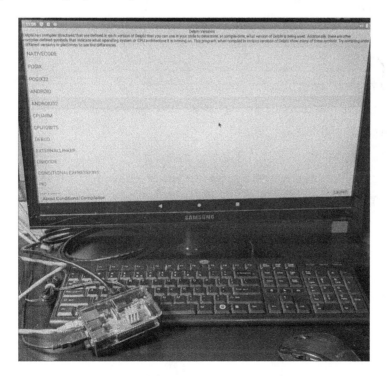

Figure 11.11 – Delphi app running on a Raspberry Pi 3

The **DelphiVersions** project can be downloaded from GitHub:

`https://github.com/corneliusdavid/DelphiVersions`

So, now your world has been expanded!

Summary

This chapter trudged through the complexities of Bluetooth, starting with its origins of discovery, pairing, and connection, then moving to the more efficient low-energy version that allowed smaller devices to take advantage of the technology. Newer still, beacons allow connectionless identity querying and proximity calculations for tracking the movement of devices and providing apps with the ability to be contextually aware. Finally, we showed how small, credit card-sized computers can help with budget and space concerns, without having to sacrifice features or switch to a different development tool—Delphi shines through all these scenarios in providing a rich development tool.

Now that we've had our fun with small, mobile devices, it's time to turn our attention to powerhouse servers where Delphi can just as effortlessly crunch monstrous SQL queries against massive databases as it can manage multiple requests from a variety of web servers. Read on!

Questions

1. How do you start searching for Bluetooth devices?
2. What are some differences between Classic Bluetooth and Bluetooth LE?
3. What are BLE Profiles?
4. What is a beacon?
5. Do beacon servers use the same component as beacon scanners?
6. What is **ThingConnect**?
7. What OS can you put on a Raspberry Pi to allow Delphi apps to run?

Further reading

- Learn About Bluetooth Versions: `https://www.bluetooth.com/learn-about-bluetooth/radio-versions/`

- Using Bluetooth: `http://docwiki.embarcadero.com/RADStudio/Sydney/en/Using_Bluetooth`

- Using Classic Bluetooth: `http://docwiki.embarcadero.com/RADStudio/Sydney/en/Using_Classic_Bluetooth`

- Microsoft's Bluetooth Low Energy Overview: `https://docs.microsoft.com/en-us/windows/uwp/devices-sensors/bluetooth-low-energy-overview`

- Using Bluetooth Low Energy: `http://docwiki.embarcadero.com/RADStudio/Sydney/en/Using_Bluetooth_Low_Energy`

- Bluetooth Specifications List: `https://www.bluetooth.com/specifications/specs/`

- Bluetooth Low Energy Applications: `https://en.wikipedia.org/wiki/Bluetooth_Low_Energy#Applications`

- Beacons and Delphi in Amsterdam – a blog from CodeRage XI: `https://community.embarcadero.com/blogs/entry/beacons-and-delphi-in-amsterdam`

- Beacons and Delphi: `https://www.youtube.com/watch?v=ieoIZ40MOjM`

- Understanding the Different Types of BLE Beacons: `https://os.mbed.com/blog/entry/BLE-Beacons-URIBeacon-AltBeacons-iBeacon/`

- A Museum Adventure with a Delphi Integration: `https://blogs.embarcadero.com/plunge-into-a-museum-adventure-with-this-delphi-integration`

- What the Internet of Things (IoT) Is and How It Works: `https://appinventiv.com/blog/what-is-internet-of-things/`

- BeaconFence: `http://docwiki.embarcadero.com/IoT/en/BeaconFence`

- Working with ThingConnect Devices: `http://docwiki.embarcadero.com/IoT/en/Working_with_ThingConnect_Devices`

- How to FULLY install Android 9 on a Raspberry Pi 3: `https://www.youtube.com/watch?v=EZxbQ5ive04`

- Android Debug Bridge (adb): `https://developer.android.com/studio/command-line/adb`

- Delphi and C++Builder on Raspberry Pi and SBC: `https://youtu.be/f_Wjqin9SXU`

- WebBroker on Android and Raspberry Pi 3: `http://delphi.org/2019/04/webbroker-on-android-and-raspberry-pi-3/`

Section 4: Server Power

The last section of the book provides you with many options for writing backend servers: Windows services, Linux daemons web modules under IIS for Windows, or Apache for Windows or Linux, or by using the power of a pre-built REST server by building modules for the RAD Server with the modifications necessary for a client app with it. The book concludes by tying everything you've learned together into a finished and deployed application suite, both server and client, to your end customers and app stores.

This section comprises the following chapters:

- *Chapter 12, Console-Based Server Apps and Services*
- *Chapter 13, Web Modules for IIS and Apache*
- *Chapter 14, Using the RAD Server*
- *Chapter 15, Deploying an Application Suite*

12
Console-Based Server Apps and Services

At the opposite end of the spectrum from microdevices, embedded electronics, beacons, and **single-board computers (SBCs)**, you will find large data centers with racks of servers processing terabytes of data at lightning speed to millions of users across the internet. These machines run 24 hours a day and, except for the initial installation process and occasional maintenance, are seldom seen by humans. These machines are called "headless" computers because they run unattended without monitors, keyboards, or mice.

You might immediately point out that Raspberry Pi and other SBCs are often used as headless mini servers and work quite well. Additionally, some small offices often use one of their more powerful workstations as a server machine with shared access to other workers. There is definitely some crossover in the variety of hardware that is available and how it's put to use. However, in this chapter, we'll focus on Windows and Linux operating systems that run on typical enterprise-class, data-center hardware, which is based on the x86 architecture.

Most server applications are built to automatically start up when the machine starts, listen and respond to multiple client requests simultaneously, handle errors gracefully, and, typically, log all activity. In the following sections, we will cover these topics:

- Starting with console apps on Windows and Linux

- Providing server connectivity for clients

- Logging activity

- Creating a Windows service

- Adopting a Linux daemon

- Exposing your server to the world

Get ready to get serious on the backend!

Technical requirements

This chapter, which focuses on server technology, will primarily require a Windows machine—your Delphi development machine will suffice quite nicely for testing and debugging Windows services.

If you have Delphi Enterprise, you can also develop and run Linux console applications and server apps. They can be deployed over your network to a standalone Linux server or onto a virtual machine. Only Ubuntu 16.08 or 18.04 or RedHat Enterprise 7 are currently supported by Embarcadero.

You can find the code for this chapter on GitHub at `https://github.com/PacktPublishing/Fearless-Cross-Platform-Development-with-Delphi/tree/master/Chapter12`.

We'll start simply by going back to square one: console apps.

Starting with console apps on Windows and Linux

Unarguably, console apps are the simplest type of application you can create in Delphi. All the code for an entire project can be contained in one unit; you don't have to deal with the nuances of any graphics engine, display framework, or even style. Of course, your user interface is quite rudimentary, but if your end goal is a server app deployed to a headless machine such as a remote Windows or Linux server, then this is the perfect starting point.

As a quick review, to create a console app in Delphi, select **File | New | Other** from the main menu and select **Console Application**. This creates a simple program that does nothing more than trap an error. Add a couple of lines of code between the `try` and `except` lines that were provided for us:

```
begin
  try
    Writeln('Hello from Delphi!');
    Readln;
  except
    on E: Exception do
      Writeln(E.ClassName, ': ', E.Message);
  end;
end.
```

These simply write out a line of text to the console. Next, pause and wait for the user to press *Enter*. Now, let's make it cross-platform, but instead of adding iOS and Android, we're going to add Linux.

First, we need to set up a simple testing environment that can be done right on your Windows 10 desktop.

Installing the Windows subsystem for Linux

There are many ways to get a Linux machine built and running on your network—along with many books, blogs, and videos to help you get there. Additionally, you can put Linux on a virtual machine on your computer (using VMWare, VirtualBox, Hyper-V, and more). You are free to use any of these methods as your experience and time allow.

However, if you don't want to wade through that learning curve right now, there's a pretty simple way to get Linux on your Windows 10 machine that takes care of a lot of details for you. Just follow these steps:

1. Go to the **Start** menu from Windows and type in `Features`. Then, select **Turn Windows features on or off**.
2. When the list of selectable features comes up, scroll to near the bottom and check **Windows Subsystem for Linux**. Then, click on **OK**. It is likely that your computer will need to reboot during this process.

3. Pull up the **Microsoft Store**, search for `Linux`, and install Ubuntu 18.04 LTS. (18.04 is the first stable release of Ubuntu in 2018; a newer version of Ubuntu might be supported by the time you read this.)

   Once it has been installed, you can launch it just like any other app, and you'll get a Linux prompt! (This is real Linux, not an emulator.) What's really nice is you can access your Windows filesystem from Linux through the `/mnt` folder and your Linux filesystem from Windows through a virtual network device exposed through `\\wsl$` (which is made from the abbreviation for **Windows Subsystem for Linux**). This makes it very simple to share files between the two systems. The first file we need to share is the **Platform Assistant Server** (**PAServer**).

   With Linux running, copy and paste `LinuxPAServer21.0.tar.gz` from the `PAServer` folder, which is inside your installed Delphi folder, to the `/home/<username>` folder inside your Linux subsystem. You can get to this from Windows easily enough with **Explorer** by navigating to `\\wsl$\Ubuntu-18.04\home` or `\\wsl$\home`, depending on your system. Then, navigate to your user subfolder.

4. From the Linux prompt, make sure you're in the same folder where the compressed LinuxPAServer was just placed (for example, `/home/david`). If not, you can get there by typing in `cd /home/david` or simply `cd ~/` from the Linux command line; then, extract the platform assistant server using the `tar xvf LinuxPAServer19.0.tar` command.

The PAServer is now ready on Linux for Delphi to connect to and send console applications.

Running our first Linux app

If you extracted the LinuxPAServer from the home folder in Linux, you can type in the following command to start it from the Linux prompt: `~/PAServer-21.0/paserver`. Similar to the PAServer running on the Mac, which we learned about in *Chapter 4*, *Multiple Platforms, One Code Base*, it prompts you for a password, which you can leave blank.

Now, switch back to Delphi and set up the connection profile to the Linux PAServer:

1. Add the **Linux 64-bit** platform to our project.

2. Create a connection to the PAServer; right-click on the platform and select **Properties**.

3. Underneath **SDK**, select **Add New...** and then click on **Select a profile to connect:**.

4. Give this new profile a name and then click on **Next >>**.

5. Enter `127.0.0.1` for the **Remote machine** (if you're using Linux on a separate computer, you will need to use the `i` command from the PAServer to list the IP addresses and then enter one of them onto your Linux connection profile instead).

6. Make sure that the **Port number** and **Password** match what you have in the PAServer running on Linux (if you accepted the defaults when you started the PAServer, you shouldn't have to change them here). The port number is listed when PAServer starts; you can use the `p` command to list it again. Note that it is usually `64211`.

7. Click on **Test Connection** and **OK**. Then, select **Finish** and **Save** and Delphi will start updating the local cache of Linux files.

   The project can now be compiled and run. If simply debugging with the *F9* key doesn't work for some reason, try one of the following ways to launch the app:

   A. Click on **Run Without Debugging** (*Shift + Ctrl + F9*).

   B. Within the Linux PAServer prompt, you might need to "wake up" the process by typing in a key or hitting *Enter* for the app to actually start.

   C. You can also manually launch the app from the Linux prompt by going to the `~/PAServer/scratch-dir/` folder, then navigating into the folder that matches Delphi's profile name for this Linux system. Finally, navigate to the folder for the project and run the app that has been deployed there.

This simple, cross-platform console app can be downloaded from GitHub at `https://github.com/PacktPublishing/Fearless-Cross-Platform-Development-with-Delphi/tree/master/Chapter12/00_SimpleConsole`.

Now that we can write a console application for Linux, let's make it do something useful.

Adding a simple database lookup module

In *Chapter 9*, *Mobile Data Storage*, we created a mobile app called *MyParks*. Instead of forcing the user to figure out the name of the park and type it in, what if we could send the coordinates captured by the phone to a server and have the server tell us which park we're at? There are GPS converter APIs that can convert longitude and latitude coordinates into an address; however, typically, they cost money and don't necessarily have the park names attached. Besides, we're here to learn how to write servers in order to extend these capabilities to other scenarios. So, we'll construct a small database and learn how to provide this service to our client app.

> **Note**
>
> I won't take up space to detail how you should set up your database. You can use whatever database product you're familiar with or to which you have access. I'll use InterBase since it's an Embarcadero product and is the default database that is used with RAD Server (we'll be exploring this in *Chapter 14, Using the RAD Server*). Therefore, the database syntax shown in the code examples will be that of InterBase.

Your database should have a table, named `Parks`, with the following fields:

- `PARK_ID – INTEGER`
- `PARK_NAME – VARCHAR(50)`
- `LONGITUDE – NUMERIC(9,6)`
- `LATITUDE – NUMERIC(9,6)`

The end goal is to create a server that listens for coordinates, looks up a matching park in the database, and sends the results to our *MyParks* app, which will update its local database. However, first, we want to test the basic functionality of the database lookup.

So, create a new console application, add a data module with the FireDAC connection and query components, hook them up to the database, and test the connection. Once we've proven this works, we'll continue building our real server.

> **Note**
>
> For database connectivity with FireDAC, make sure that you use the **Connection Editor** of the `TFDConnection` to set up your database and test the connection. Remember that this data module will be used with both Windows and Linux projects, so if your database is on your Windows machine, the InterBase protocol will need to be set to `Remote` for it to connect to Linux.

GPS coordinates can vary depending on the accuracy of the device and wherein the park you're at, especially if it's a big park. So, when we construct the lookup query, we should write it in such a way that the location could vary a little from what is stored in the database. For instance, you could make that variance a setting in your application, provide a nice configuration utility, and perhaps even provide a way to change the lookup precision in the mobile app. For demonstration purposes, we'll hardcode this variance into the query by searching for our longitude and latitude values between the stored value plus or minus 0.002:

```
select PARK_ID, PARK_NAME, LONGITUDE, LATITUDE
from Parks
```

```
where :long between (LONGITUDE - 0.002) and (LONGITUDE +
   0.002)
   and :lat  between (LATITUDE  - 0.002) and (LATITUDE +
      0.002)
```

The identifiers that start with a colon create parameters to the query. These are the two values that the MyParks app will send. We'll need to add a public function inside the data module that takes two double parameters and returns a record—let's define the record first:

```
public
  type
    TParkDataRec = record
      ParkID: Integer;
      ParkName: string;
      Longitude: Double;
      Latitude: Double;
      procedure Clear;
    end;
```

The record type will be what the function returns after querying the database:

```
function TdmParksDB.LookupParkByLocation(
    const ALongitude, ALatitude: Double): TParkDataRec;
begin
  Result.Clear; // procedure to clear the record's fields
  try
    qryParkLookup.ParamByName('long').AsFloat := ALongitude;
    qryParkLookup.ParamByName('lat').AsFloat  := ALatitude;
    qryParkLookup.Open;

    if qryParkLookup.RecordCount > 0 then begin
      Result.ParkID := qryParkLookupPARK_ID.AsInteger;
      Result.ParkName := qryParkLookupPARK_NAME.AsString;
      Result.Longitude := qryParkLookupLONGITUDE.AsFloat;
      Result.Latitude  := qryParkLookupLATITUDE.AsFloat;
    end;
  finally
```

```
    qryParkLookup.Close;
  end;
end;
```

Before we leave the data module, you need to remember an important difference between the console apps and the GUI apps you're used to writing. When writing either VCL or FireMonkey windowed apps, the form units are automatically added to the project file in the main `begin-end` loop with `Application.CreateForm()` statements so that the forms are readily accessible in your application code (you can turn this off by navigating to **Tools | Options | User Interface | Form Designer** and then unchecking **Auto create forms & data modules**). No data modules are automatically created in the console applications—you have to write the code to create them yourself before you try to access them; otherwise, you'll get an access violation error.

Since this data module could be used by a console app or a different kind of server app that we'll create later in this chapter, it'll be more convenient to create and free the data module class in the `initialization` and `finalization` sections of the unit. That way, it's created when the app starts, automatically freed, and accessible to any program or module that uses this data module. I named the data module `TdmParksDB`, so the autogenerated global variable is `dmParksDB` and the creation and freeing code looks like this:

```
initialization
  dmParksDB := TdmParksDB.Create(nil);
finalization
  dmParksDB.Free;
```

Now our data module is ready to be tested.

Testing the data module with a console app

As we touched upon a moment ago, the console app we're building right now is just to test the functionality that the eventual server we're building will provide. A console app is a really easy way to do that. We put all the common functionality for the "real" server into data modules and other units. Then, we write a simple interface that collects and displays data in the main part of the test console app, which calls methods from the used units. Our strategy will become clearer in later sections when we build a Windows service and a Linux daemon.

Our console app for testing the new database lookup function needs to write some code to read values from the console, call the lookup function from the data module, and display the results back to the console. Here's the main project code:

```
var
  long, lat: Double;
  ParkRec: TdmParksDB.TParkDataRec;
begin
  Writeln('--MyParks Lookup Test--');
  Write('Enter the longitude: ');
  Readln(long);
  Write('Enter the latitude: ');
  Readln(lat);

  ParkRec := dmParksDB.LookupParkByLocation(long, lat);
  if ParkRec.ParkID > -1 then
    Writeln(Format('Park %d: "%s" (%0.3f, %0.3f)',
      [ParkRec.ParkID, ParkRec.ParkName,
        ParkRec.Longitude, ParkRec.Latitude]))
  else
    Writeln(Format('Coordinates (%0.3f, %0.3f) not found.',
      [long, lat]));

  Write('Press Enter...');
  Readln;
end.
```

Running it looks like this:

```
--MyParks Lookup Test--
Enter the longitude: -103.424
Enter the latitude: 44.475
Park 122: "Bear Butte State Park" (-103.424, 44.475)
Press Enter...
```

Figure 12.1 – The console app on Windows looking up data from InterBase

Everything is set up for you on the Windows platform. To get this to run on the Linux platform, we have to tell Linux about some extra files that will need to deploy in order to access InterBase (note that if you're using a different database product, your file selection will be different):

1. Add the **Linux 64-bit** platform to the **Target Platforms** in your project manager.
2. Open up the **Deployment** tab (that is, navigate to **Project | Deployment** from the main menu).
3. Make sure **Linux 64-bit platform** under **All configurations** has been selected.
4. Click on the **Add Featured Files** button (this is the folder icon with a plus sign; it is the fourth from the left-hand side)
5. Expand **InterBase Client** and select **Linux64**. Then, click on **OK**.

With the **Linux 64-bit** target selected as the active platform, and the PAServer running on your Linux system, run the app and switch to your Linux shell. You should see the same interface show up and run exactly the same way as it does on Windows.

You can download the source for the `MyParksConsoleApp` project on GitHub at `https://github.com/PacktPublishing/Fearless-Cross-Platform-Development-with-Delphi/tree/master/Chapter12/01_ParksConsole`.

Now that the data module works with a lookup function, it's time to turn our app into a real server by actually serving data outside of itself.

Providing remote server connectivity for clients

Our console app isn't really a true server yet because it has a user interface (albeit a crude one) that accepts input and displays output and then exits—no client app interacts with it. Our next step, then, is to replace the `Readln` and `Writeln` statements in the body of the main program loop, which waits for a human to type something into the console, with a way to start a listening and response mechanism that can be used by a client app.

Copy the data module (both the `.pas` and `.dfm` files) into a new folder and create a new app with it added; call this new app `MyParksTCPServerConsole`.

There are many ways for server and client applications to talk to one another. We discussed how Bluetooth was used in the previous chapter, which works when two devices are close together. In this chapter, we will use a very common method of communication used across the internet, which works anywhere a network connection is available, that is, through a port over TCP/IP. One popular set of components that support TCP/IP communication is known as **Internet Direct (Indy)**.

Indy is a set of tried-and-true components for handling both the server and client sides of a wide assortment of internet protocols such as FTP, SMTP, Echo, TraceRoute, SSH, Telnet, and even HTTP (yes, you can write your own web server with these components). The one we'll use is the TCP server, which listens on a port for requests and sends back responses. Our server will listen for coordinates from a client app and respond with park information.

Let's add another data module to our new server app and place `TIdTCPServer` on it. It listens on a port that we define for incoming requests; we will parse the request, look up the data, format a response, and then send it out on that same port back to the client. Therefore, our client and server apps need to coordinate regarding the port and data formats to use in order to communicate effectively. Here's what we'll use:

- Port: 8081
- Request: `longitude=x.xxxx,latitude=y.yyyy`
- Response: `ParkName=<park_name>`

The requests, expected on port 8081, take the form of a string with two coordinates, separated by a comma; the response back to the client will simply be the park name. For example, if the server receives `longitude=-122.813,latitude=45.517`, we would parse the coordinates, pass them to the database module's lookup function that we wrote earlier, and respond with the park name stored in the database for those coordinates. In my database, as we saw earlier, it should respond with `ParkName=Commonwealth Lake Park`.

> **Note**
> Why did I choose port 8081? There are 65,535 TCP ports to choose from. Many of them are standard ports, which are defined and used in public contexts. For example, port 80 is used for unsecured web pages or HTTP, 443 is for HTTPS, 25 is for SMTP, and so on. Many sample web applications or local servers use 8080 as an example or as a temporary port. I guess this is because it's easy to remember by thinking of it as the common port "80" twice. I simply incremented that port number by one to avoid possible conflicts, but you can use a different one if you wish—as long as it isn't already in use.

Setting the port is easy; there's a property on the component for that, which is
DefaultPort. Detecting requests and sending responses requires writing some code in
the event handler, OnExecute. Since there are two values we're expecting, separated by a
comma, we can use TStringList to read them in and conveniently parse them for us.
Additionally, we'll need a couple of variables to hold the longitude and latitude values we
parse and a response string. This is how we'll set up the initial part of the procedure:

```
procedure TdmTCPParksServer.IdTCPMyParksServerExecute(AContext:
TIdContext);
var
  ValidRequest: Boolean;
  Requests: TStringList;
  ReqLong, ReqLat: Double;
  ResponseStr: string;
begin
  ValidRequest := False;
  ResponseStr := EmptyStr;
```

Next, we'll parse the string and do some error checking:

```
  Requests := TStringList.Create;
  try
    Requests.CommaText := Trim(AContext.Connection.Socket.
      ReadLn);
    if Requests.Count = 2 then begin
      if TryStrToFloat(Requests.Values['longitude'], ReqLong)
    and
        TryStrToFloat(Requests.Values['latitude'], ReqLat)
          then begin
          var ParkInfo := dmParksDB.
            LookupParkByLocation(ReqLong, ReqLat);
          if ParkInfo.ParkName.Length > 0 then begin
            ResponseStr := ParkInfo.ParkName;
            ValidRequest := True;
          end else
            ResponseStr := '<Unknown Park>';
        end;
```

```
      end;
  finally
    Requests.Free;
  end;
```

Finally, we'll check to see whether we have had any errors. Then, we'll set the response accordingly and return it:

```
  if (not ValidRequest) and (ResponseStr.Length = 0) then
    ResponseStr := 'ERROR: Invalid request';

  AContext.Connection.Socket.WriteLn(ResponseStr);
end;
```

We'll also provide the public `Start` and `Stop` procedures to hide the actual component in use from the calling module. This is in case we want to change to a different TCP server component in the future:

```
procedure TdmTCPParksServer.Start;
begin
  IdTCPMyParksServer.Active := True;
end;
procedure TdmTCPParksServer.Stop;
begin
  IdTCPMyParksServer.Active := False;
end;
```

The last thing we need is a way for the calling module to tell whether the server component is still active. I created a public property, `Running`, and wrote its getter function like this:

```
function TdmTCPParksServer.IsRunning: Boolean;
begin
  Result := IdTCPMyParksServer.Active;
end;
```

Note that there are a few other events you can hook into that are useful for reporting when a client connects and disconnects or when an error occurs. It's a good idea to use them for unattended server applications running on a remote server. This is so it can log activity that can be queried periodically to check whether things are running properly. You can do this by creating event handlers in the data module that can be hooked into the calling module. The current app we're writing will just write the activity to the console; however, in the next section, we'll turn this into a service. This data module should work in either situation, so don't assume anything about the interface or have any dependencies regarding how it will be used.

To support this concept of event-driven programming, we recommend that you write public properties that are checked at key points in your code and call the appropriate notification procedures. For example, when a client connects to this server, you should provide a way to notify the calling module with the following public property:

```
public
    property OnConnect: TNotifyEvent read FOnConnect write
FOnConnect;
```

The private variable should be declared like this:

```
private
    FOnConnect: TNotifyEvent;
```

I like to write *Do* procedures to check to see whether the event handler is hooked up by the calling module and, if so, call the calling event handler:

```
procedure TdmTCPParksServer.DoOnConnect;
begin
  if Assigned(FOnConnect) then
    FOnConnect(self);
end;
```

Then, double-click on the OnConnect event in the IdTCPServer component and simply call the *Do* procedure:

```
procedure TdmTCPParksServer.IdTCPMyParksServerConnect(AContext:
TIdContext);
begin
  DoOnConnect;
end;
```

This might seem like an unnecessary simplification, but implementing this technique as a habit in all of your applications makes your code readable, consistent, and flexible. We'll now be able to use this data module, unchanged, in a variety of ways; first, in a simple Windows console app for testing, and then again in the next couple of sections when we write a Windows service and a Linux daemon.

The main loop of the project will still be a console app. However, instead of asking for coordinates and making the call ourselves, it'll simply be used to start and stop the TCP server and display some messages as clients connect and send requests. In fact, it could be as simple as this:

```
begin
  Writeln('Starting MyParks TCP Server on port ' +
  dmTCPParksServer.IdTCPMyParksServer.DefaultPort.ToString);
  dmTCPParksServer.Start;
  while dmTCPParksServer.Running do
    Sleep(100);
  Writeln('MyParks TCP Server quitting ');
end.
```

Running it simply displays the `Starting` ... message; it then silently listens and processes TCP messages until it's shut down. To utilize the event handler we wrote, we have to create a class; I called it `TConsoleParkDisplay`, and the implementation of the `OnConnect` and `OnDisconnect` procedures looks like this:

```
procedure TConsoleParkDisplay.OnConnect(Sender: TObject);
begin
  Writeln('>> client connected');
end;
procedure TConsoleParkDisplay.OnDisconnect(Sender: TObject);
begin
  Writeln('<< client disconnected');
end;
```

To add them to the main loop just requires you to create the class and assign the event handlers:

```
var
  ConsoleDisplay: TConsoleParkDisplay;
begin
```

```
  Writeln('Starting MyParks TCP Server on port ' +
    dmTCPParksServer.IdTCPMyParksServer.DefaultPort.ToString);
  ConsoleDisplay := TConsoleParkDisplay.Create;
  try
    dmTCPParksServer.OnConnect := ConsoleDisplay.OnConnect;
    dmTCPParksServer.OnDisconnect := ConsoleDisplay.
      OnDisconnect;
    dmTCPParksServer.Start;
    while dmTCPParksServer.Running do
      Sleep(100);
  finally
    ConsoleDisplay.Free;
  end;
  Writeln('MyParks TCP Server quitting ');
end.
```

I implemented a few more events in the full version of this project. Check out the MyParksTCPServerConsole project in GitHub at https://github.com/PacktPublishing/Fearless-Cross-Platform-Development-with-Delphi/tree/master/Chapter12/02_ParksServer.

> **Note**
>
> If you are downloading and running this project from GitHub, you might want to look ahead and read the *Logging activity* section before trying to compile it, so you can get the logging library installed.

To view this in action, we need to build a client.

Testing with a console client

Now, we'll build one more console app because it's quick and simple and keeps us focused on the client-server interaction. As with the server counterpart we just finished, we'll add a data module to hold the TCP component, but on this one, of course, we'll add TIdTCPClient. Additionally, instead of starting and stopping a server, we'll add the public procedures of Connect and Disconnect:

```
procedure TdmTCPParkClient.Connect;
begin
  IdTCPMyParksClient.Connect;
```

```
end;
procedure TdmTCPParkClient.Disconnect;
begin
  IdTCPMyParksClient.Disconnect;
end;
```

We'll also provide a few event handlers that the calling module can use. For example, when the TCP client's OnConnected event handler is triggered, we'll call our *Do* procedure, which checks whether an event handler we created, OnClientConnected, is hooked up and if so, calls it.

The most important method of the data module is the call to the server that actually looks up park information by passing longitude and latitude coordinates to the server:

```
function TdmTCPParkClient.GetParkName(const ALong, ALat:
Double): string;
begin
  IdTCPMyParksClient.IOHandler.WriteLn(Format(
        'longitude=%1.4f,latitude=%1.4f', [ALong, ALat]));
  Result := IdTCPMyParksClient.IOHandler.ReadLn;
end;
```

The main body of the client console project needs to ask for the coordinates, call the TCP data module for the park name, and then write the result out on the console. Variables and constants are left out of the following code snippet for brevity, but they can be implied by context; GetCoordinates, the function that asks for the coordinates, is also not shown here:

```
begin
  dmTCPParkClient.IdTCPMyParksClient.Host := '127.0.0.1';
  dmTCPParkClient.IdTCPMyParksClient.Port := 8081;
  dmTCPParkClient.Connect;

  done := False;
  repeat
    cmd := GetCoordinates(Long, Lat);
    if SameText(cmd, QUIT_CMD) then
      done := True
    else if SameText(cmd, QUERY_CMD) then
```

```
      Writeln(dmTCPParkClient.GetParkName(Long, Lat))
    else
      Writeln(cmd);
  until done;

  Writeln('Good-bye');
  dmTCPParkClient.Disconnect;
end.
```

Running the client from Windows looks like this:

```
Enter longitude or "quit" to end: -81.055
Enter latitude or "quit" to end: 27.611
Kissimmee Prairie Preserve State Park

Enter longitude or "quit" to end:
```

Figure 12.2 – The MyParks console client app on Windows requesting park information from a server

The server, running from a Linux subsystem, receives the client requests via TCP, queries the InterBase database for the park information, and returns the result:

```
david@goliathvm-pro: ~
Platform Assistant Server  Version 12.2.10.3
Copyright (c) 2009-2021 Embarcadero Technologies, Inc.

Connection Profile password <press Enter for no password>:

Starting Platform Assistant Server on port 64211

Type ? for available commands
>Starting MyParks TCP Server on port 8081
>> client connected
Response: Response: Kissimmee Prairie Preserve State Park
<< client disconnected
Response: Connection Closed Gracefully.
```

Figure 12.3 – The MyParks console server on Linux processing TCP requests for park information

The source for the console-based client app can be used to test variations of the server that we write in the rest of this chapter. Its source code is available from GitHub at https://github.com/PacktPublishing/Fearless-Cross-Platform-Development-with-Delphi/tree/master/Chapter12/03_ParkClient.

> **Note**
>
> If you are downloading and running this project from GitHub, you could encounter an error, as it can't find the logging library that we will add to the project in the *Logging activity* section.

Now we have a true server running, and it has been client-tested. However, it has to be started and stopped manually and only runs under a user account. The next step will be to create a background process that starts when the machine boots up, enabling a complete unattended operation on a headless server. The problem is that background processes can't write to the console in the same way as these test apps. So, we will have to make a side trip and implement logging.

Logging activity

In this chapter, our focus is on building robust server applications that run in the background without human intervention, providing services to clients. In a similar way to the beacons and IoT devices we learned about in *Chapter 11*, *Extending Delphi with Bluetooth, IoT, and Raspberry Pi*, addresses and data formats must be agreed upon by both the client and server for there to be useful communication. However, one big conceptual difference is that IoT devices don't wait for clients to request data; instead, they just continually spew forth information, including the services they provide and what format to use, over Bluetooth for anyone close enough to pick it up. The servers we're building in this chapter listen for specific requests on a specific port and then respond. This requires a little more coordination to get right because there's no advertisement mode for the service that is provided.

Because of the "silent waiting" mode that these server applications are in, and because servers are often managed en masse by only a few IT people, logging is a very useful and commonly expected feature for server apps. Disk space is cheap compared to knowing what an application you can't see is doing.

Application logging can be as simple as adding a line to a text file every time something happens or as complex as capturing detailed diagnostics about the running process, which requires specialized log-viewing tools to parse compressed data. Some logging data is encrypted to protect it from snooping by unauthorized users; some is automatically sent to tech support to help fix a problem before it is detected by the general public.

A good logging library will provide options for setting file size limits, automatically purging old log files, and sending log messages to multiple places. One such library for Delphi is `LoggerPro`, which is a well-supported open source project on GitHub. You can find it at `https://github.com/danieleteti/loggerpro`. We will add this library to our project, test it in our console app, and then demonstrate how it can be used on both Windows and Linux.

After downloading LoggerPro to a folder of your choosing, building the project for all of the platforms we'll be working with (that is, Windows 32-bit, Windows 64-bit, and Linux 64-bit), and, finally, making sure the compiled package and units are in the Delphi library path for each platform, reopen the `MyParksTCPServerConsole` project and add `LoggerPro.GlobalLogger` to the `uses` clause. That unit establishes some good default options so that you can use the library right away and in the simplest way possible by creating a global `Log` variable.

LoggerPro implements four types of log messages by calling one of the following methods, `Info`, `Debug`, `Warning`, or `Error`. Each method takes two string parameters: the first one is the message that is to be logged, and the second is a tag that puts the message into a separate log file for additional categorization.

Just underneath the `uses` clause, create a constant to define the tag we'll use for each log call:

```
const
  LOG_TAG = 'console';
```

Then, wherever there is a `Writeln` statement, duplicate the line to call `Log` and one of its log message methods. For example, in the `OnConnect` procedure, add the following line:

```
Log.Info('Client connected', LOG_TAG);
```

The `OnException` procedure would use this:

```
Log.Error(s, LOG_TAG);
```

Before we run this, do the same thing to the two data modules we added so that we can get a full report of activity. In both cases, add `LoggerPro.GlobalLogger` to the `uses` clause, in the `implementation` section. Then, create a string constant defining the log tag for that unit. Afterward, go through the important methods where you might want to see activity logged and add `Log.Info` calls (or other log methods, as appropriate).

Now, after you run the server, connect to a client, run a park lookup, and disconnect, you'll find one or more log files in your project folder. For my database module, udmParksDB, I set my LOG_TAG constant to 'database'. And in my TCP module, udmTCPParksServer, I set my LOG_TAG constant to 'tcp'. Therefore, I had three log files, as follows:

- MyParksTCPServerConsole.00.console.log

- MyParksTCPServerConsole.00.database.log

- MyParksTCPServerConsole.00.tcp.log

Each of these filenames starts with the project name, followed by a number, and ends with the tag as the base filename for the log file. This helps us to keep the files together, but it also separates them based on the log tag you used. As log entries are added and the files reach a maximum size, the number is incremented, the file is kept as a backup, and a new file with '00' is started.

Here are the contents of the first file, which have been generated from the main body of the project:

```
2021-05-24 21:41:04:548  [TID      2812][INFO   ] Starting MyParks TCP Server on port 8081 [console]
2021-05-24 21:41:07:990  [TID      744][INFO   ] Client connected [console]
2021-05-24 21:41:39:224  [TID      744][INFO   ] Client disconnected [console]
```

Figure 12.4 – The sample log entries of the console server

There's a lot of information in these lines to just add a few log lines. LoggerPro adds a lot of useful information such as the exact date and time of the entry, a thread ID, the type of log entry (such as info, debug, warning, or error), the message, and the tag used.

Let's take a look at the TCP log (note that I cut off the date/time and thread ID columns):

```
[INFO    ] START [tcp]
[DEBUG   ] OnConnect [tcp]
[INFO    ] Received Request [tcp]
[DEBUG   ]     Request parameters: longitude=-115.90, latitude=33.88 [tcp]
[DEBUG   ]     Returning park name: Joshua Tree National Park [tcp]
[DEBUG   ] OnExecute [tcp]
[INFO    ] Received Request [tcp]
[DEBUG   ]     Request parameters: longitude=-109.59, latitude=38.73 [tcp]
[WARNING ]     Returning: <Unknown Park> [tcp]
[DEBUG   ] OnExecute [tcp]
[INFO    ] Received Request [tcp]
[DEBUG   ]     Request parameters: longitude=-109.59, latitude=38.73 [tcp]
[DEBUG   ]     Returning park name: Arches National Park [tcp]
[DEBUG   ] OnExecute [tcp]
[INFO    ] Received Request [tcp]
[DEBUG   ] OnDisconnect [tcp]
[ERROR   ] OnException: Connection Closed Gracefully. [tcp]
```

Figure 12.5 – Sample log entries from the TCP data module of the server

These entries show the requests and responses processed in the TCP communications of the server using several different log types. One technique that I like to use is indenting the lines to indicate multiple log messages within a procedure. You can see this in the preceding screenshot after each entry containing `Received Request`.

Finally, let's take a look at the database log file:

```
[INFO    ] LookupParkByLocation(-115.90, 33.88) [database]
[INFO    ]    returning ParkID=108, ParkName=Joshua Tree National Park [database]
[INFO    ] LookupParkByLocation(-109.59, 38.73) [database]
[INFO    ]    returning ParkID=-1, ParkName= [database]
[INFO    ] LookupParkByLocation(-109.59, 38.73) [database]
[INFO    ]    returning ParkID=132, ParkName=Arches National Park [database]
```

Figure 12.6 – Sample log entries from the database module of the server

There are two things I dislike about how we implemented this: first, the main body of the server duplicates each activity message, once for the log file and once for the console; and second, the data modules don't log into the console in the same way as the main body. It would be easier to simply remove the `Writeln` statements from the project file, but then the console application wouldn't display anything while it's running. This can easily be solved.

Sending logs in two directions

Earlier in this section, I mentioned that a good logging library could send log entries to multiple places. LoggerPro supports this—and without much difficulty, we can satisfy both of my complaints.

The `LoggerPro.GlobalLogger` unit that is included in each of the data modules and the main project defines a global variable, `Log`, that we have used for logging. If you look within that unit, you can see the declaration and that it is initialized by calling `BuildLogWriter([TLoggerProFileAppender.Create])`. This writes the default log files that we've seen. However, note that the parameter is an array. That means we could add other `ILogAppender` instances. LoggerPro comes with some of these supplied in its directory, which you can take advantage of. There's one for Delphi's `OutputDebugString`, one to send emails, one to send a VCL memo, and more. The one we're interested in is `ConsoleAppender`.

To use this, we'll replace `LoggerPro.GlobalLogger` with our own unit that houses our global `Log` variable and includes that in the project and data modules, and then we'll initialize it the way we want in the project. With the `MyParksTCPServerConsole` project still open, create a new unit called `uMyParksLogging` that uses the LoggerPro unit and declares our global `Log` variable:

```
unit uMyParksLogging;
interface
uses
  LoggerPro;
var
  Log: ILogWriter;

implementation
end.
```

It's a very tiny unit but gives us a chance to initialize it. Replace the `LoggerPro.GlobalLogger` unit with these three: `LoggerPro`, `LoggerPro.ConsoleAppender`, and `LoggerPro.FileAppender`. Then, right after the main program's `begin`, add the following:

```
Log := BuildLogWriter([TLoggerProFileAppender.Create,
                       TLoggerProConsoleAppender.Create]);
```

The only significant difference between our initialization of the `Log` variable and the one in LoggerPro's `GlobalLogger` unit is that ours will log to both a file and to the console. You can add more log destinations by simply adding another "appender" unit to the `uses` clause and another `ILogAppender` instance to the array parameter of `BuildLogWriter`. When I write VCL applications, I like to use `TVCLListBoxAppender` in addition to the file appender. You can also write your own appenders for your own custom needs.

Now you can remove all of the `Writeln` statements from the project and data modules we used and simply let the `Log` statements send information where they need to. Once that has been done and you have run the server with a client connected, you will be able to view all of the log statements from the project and each of the data modules inside the console without needing to dig through the log file:

```
[INFO    ]  Starting MyParks TCP Server on port 8081 [console]
[INFO    ]  START [tcp]
[DEBUG   ]  OnConnect [tcp]
[INFO    ]  Client connected [console]
[INFO    ]  Received Request [tcp]
[DEBUG   ]    Request parameters: longitude=-78.30, latitude=38.77 [tcp]
[INFO    ]  LookupParkByLocation(-78.30, 38.77) [database]
[INFO    ]    returning ParkID=-1, ParkName= [database]
[WARNING ]    Returning: <Unknown Park> [tcp]
[DEBUG   ]  OnExecute [tcp]
[INFO    ]  Response: <Unknown Park> [console]
[INFO    ]  Received Request [tcp]
[DEBUG   ]    Request parameters: longitude=-78.29, latitude=38.70 [tcp]
[INFO    ]  LookupParkByLocation(-78.29, 38.70) [database]
[INFO    ]    returning ParkID=174, ParkName=Shenandoah National Park [database]
[DEBUG   ]    Returning park name: Shenandoah National Park [tcp]
[DEBUG   ]  OnExecute [tcp]
[INFO    ]  Response: Shenandoah National Park [console]
[INFO    ]  Received Request [tcp]
[DEBUG   ]  OnDisconnect [tcp]
[INFO    ]  Client disconnected [console]
[ERROR   ]  OnException: Connection Closed Gracefully. [tcp]
[INFO    ]  Connection Closed Gracefully. [console]
```

Figure 12.7 – Sample log entries from multiple units of our server (with the date/time and thread ID columns removed)

This works great for Windows, but unfortunately, the console appender that comes with LoggerPro only works on Windows, so we can't compile this for the Linux platform. However, this does present us with an opportunity to write our own custom appender that *does* work on Linux. And it's not as hard as you might think.

Adding a custom logging mechanism

We can create a new, custom appender right inside our project folder that depends on the LoggerPro library. So, create a new unit in this project and call it `LoggerPro.SimpleConsoleAppender`. All LoggerPro appenders must descend from `TLoggerProAppenderBase`, which is found in the `LoggerPro` unit. So, include that unit and create a class that descends from it, overriding the three abstract procedures in the base class and adding a `TFormatSettings` field that we'll use when writing the log entry:

```
TLoggerProSimpleConsoleAppender =
  class(TLoggerProAppenderBase)
private
```

```
  FFormatSettings: TFormatSettings;
public
  procedure Setup; override;
  procedure TearDown; override;
  procedure WriteLog(const aLogItem: TLogItem); override;
end;
```

The Setup procedure simply initializes the format field:

```
procedure TLoggerProSimpleConsoleAppender.Setup;
begin
  FFormatSettings := LoggerPro.GetDefaultFormatSettings;
end;
```

We don't need to do anything in the TearDown procedure, but it must be overridden since the base class declares it as abstract:

```
procedure TLoggerProSimpleConsoleAppender.TearDown;
begin
  // do nothing
end;
```

The only other thing we need to do is actually write out the log entry, formatting the data in the way we'd like to see it:

```
procedure TLoggerProSimpleConsoleAppender.WriteLog(const
aLogItem: TLogItem);
var
  lText: string;
  ds: string;
begin
  ds := DateTimeToStr(aLogItem.TimeStamp, FFormatSettings);
  lText := Format('[%-8s] %s [%2:-10s] %s', [aLogItem.LogTag,
    ds, aLogItem.LogTypeAsString, aLogItem.LogMessage]);
  Writeln(lText);
end;
```

In the main project file, change the `uses` clause to include `LoggerPro.SimpleConsoleAppender`, change the `BuildLogWriter` function to use `TLoggerProSimpleConsoleAppender`, and run this on Linux. The output isn't as colorful, but it is now cross-platform and still very functional:

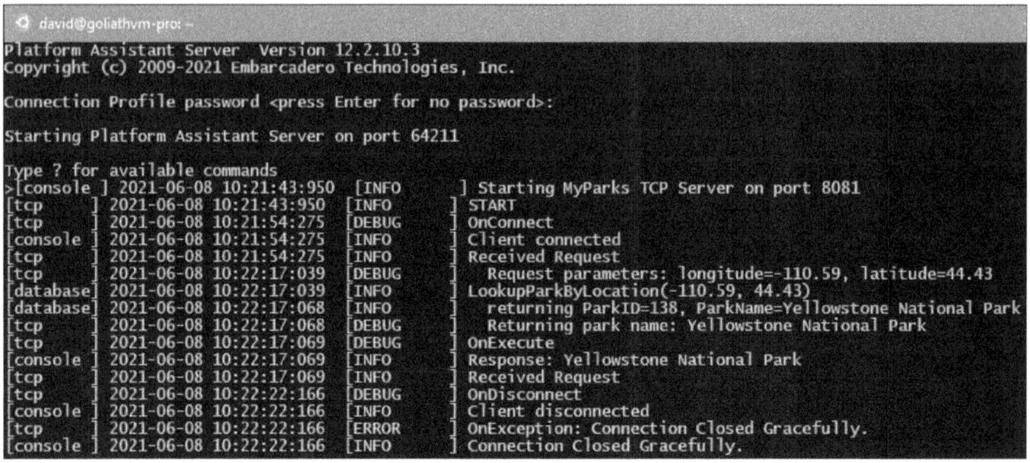

Figure 12.8 – Sample log entries sent to the console from a Linux server

LoggerPro is very configurable; from the parts of the filenames to the order and selection of fields in each log line, comments in the accompanying sample projects demonstrate how you can customize everything to your liking.

It's useful to list parameters, return values, sometimes SQL queries, and even byte streams, as they can become extremely useful to diagnose obscure problems. By using the proper log type, log-viewing tools can filter out information that will get in the way of finding what you need quickly.

Once again, the full `MyParksTCPServerConsole` project is on GitHub at `https://github.com/PacktPublishing/Fearless-Cross-Platform-Development-with-Delphi/tree/master/Chapter12/02_ParksServer`.

> **Note**
>
> LoggerPro is by no means the only logging option out there. A recent addition in the **GetIt Package Manager** is **QuickLogger** by **Exilon Soft**. There are undoubtedly others as well. Do some research, write some tests, and see what works for you.

With flexible logging in place in addition to the data modules for real functionality, we now have everything ready to move our server to the background and still have visibility of what is happening. Both of the two headless server applications that we'll build in this section are very specific to their respective platforms. The other application types that we cover in this book include conditional compilation to allow the application to work on different platforms by simply switching the target and recompiling. There's so much "scaffolding" code for Windows and Linux background processes that the best approach here is to put the bulk of the functionality into data modules, as we've done, and then include those data modules in separate, platform-specific applications.

Let's start this discussion by building a Windows service.

Creating a Windows service

There are dozens of services running on every Windows machine all of the time: some to keep your machine updated or secure, some to provide access to installed databases, others to watch for malicious activity, and many others to provide myriad functionality. Delphi has supported building Windows services for a long time. In fact, it's pretty simple to get one up and running—just follow these steps to bring the data modules from our console app into a new service app:

1. Create a new **Windows Service** application (navigate to **File | New | Other** from the menu, and then expand **Delphi** and **Windows**).

2. Answer "Yes" to enable the **Visual Component Library** framework.

3. Add the two data modules we created for the server: udmParksDB and udmTCPParksServer.

4. Add the logging units of LoggerPro, LoggerPro.FileAppender, and uMyParksLogging (these are the same ones from the previous server project except for the console logging unit).

5. Initialize the Log variable right after the service has been created and just before the Application.Run line:

```
Log := BuildLogWriter([TLoggerProFileAppender.Create]);
```

6. Save the automatically created data module as udmMyParksService and the project as MyParksTCPServerService.

7. In the new udmMyParksService unit, set the Name and DisplayName properties to what you'd like to see listed in the Windows services list. The Name property should be short but unique and not have any spaces—I named mine as MyParksIBService. You can set the DisplayName property to be something more descriptive, and it can include spaces. However, don't make it much longer than the Name property—it's shown in the **Name** column in Windows Services; I set mine to MyParks InterBase Lookup Service.

8. Include the udmTCPParksServer unit in the implementation section. Then, in the **Object Inspector** for the udmMyParksService unit, double-click on both the OnStart and OnStop events to create the following event handler procedures for them:

```
procedure TMyParksIBService.ServiceStart(Sender:
TService; var Started: Boolean);
begin
  dmTCPParksServer.Start;
end;
procedure TMyParksIBService.ServiceStop(Sender: TService;
var Stopped: Boolean);
begin
  dmTCPParksServer.Stop;
end;
```

That's all that is necessary since we've already built and tested the real working parts of the service (that is, the TCP port listening and the database lookup). After compiling it, start **Command Prompt** as an administrator, go to the directory where the .exe file is generated, and type in the following command:

```
MyParksTCPServerService.exe /install
```

A message should pop up telling you the **Service installed successfully**.

You can go to **Services** and see it listed as follows:

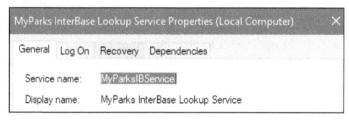

Figure 12.9 – First version of MyParks Windows service installed

Pay attention to a couple of things here. The Name property we set cannot be seen anywhere, and the Description column is blank. If you right-click on the service and select **Properties**, you'll see the **Service Name** set to our Name property and the value under the **Name** column in the Windows Services list is now shown as the **Display name**, matching our DisplayName property:

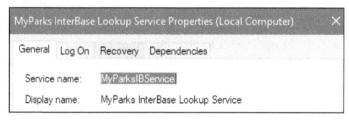

Figure 12.10 – The MyParksIBService properties

So, how do we set the description for the service? Windows services are listed and described in the Windows registry under HKEY_LOCAL_MACHINE\SYSTEM\ CurrentControlSet\Services. You can access a specific service by its unique name (this is where the Name property of our service data module comes in). A perfect time to do that is right after the service has been installed. So, create an event handler for the AfterInstall event and write the following code:

```
procedure TMyParksIBService.ServiceAfterInstall(Sender:
TService);
var
  Reg: TRegistry;
begin
  Reg := TRegistry.Create(KEY_READ or KEY_WRITE);
  try
    Reg.RootKey := HKEY_LOCAL_MACHINE;
```

```
  if Reg.OpenKey('\SYSTEM\CurrentControlSet\Services\' +
        Name, False) then begin
    Reg.WriteString('Description', 'Listens for Longitude and
      Latitude data, returns the matching park name.');
    Reg.CloseKey;
  end;
finally
  Reg.Free;
end;
end;
```

Navigate back to **Command Prompt** as an administrator, uninstall the service, and then reinstall it:

```
MyParksTCPServerService.exe /uninstall
MyParksTCPServerService.exe /install
```

Now refresh the Windows Services list and you'll see the description show up. Start the service either with the **Start Service** toolbar button or by selecting **Start** from the right-click pop-up menu; if you have the **Extended** tab view selected, you can also simply click on the **Start** link.

Let's check our logging.

Logging to the Windows Event Log

Remember the logging that we added in the previous section? Since it was embedded in the data modules we added to this project, the file logging is automatically part of this Windows service. However, we also added a module, udmMyParksService, to handle the Windows service management. If we take a look at that unit, there are several obvious places in which you can add logging.

By default, LoggerPro puts the log files for Windows services in the same folder as the executable for the service itself, so we don't need to change anything to get our data modules to work equally well from a console app or a Windows service. However, service apps often send messages to the Windows Event Log, not files. We can keep our file logging, but for best practices, we should also learn how to do it the preferred way—especially since it's so simple.

The service data module that was automatically created for us descends from TService. By simply calling its LogMessage method and passing in a message string and an event type (such as the integer constants defined in Delphi's Winapi.Windows unit), we can log to the Windows Event Log—yes, it's that simple! Try it out by adding a few in our service data module:

```
procedure TMyParksIBService.ServiceStart(Sender: TService; var
Started: Boolean);
begin
  dmTCPParksServer.Start;
  Log.Info('Started', LOG_TAG);
  LogMessage('Started', EVENTLOG_INFORMATION_TYPE);
end;
procedure TMyParksIBService.ServiceStop(Sender: TService; var
Stopped: Boolean);
begin
  dmTCPParksServer.Stop;
  Log.Info('Stopped', LOG_TAG);
  LogMessage('Stopped', EVENTLOG_WARNING_TYPE);
end;
```

I added ones in the AfterInstall and AfterUninstall events, too. Then, after installing, starting, stopping, and uninstalling, I pulled up the Windows **Event Viewer**, expanded the **Windows Logs** to **Application**, looked for the name of my service, **MyParksIBService**, and saw the events:

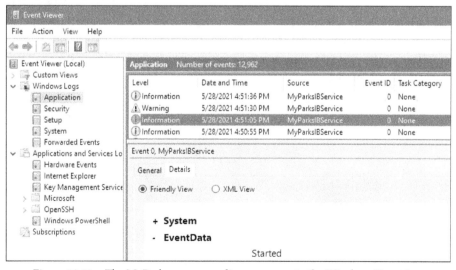

Figure 12.11 – The MyParks server sending messages to the Windows Event Log

All of them sent informational messages except for the `ServiceStop` procedure; I used `EVENTLOG_WARNING_TYPE` for that one to simply show how it looks in the **Event Viewer**. The **Started** word, which is displayed in the screenshot, is what I sent as the `Message` parameter. I left the `Category` and `ID` parameters off, which kept them at their default settings.

So, now, of course, we would like LoggerPro to take care of that for us since we're already using it in the data modules, which are still logging only to files. It's time for another custom log appender!

Again, it's really simple. Create another unit called `LoggerPro.WindowsEventLogAppender`, and use the `Vcl.SvcMgr` and `LoggerPro` units. Create a class that descends from `TloggerProAppenderBase`, just as we did with the last one. However, this time, add a reference to the Windows service and provide a new constructor:

```
TLoggerProWindowsEventLogAppender =
  class(TLoggerProAppenderBase)
private
  FService: TService;
public
  constructor Create(AService: TService); reintroduce;
  procedure Setup; override;
  procedure TearDown; override;
  procedure WriteLog(const aLogItem: TLogItem); override;
end;
```

The Windows event log will record the date and time of each entry, so we don't need to format the date and time as part of our log message. This means that we won't need anything in the `Setup` procedure; the `TearDown` procedure will, once again, be unused. Here's the short constructor:

```
constructor TLoggerProWindowsEventLogAppender.Create(AService:
TService);
begin
  inherited Create;
  FService := AService;
end;
```

The `WriteLog` procedure is where we'll log different types of Event Log messages based on LoggerPro's `TLogItem.LogType`:

```
procedure TLoggerProWindowsEventLogAppender.WriteLog(const
aLogItem: TLogItem);
begin
  case aLogItem.LogType of
    TLogType.Debug,
    TLogType.Info:
      FService.LogMessage(aLogItem.LogMessage, EVENTLOG_
        INFORMATION_TYPE);
    TLogType.Warning:
      FService.LogMessage(aLogItem.LogMessage, EVENTLOG_
        WARNING_TYPE);
    TLogType.Error:
      FService.LogMessage(aLogItem.LogMessage, EVENTLOG_ERROR_
        TYPE);
  end;
end;
```

Note that there's no separate event type for debug messages. You can pass an `ID` and `Category` (which are both numeric) to the Windows event log if you wish to.

With this new custom appender sending messages to the Windows Event Log, we can take out our tests, leave the LoggerPro calls, and implement standard logging from our completed Windows service.

You can view my version of the `MyParksTCPServerService` project on GitHub at `https://github.com/PacktPublishing/Fearless-Cross-Platform-Development-with-Delphi/tree/master/Chapter12/02_ParksServer`.

Test out the new service by running the `MyParksTCPClientConsole` app we built earlier. It should work exactly the same as when we started the TCP server manually from the console server.

Now, let's examine how background processes work on Linux.

Adopting a Linux daemon

Getting a background process up and running in Linux isn't straightforward. It involves a deep understanding of Linux processes that most Windows developers (and even many who work on Linux) never worry about.

A Linux **daemon** is a background process owned by the system initialization process (**Init**). Usually, they are spawned (or created) by Init, but they can be started by applications that then **fork** a child process (somewhat like creating a separate thread) and then exit. When that happens, the spawned child process becomes orphaned—at which point Init adopts it. You can find a very brief overview of what takes place, with many of the details left out, in the *Building a (real) Linux daemon in Delphi* blog at `http://blog.paolorossi.net/2017/09/04/building-a-real-linux-daemon-with-delphi-part-2`.

> **Note**
>
> If you know a thing or two about Linux, you might be aware that you can start an app and send it immediately to the background by simply appending an ampersand after the application on the command line. While this does return control to the user at the command line with the application running in the background, it is not a true Linux daemon; if the terminal from which the process was started is logged out, it kills the background processes along with it. This is not what we want.

Note that there is no "Linux daemon" type of application when starting a new project; you have to create a regular Linux console app and get it to spin off a daemon process. Instead of walking through the details of how to build this and trying to explain the nuances of processes, signals, and filehandles, I'd like to invite you to simply download the completed project from GitHub at `https://github.com/PacktPublishing/Fearless-Cross-Platform-Development-with-Delphi/tree/master/Chapter12/04_ParksLinuxDaemon`.

Looking through the code of this project is quite different from anything else in this book. It diverges significantly from what you typically find in a Delphi program. However, it shows that given enough knowledge about a system and an understanding of how to hook into the right libraries, you can build any type of application you need.

This project has been modified only slightly from the GitHub project, which was referenced in the blog that was mentioned earlier in this section. There were a few things not needed, and, of course, the data modules were added and the TCP server was started.

You'll notice that LoggerPro is not used in this project; instead, calls to **syslog** from the original code are used because that is the preferred way a Linux daemon should log. However, I put the calls to syslog in conditional compilation directives, so this modified data module is still quite capable of being compiled for any of the other project types we've discussed in this chapter. After connecting to it with our test console client, making a typical query, and then shutting it down, I listed the contents of the syslog:

Figure 12.12 – Syslog entries generated by our Linux daemon

I also had to make a minor change to the park lookup data module. With the other platforms, I've been lazy in terms of leaving the database connected at runtime, but the Linux daemon only encountered database problems once I ran the first query and had activated it —that's very likely due to the main process being forked and killed when it starts up! With these small changes, however, the data modules can be ported back to be used in our previous Windows or Linux console apps, Windows services, or Linux daemons.

With the server ready and having been tested with a simple console client, it's time to make it available outside of our small test environment.

Exposing your server to the world

If you've been following along with these examples and testing the servers on your own machine, the connection is practically guaranteed. The IP address we've used, 127.0.0.1, always points to the local machine, which typically allows connections from itself without question. However, the purpose behind building a server is to provide information to a wider audience rather than just ourselves. This final section of the chapter will briefly discuss some methods and considerations to bear in mind when you are making a server app publicly available. Additionally, we will test them by modifying the *MyParks* app from *Chapter 10*, *Cameras, the GPS, and More*, and actually using our new park lookup feature at an actual park.

First, if you have one available, run the server of your choice from this chapter on another computer on your network and try to connect to it. Any Windows or supported Linux computer will do; I have a Windows server on my network and have copied the Windows Service app we built over to it. Additionally, I have installed and started it. After checking the event log to confirm it was running, I pulled up the console client app we built, changed the IP address in the body of the program to point to the machine on which the server app was running, and ran it. By doing so, I successfully connected to it and was able to get the name of the park back as expected.

Typically, local networks are shielded from the internet by firewalls. Additionally, IP addresses on a local network are usually configured to use a local range of addresses that are not exposed to the internet. When a request comes in from outside of your home or organization, it tries to connect to the specified port on the router, and if that router isn't configured to handle that port, the request will either time out or get bounced back with an error. Depending on your level of expertise, your familiarity with configuring routers, and your security concerns, you can configure your router to forward a port to the computer running the server app. It's a good idea to forward ports to computers with static IP addresses; otherwise, every time it changes, you'll have to reconfigure your router.

> **Note**
>
> The information provided here concerning IP address ranges, port forwarding, and security is an extremely brief list of the considerations you can face when deciding whether to expose your server to the outside world. There are many issues at stake and further study and/or consulting with security and networking experts is highly recommended.

Instead of installing your server app on your own hardware and dealing with the infrastructure yourself, you could rent server space. Microsoft Azure, Amazon Web Services, Rackspace, and many other service providers offer many such options. It's also possible that if have you have a website, your hosting provider might allow custom-written applications to run. Any of these scenarios will require a little bit of effort to research and configure, but since you can now write applications that work on either Linux or Windows as console apps or background processes, your options are much broader.

After your server has been established with an open port to the world, we'd like to actually make use of it in our *MyParks* app, which we built in *Chapter 10, Cameras, the GPS, and More*.

Modifying our client app to use the new server

I've made a copy of the MyParks app inside a new folder on GitHub at `https://github.com/PacktPublishing/Fearless-Cross-Platform-Development-with-Delphi/tree/master/Chapter12/05_MyParksAppClient`.

We need to add a button that grabs the current location and sends a request to our server at an IP address and port. Ideally, we would add a screen to allow the configuration of these values, but we'll save that for a later chapter. For now, we'll just hardcode the address and port.

To begin, first address the user interface. On the main form, add an image, a button to the panel at the bottom of `tabParkList`, and an action to the `ActionList`, called `actParkNearMe`. This new action needs to grab the coordinates and make the call to our server.

Calling the server can be done using a `TIdTCPClient` component. We already built a data module to do this when we wrote the client console app earlier in this chapter, so we can simply copy and paste the two files comprising that module (`udmTCPParkClient.pas` and `udmTCPParkClient.dfm`) into this project folder and add the unit name to the project. In addition to this, add the unit to the `uses` clause in the `implementation` section of the main form.

> **Note**
>
> It is perfectly reasonable to share files between projects, just as we did with multiple servers sharing the data modules. The `TCPParkClient` data module could be in a shared folder or the MyParks app could've been copied to the folder where we created the client console app. I've put them in a separate folder simply for the organizational purposes of this chapter.

When we originally wrote the MyParks app, we left the location services option off until we entered the park edit tab. By adding the park query button to the first tab, which is displayed as soon as the app starts, we'll need location services immediately rather than waiting for an event to arise. So, before we go further, we need to move the section of code that asks for permission to use location services and subsequently turns on the location sensor, if permission is granted, to the `OnCreate` event of the app. That code was in the `OnItemClick` event of the ListView. After moving that code, only one line is left in the event handler:

```
procedure TfrmMyParksMain.lvParksItemClick(const Sender:
TObject; const AItem: TListViewItem);
begin
```

```
   NextParkTabAction.Execute;
 end;
```

The updated `OnCreate` event of the form is a little fuller now:

```
procedure TfrmMyParksMain.FormCreate(Sender: TObject);
const
  PermissionAccessFineLocation =
      'android.permission.ACCESS_FINE_LOCATION';
begin
  (* other FormCreate code ... *)

  PermissionsService.RequestPermissions(
    [PermissionAccessFineLocation],
    procedure(const APermissions: TArray<string>;
              const AGrantResults: TArray<TPermissionStatus>)
    begin
      if (Length(AGrantResults) = 1) and
         (AGrantResults[0] = TPermissionStatus.Granted) then
        LocationSensor.Active := True
      else begin
        actParkNearMe.Enabled := False;
        TDialogServiceAsync.ShowMessage(
            'Park location data will not be available.');
      end;
    end);
 end;
```

With our location sensor continuously grabbing coordinates when the app starts, we can simply snag those values whenever the park query button is clicked on, just like we did when saving the coordinates for the new park we entered. This time, however, we'll send the coordinates to the remote server and let it tell us which park we're at.

The servers we've built in this chapter using the `TIdTCPServer` component are multithreaded. This means that they can handle multiple requests with ease. Additionally, InterBase is very capable of handling many connections and queries simultaneously. Still, it's always a good idea to connect, query, and disconnect as quickly as possible. We do this in the new action event handler that is assigned to the park query button:

```
procedure TfrmMyParksMain.actParkNearMeExecute(Sender:
TObject);
var
  QueriedParkName: string;
begin
  // replace the Host IP address with your external internet
address
  dmTCPParkClient.IdTCPMyParksClient.Host := '192.168.1.15';
  dmTCPParkClient.IdTCPMyParksClient.Port := 8081;
  dmTCPParkClient.Connect;

  try
    QueriedParkName := dmTCPParkClient.GetParkName(
            FParkLongitude, FParkLatitude);
  finally
    dmTCPParkClient.Disconnect;
  end;

  TDialogServiceAsync.ShowMessage(
      Format('The park at location (%6.3f, %6.3f) is "%s"',
            [FParkLongitude, FParkLatitude, QueriedParkName]));
end;
```

The assignment of the host address and port would normally be done elsewhere in the program and, as mentioned earlier, be saved in a configuration file. Another obvious feature is automatically creating a new park entry in your database for you to use once a park name has been successfully created. We'll address these and more as we continue to expand on the usefulness of connecting to servers in the next couple of chapters. For now, the app simply displays the name of the park found in the database matching the current coordinates:

Figure 12.13 – The coordinates sent to a custom server returning with the name of the park

The final test of our server shows the app connected across the internet, through the port that has been forwarded from my router to my Windows server. It has queried the database successfully and returned the name of the park back to my Android phone.

Summary

We have begun building server applications that can provide data to a client app that we wrote in earlier chapters. We demonstrated how you can simply start with a console application, putting the bulk of the functionality into data modules, testing to make sure they work well on both Windows and Linux, and then creating the platform-specific project that utilizes the data modules. Logging data is an important element of background processes that are not managed directly and work without interaction. We learned how to implement this feature into our server applications in a way that is appropriate for each platform. Finally, we tested our new servers with both a simple console app locally and then with a real app over the internet, completing our goal of producing a usable server in a progressively more complex but completely manageable fashion. Your skillset has now expanded to include the ability to create true native servers for both Windows and Linux!

In the next couple of chapters, we will give you more options for server-based applications: first, by integrating them with web servers, and second, by utilizing RAD Server.

Questions

1. What edition of Delphi is required to build Linux server applications?

2. How do you access files on a Windows subsystem for Linux?

3. Why is testing a new server as a console app recommended?

4. Which common component set that is included with Delphi is often used for TCP/IP communication?

5. What is a common method of logging from Windows services?

6. What is a common method of logging from a Linux daemon?

7. How is the description for a Windows service set?

8. What is the IP address of the local machine on any computer?

9. What does port forwarding mean?

Further reading

- *Configure Delphi and RedHat or Ubuntu for Linux development*: `https://chapmanworld.com/2016/12/29/configure-delphi-and-redhat-or-ubuntu-for-linux-development/`

- *What is the Windows Subsystem for Linux?*: `https://docs.microsoft.com/en-us/windows/wsl/about`

- *Developing on Windows with WSL2 (Subsystem for Linux), VS Code, Docker, and the Terminal*: `https://www.youtube.com/watch?v=A0eqZujVfYU`

- *List of TCP and UDP port numbers*: `https://en.wikipedia.org/wiki/List_of_TCP_and_UDP_port_numbers`

- *Internet Protocol*: `https://en.wikipedia.org/wiki/Internet_Protocol`

- *Service Applications*: `http://docwiki.embarcadero.com/RADStudio/Sydney/en/Service_Applications`

- *Writing to the Windows Event Log using Delphi*: `https://stackoverflow.com/questions/30229826/writing-to-the-windows-event-log-using-delphi`

- *Linux Jargon Buster: What are Daemons in Linux?*: `https://itsfoss.com/linux-daemons/`

- *Building a (real) Linux daemon with Delphi*: `http://blog.paolorossi.net/2017/09/04/building-a-real-linux-daemon-with-delphi-part-2`

13
Web Modules for IIS and Apache

Some people are surprised when they find out Delphi has support for building web servers. They are even more surprised to find out that support has been there for over 20 years! Delphi and its surrounding ecosystem of tools, open source frameworks, and third-party libraries can be used to build backend web servers based on SOAP or REST protocols, frontend JavaScript clients, or complete full stack web solutions.

In this chapter, we'll take a brief look at the options for building web server applications that come with Delphi, walk through the wizard interfaces that create them, and then show you how to get them up and running under both Microsoft's IIS on Windows and Apache on Windows and Linux.

In this chapter, we will cover the following main topics:

- Surveying website-building options in Delphi
- Getting comfortable with the underlying framework
- Building an ISAPI web module for IIS on Windows
- Getting started with the Apache HTTP server
- Writing cross-platform Apache web modules

Let's get our feet wet in the World Wide Web!

Technical requirements

Like the previous chapter, we'll focus on the two major platforms in the server arena: Windows and Linux. The former can be done with Delphi Professional and can be tested on any version of Windows. Apache runs on both Windows and Linux (and other operating systems), so we'll cover both of those, with the latter requiring Delphi Enterprise or higher. The complete code for this chapter can be found online on GitHub: `https://github.com/PacktPublishing/Fearless-Cross-Platform-Development-with-Delphi/tree/master/Chapter13`.

Surveying website-building options in Delphi

There are many ways to build content-rich, dynamic websites in Delphi. Some of these are server-side only, producing SOAP or REST APIs to be consumed by client applications (WebBroker, DataSnap, and RAD Server, to name a few). Other tools build end-to-end solutions from the ground up, encompassing databases, grids, a rich set of user interface components, and themed web views (IntraWeb and UniGUI). Still, others are custom IDEs that use a Pascal-like language to generate HTML/JavaScript code that works only in the browser and connects to web servers for their data (Elevate Web Builder, TMS Software Web Core, and Smart Mobile Studio). There are also several open source frameworks with a wide following (DelphiMVCFramework, mORMot, and Mars Curiosity).

To see an excellent overview of these options, visit the first link in the *Further reading* section at the end of this chapter, called *Ultimate Web Frameworks for Ultra-Fast Web Application Development Using Delphi/C++ Builder*. It's an Embarcadero blog containing links and a YouTube video and is a great starting point for exploration.

This book focuses on cross-platform development with Delphi, so we won't be looking at third-party products that concentrate on the frontend. Instead, we'll look at the foundation of several web technologies Delphi has supported for many years and how to get them working in three different environments.

Understanding the Web Server Application wizard

In the previous chapter, we used one of the Indy components to provide generic TCP/IP communication between a client and a server. When implementing a server from scratch, we have a wide variety of internet protocols to choose from, and we could have selected a different one, such as FTP or Telnet, just as easily. Web servers use **Hypertext Transfer Protocol** (**HTTP**), and the Indy component that supports this protocol is `TIdHTTPServer`. But instead of creating our own web server, we can write a module that works inside an existing one. Delphi has a built-in wizard to automate the creation of both standalone web servers and modules that integrate with IIS or Apache.

When you select **File | New | Other,** jump down to the **Web** section, and select **Web Server Application**. The wizard starts. All types of web servers support **Windows** and if you have Delphi Enterprise, you can also select **Linux** as a target platform for the new application. Leave Linux unselected for now – we'll come back and add another server later for that platform.

Five options are presented by the **New Web Server Application** wizard (if Linux is not checked):

- **Apache dynamic link module**: This creates a DLL that works inside the Apache web server on Windows.

- **Standalone console application**: This creates a standalone, console-based web server application that uses the IdHTTPServer component, so it does not require IIS or Apache.

- **Standalone GUI application**: This creates a standalone, VCL, or Firemonkey web server application that uses the IdHTTPServer component, so it does not require IIS or Apache. This option is only available on Windows.

- **ISAPI dynamic library**: This uses the Internet Server API to provide extensions and filters as Windows DLLs that get registered in Microsoft's IIS web server; modules have been added to other web servers, such as Apache, to support ISAPI DLLs, but only on Windows.

- **CGI standalone executable**: This creates a standalone, console-based application where web requests and responses are passed through standard input and output file handles.

All of these project types use **WebBroker,** a cross-platform, multithreaded, event-driven framework for producing HTML pages. It first appeared with Delphi 3 and is the foundation for many other web server products for Delphi. Before we build a module for IIS or Apache, we need to review how WebBroker works in a standalone web server app.

Getting comfortable with the underlying framework

The simplest way to set up a WebBroker module is to create a standalone web server application using Delphi's wizard and test it there first. Here are the steps:

1. From the **New Web Server Application** wizard, select **Stand-alone GUI application** and click **Next**.

2. Choose either **VCL application** or **FireMonkey application** (it doesn't matter which) and click **Next**.

3. The last page of the wizard asks for an **HTTP Port**. If your server from *Chapter 12, Console-Based Server Apps and Services*, is still running and you selected 8081 for that one, simply put in 8082 for this one. Click **Test Port** to make sure it's available, then click **Finish**.

This creates a simple web server app that you can launch right from Delphi, start and stop, and, with the click of a button, open a browser that conveniently loads the page for your web server. This provides a great way to quickly make changes and test your server.

Select the web module of your new project. Then, in the **Object Inspector** window, click the ellipses button for the `Actions` property. You'll see that one item has already been entered for you and that its `Default` property is checked. This means this web module will handle one default web page, returning the HTML document (expressed as a constant string) assigned to the `Content` property of the `TWebResponse` object referenced by the `Response` parameter. You can customize this HTML all you want.

> **Note**
>
> The `Response` object can return any type of data, not just HTML. By setting its `ContentType`, `CustomHeaders`, and other properties, a WebBroker server can look and act just like any modern REST or SOAP server. Indeed, both the DataSnap and RAD Server products, which can be found in the Enterprise and higher editions of Delphi, are based on WebBroker.

To handle a different page, create a new action item and set the `PathInfo` property to the subpage you want to handle. Then, set its `Response.Content` parameter property in a similar fashion as the default handler.

For example, let's say the main page of a company website lists the company name, some brief information, and marketing material, and includes links to various subpages – perhaps one of them is an About page that gives more detailed information about the company. In this case, the default handler (the first action item) would return some HTML for the main company page, with one of the lines containing `<a href="/about">About</a>` to take the user to the About page. The second action item of this example would handle the About page by setting the `PathInfo` property to `/about` and returning some HTML containing the historical information in `Response.Content`.

Try this out in your project by changing the default action item's `OnAction` event handler:

```
procedure TwmMyParks.wmMyParksDefaultHandlerAction(Sender:
TObject;
  Request: TWebRequest; Response: TWebResponse;
```

```
  var Handled: Boolean);
begin
  Response.Content :=
    '<html><head><title>MyParks</title></head>' +
    '<body><p>MyParks</p>' +
    '<a href="about">About</a>' +
    '</body></html>';
end;
```

Now, add the second web action item for the About page by setting its `OnAction` event handler to this:

```
procedure TwmMyParks.wmMyParksAboutAction(Sender: TObject;
  Request: TWebRequest; Response: TWebResponse;
  var Handled: Boolean);
begin
  Response.Content :=
    '<html>' +
    '<head><title>MyParks - About</title></head>' +
    '<body>This is a simple test site for the MyParks app
    </body>' +
    '</html>';
end;
```

With the `PathInfo` property for the default action item set to / and the `PathInfo` property for the about action item set to /about, run the app, start the server, and then launch the browser.

You'll see **MyParks** on the first line with a link to **About** on the second. When you click that, you'll see the simple **About** page we coded in its web action event:

Figure 13.1 – The MyParks "About" page in a web browser powered by WebBroker

If you had to build your entire website by hand this way, it would get tedious very quickly, like going back to the early 1990s when website-building tools were primitive. Fortunately, there's a lot of power that builds from here and the first of these are **page producers**.

Place a TPageProducer from the **Internet** section of the tool palette on the web module. It has a couple of interesting properties, such as HTMLDoc, which allows you to enter the HTML that will be returned to the component itself, such as lines in a ListBox; and HTMLFile, which takes a filename and loads the HTML from the file at runtime. This immediately separates building the web server from building the web pages, since you could have someone else with HTML design skills create individual pages that you load at runtime.

For educational purposes, simply take the HTML from the about web action item we just wrote and put that into the HTMLDoc property. Since the HTML is now stored in the page producer, we can eliminate the event handler and code and simply assign the page producer to the web action item's Producer property. Run the application again, click on the **About** link, and you'll see the same **About** page you saw earlier (assuming you took out the quotes and operators that were used to assign the text value in the Delphi code).

This is still all just static HTML, though. We need to add some dynamic capabilities.

Templating your HTML

Page producers have one event to hook into: OnHTMLTag. This allows us to use the HTML as a template, substituting various tags at runtime in code. A tag is embedded in the HTML with a special XML format that gets replaced at runtime: <#TagName Param1=Value1 Param2=Value2 ...>.

You could do something such as replace the application name, currently hardcoded in your HTML, with a tag such as <#AppName> (leave the optional parameters out for now) and then set a constant in the web module for the application name and replace every instance of the MyParks string in the HTML with <#AppName>. The **About** page's HTML would now look like this:

```
<html>
<body>
<h1>About <#AppName></h1>
<body>This is a simple test site for the <#AppName> app</body>
</body>
</html>
```

To handle this tag, the `OnHTMLTag` event handler should look like this:

```
const
  APP_NAME = 'MyParks';

procedure TwmMyParks.ppAboutHTMLTag(Sender: TObject; Tag: TTag;
  const TagString: string; TagParams: TStrings;
  var ReplaceText: string);
begin
  if SameText(TagString, 'AppName') then
    ReplaceText := APP_NAME;
end;
```

By simply changing the `APP_NAME` constant, you can change the name of the application that's displayed in the produced HTML everywhere it appears in the template. You can refactor that to a separate function and call it from the `OnHTMLTag` events of other page producers as well. This will ensure you have consistent template support throughout your web server.

> **Tip**
>
> Even when splitting various parts of the web page among separate page producers, all the page text will still be in your Delphi code if you're using the `HTMLDoc` property. Delphi is great for building compiled applications and web modules but is not the best tool for editing the layout of a modern website built with a rich mix of HTML, CSS, and JavaScript. As we mentioned previously, using the `HTMLFile` property of the page producers allows you or someone else to use any of several powerful website-editing tools on the market today that specialize in this job, saving files and templates for runtime loading. What's more is that these external files can be changed after the compiled web module is in place, allowing the look and feel of the pages to change without us having to rebuild and redeploy the Delphi web module, which is a huge consideration.

Now, let's go a step further and show you how another page producer makes it easy to display a list of parks from our database. Follow these steps to get started:

1. Copy over the MyParks data module from the previous chapter and add it to the project. Then, remove it from the list of autocreated forms in the project manager as it creates and frees itself in the `initialization` and `finalization` sections of the unit. (If you included Linux support in those data modules, remove the used unit, `Posix.Syslog`, and all calls to `syslog` as we won't be using that type of logging in this chapter.)

2. Copy the logging unit that it uses and add it to the project.

3. Add the data module to the `uses` clause in the `implementation` section of the web module.

4. Add a `TFDQuery` component to the data module so that it's hooked up to `TFDConnection` on the data module.

5. Add `SELECT * FROM Parks ORDER BY PARK_NAME` to the FDQuery and test it by clicking the **Execute** button in the query editor.

6. Add all the fields to the query's **Field Editor**.

7. In the web module, place a `TDataSetTableProducer` on the web module; set the `DataSet` property to the new query we just added to the data module.

8. Click the ellipses button on the `Columns` property and add each of the fields from the associated query.

9. Create a new web action item in the web module, set its `PathInfo` property to `/parklist`, and assign the new `TDataSetTableProducer` we set up to the `Producer` property.

10. Create an `OnCreate` event handler for the web module and add the following lines to the standard file logging:

    ```
    Log := BuildLogWriter([TLoggerProFileAppender.Create(
        5, 1000, TPath.Combine(WebApplicationDirectory,
          'logs'))]);
    Log.Info('web module created', LOG_TAG);
    ```

11. Finally, modify the HTML for the main page by adding a link to the new `parklist` page.

You can make further modifications to the HTML table that's been produced by looking through the various properties of `TDataSetTableProducer`. I set the table background to `Lime` and left-justified the park names; with several imported parks, my version looks like this:

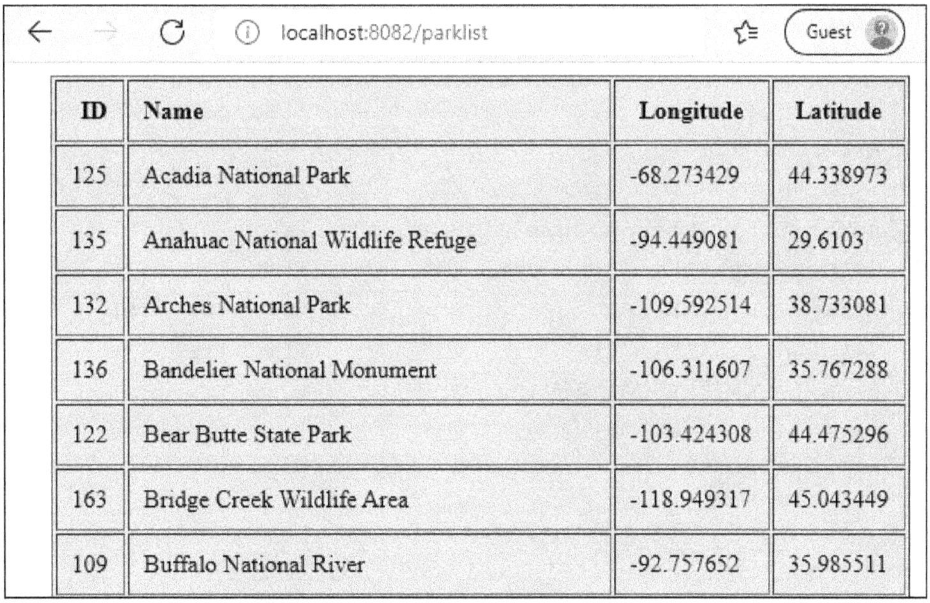

ID	Name	Longitude	Latitude
125	Acadia National Park	-68.273429	44.338973
135	Anahuac National Wildlife Refuge	-94.449081	29.6103
132	Arches National Park	-109.592514	38.733081
136	Bandelier National Monument	-106.311607	35.767288
122	Bear Butte State Park	-103.424308	44.475296
163	Bridge Creek Wildlife Area	-118.949317	45.043449
109	Buffalo National River	-92.757652	35.985511

Figure 13.2 – HTML list of parks generated by WebBroker's DataSetTableProducer

This is crude HTML, but you can see the power quickly escalating through more sophisticated components built on top of WebBroker. We have one more aspect of these components to show before converting this into a true web module.

The initial functionality of the `MyParks` data module was to look up a park based on its coordinates. We should provide that functionality on the web page, partially to surface already existing functionality, but also to show how query parameters can be utilized in WebBroker.

Let's add another link to the main HTML. When you do this, consider using one of the predefined tags; that is, `Link`. It doesn't save a lot of code or build the link for you, but it is a standard way you can add links to your HTML template. Here's the body of my default page now:

```
<h1><#AppName></h2>
Links:<br>
<ul>
<li><#Link Link="parklist" title="List of Parks"></li>
```

```
<li><#Link Link="getpark" title="Get park by location"></li>
<li><#Link Link="about" title="About"></li>
</ul>
```

Here's the OnHTMLTag event handler:

```
procedure TwmMyParks.ppMainPageHTMLTag(Sender: TObject; Tag:
TTag; const TagString: string; TagParams: TStrings;
  var ReplaceText: string);
begin
  ReplaceText := CheckAppName(TagString);
  if ReplaceText.IsEmpty then
    if (Tag = tgLink) and (TagParams.Count > 0) then begin
      var Link, Title: string;
      Link := TagParams.Values['Link'];
      Title := TagParams.Values['Title'];
      ReplaceText := MakeLink(Link, IfThen(Title.IsEmpty, Link,
        Title));
    end;
end;
```

There are several things to notice from the preceding code blocks:

- I factored out the #AppName tag handling to a function called CheckAppName (if the tag being handled in this instance isn't #AppName, the string that's returned is blank).

- I don't have to check the TagString property to see whether it's a Link because, as one of the predefined tags, it sets the Tag property to one of the TTag enumerated constants, tgLink. Otherwise, it sets Tag to tgCustom, which is the only type we've handled until now.

- I used the parameters in the Link tag to pass in the values for the HTTP link and title that get passed to the new MakeLink function I wrote:

```
function TwmMyParks.MakeLink(const LinkDest, LinkTitle:
string): string;
begin
  Result := '<a href="' + LinkDest + '">' + LinkTitle +
    '</a>';
end;
```

The result is a simple main page:

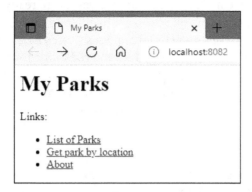

Figure 13.3 – Main page of the MyParks web server built by WebBroker

Now, build the page that asks for coordinates by creating a new page producer and adding an HTML `form` element with a few `label` and `input` elements to get the information from the user:

```
<form action="/showpark">
  <label for="long">Longitude:</label><br>
  <input type="text" id="long" name="long"><br><br>
  <label for="lat">Latitude:</label><br>
  <input type="text" id="lat" name="lat"><br><br>
  <input type="submit" value="Submit">
</form>
```

Here, the form's `action` sends us to yet another new page, `showpark`, which will display the results from the park lookup function in the database. The HTML should contain a line like this:

```
The park at <#longitude>, <#latitude> is: <#ParkName>
```

Its three tags will be substituted in the OnHTMLTag event handler, as you would expect, but before that gets called, we need to make sure the values are ready to plug in. We can do that in the web action item's OnAction event handler. In there, the parameters that are passed in the URL are put into the HTTP request's QueryFields property, which is an array of the two coordinate values that were entered in the form. The OnAction event handler gets called after the page producer is processed, so we have to leave the Producer property blank and manually assign it in the OnAction event handler, once the park information has been retrieved. We also need to save the values that the OnHTMLTag event handler will need in class-level private fields. Here is the event handler:

```
procedure TwmMyParks.
wmMyParkswaiShowParkFromCoordsAction(Sender: TObject; Request:
TWebRequest; Response: TWebResponse; var Handled: Boolean);
begin
    if Request.QueryFields.Count = 2 then begin
      var long, lat: Double;
      if TryStrToFloat(Request.QueryFields.Values['long'], long)
        and
        TryStrToFloat(Request.QueryFields.Values['lat'], lat)
          then begin
        FLongitude := long;
        FLatitude := lat;
        var ParkInfo := dmParksDB.
          LookupParkByLocation(FLongitude, FLatitude);
        FParkName := ParkInfo.ParkName;
      end;
    end;
    Response.Content := ppShowParkFromCoords.Content;
    Handled := True;
end;
```

Here's the OnHTMLTag event handler that uses those fields:

```
procedure TwmMyParks.ppShowParkFromCoordsHTMLTag(Sender:
TObject; Tag: TTag; const TagString: string;
    TagParams: TStrings; var ReplaceText: string);
var
    plong, plat: Double;
    ParkInfo: TdmParksDB.TParkDataRec;
```

```
begin
  ReplaceText := CheckAppName(TagString);
  if ReplaceText.IsEmpty then
    if SameText(TagString, 'longitude') then
      ReplaceText := FLongitude.ToString
    else if SameText(TagString, 'latitude') then
      ReplaceText := FLatitude.ToString
    else if SameText(TagString, 'ParkName') then
      ReplaceText := FParkName;
end;
```

Compile and run this code, and then enter some coordinates on the form in the browser:

Figure 13.4 – Entering the coordinates for a park on a form in the MyParks web server

After clicking **Submit**, the results will appear on the subsequent page:

Figure 13.5 – Results of a database lookup by the standalone MyParks web server

Notice the URL in the web browser address line. The longitude and latitude parameters are passed in using a typical URL parameter style.

Several open source and commercial packages build on these simple concepts to provide powerful web building packages that include JavaScript libraries and CSS for creating beautiful, modern web solutions. Expanding on the simple tag replacement concept of WebBroker, it's not hard to do this yourself. I inserted a couple of links to add the Bootstrap library (documented at `https://getbootstrap.com`) and created some header and footer page producers that are included by tag replacements in the main page producers, enhancing the look and feel of the produced HTML significantly. I went so far as to turn the links into tabs, eliminating the need for a "home" page with links to the subpage, by using the **About** page as the default page. Here's how my **Find a Park** page looks now:

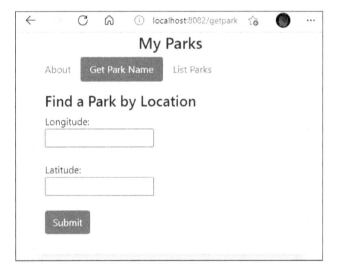

Figure 13.6 – Updated Find a Park by Location page from WebBroker with Bootstrap

The full source, including the embedded links to the Bootstrap library, can be found on GitHub at `https://github.com/PacktPublishing/Fearless-Cross-Platform-Development-with-Delphi/tree/master/Chapter13/01_MyParksStandAloneGUI`.

That's enough about getting a standalone web server up and running using WebBroker. Now, it's time to turn this into an actual web module that works inside the two most popular web servers. The first one we'll tackle is an ISAPI module for IIS on Windows.

Building an ISAPI web module for IIS on Windows

You don't need to have a separate computer with the server edition of Windows to use and test **Internet Information Services (IIS)**. If your local Windows machine doesn't currently have it installed, all you have to do is enable it by performing these steps:

1. Click your Windows **Start** button and type in the word `features` to launch **Turn Windows features on or off** (or, from the Control Panel, select **Programs and Features** and click on **Turn Windows features on or off**).

2. Scroll down the list of features to find **Internet Information Services**; check its box.

3. Open the IIS feature, and then expand **World Wide Web Services** and **Application Development Features**.

4. Check the box for **ISAPI Extensions** (and optionally **CGI**) and click **OK**:

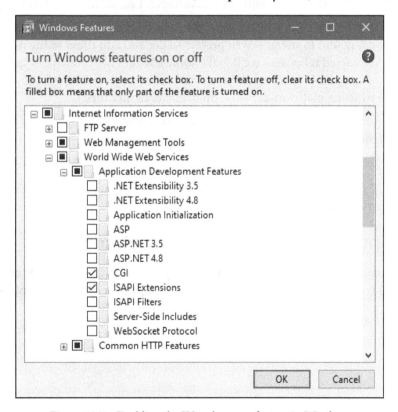

Figure 13.7 – Enabling the IIS web server feature in Windows

It only takes a few moments to install the web server. To verify it is working, open a new tab in any web browser and type localhost into the address line; you should see the default IIS web page, which has a blue background with the word **Welcome** in several different languages spread across the page.

> **Note**
>
> We will not cover building a CGI app in this book; include it if you want to do your own research and testing.

With our web server working and ready-to-load web modules and host websites, switch back to Delphi and follow these steps:

1. Create a new **Web Server Application**, check only **Windows**, select **ISAPI dynamic link library** in the creation wizard, and click **Finish**.

2. Save the project with a good name in a new folder. I called mine MyParks.

3. From the previous project's folder, copy the web module, the MyParks data module, and the logging unit to the new web project folder and add them to the new Delphi project. (We copied it because we'll make some changes but often, you can share code between multiple kinds of projects, such as Console, ISAPI, Windows Service, and more by using platform-agnostic units and data modules.)

4. View the source of the main project and remove the line that added the creation of the data module.

5. Compile the project.

You now have an ISAPI DLL ready to deploy to IIS – yes, it's that simple! Now, we just have to prepare IIS for our web module.

Configuring IIS to support Delphi web modules

Bring up **Internet Information Services Manager** by clicking the Windows **Start** button and typing IIS to find it. It can also be found in the Control Panel under **Administrative Tools**. Here's what the IIS Manager interface looks like:

Figure 13.8 – IIS Manager

Follow these steps to get IIS ready for your web module:

1. Click on your computer name listed under **Connections** in the left column.

2. Expand it, and then click on **Application Pools**.

3. You'll see that an application pool has already been created called **DefaultAppPool**. Create a custom application pool for our Delphi apps by clicking on **Add Application Pool...** in the right column, under **Actions**.

4. In the **Add Application Pool** dialog, enter a name, then switch the drop-down list for **.NET CLR version** to **No Managed Code**:

Figure 13.9 – Add Application Pool dialog

5. If your ISAPI project's target platform is `Windows 32-bit`, you need to right-click on the new application pool and select **Advanced Settings...** to switch **Enable 32-Bit Applications** from the default of `False` to `True`.

> **Note**
>
> If you need to support a mix of 32-bit and 64-bit ISAPI DLLs, you will need a different application pool for each with this option set accordingly – although you may choose to create multiple pools anyway to minimize the impact restarting one would have on other parts of your site.

6. In the left column, expand and then right-click on **Sites** and select **Add Website...**.

7. In the **Add Website** dialog, fill in the **Site Name** box, click the **Select...** button to change **Application Pool** from the default to the new application pool you just created, and select a **Physical Path** where the new DLL and any necessary supporting files will be placed. The default website's path is `C:\inetpub\ wwwroot`; your new one could be a separate subfolder or even be located elsewhere. Also, assign a unique **Port** for this website; I chose `8080` as it's easy to remember and is popular for testing:

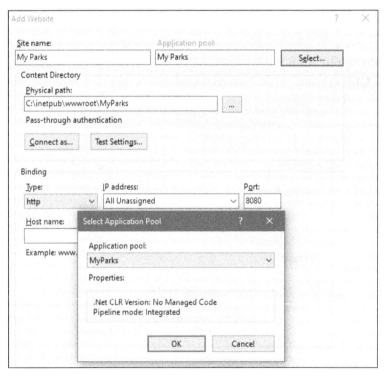

Figure 13.10 – Adding a website to IIS

8. Back under **Connections**, click on the server name (the root-level item in the tree view) to view all the feature icons for your web server; double-click on **ISAPI and CGI Restrictions**. This is where you can configure the server to allow your web module to execute. You have a few options regarding how to configure this; the easiest is to click **Edit Feature Settings...** in the right column and check **Allow unspecified ISAPI modules** (if you will never write CGI modules, you can leave this one unchecked):

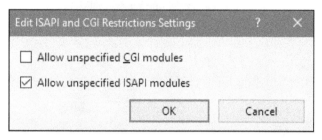

Figure 13.11 – Edit ISAPI and CGI Restrictions Settings page

9. If, instead, you want to specify every module you write explicitly, leave the checkboxes shown in the preceding screenshot unchecked, cancel the dialog, and click the **Add...** link to pull up the **Add ISAPI or CGI Restriction** dialog. Now, you can enter the specific DLL or EXE that you want to run:

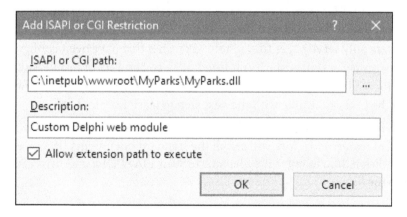

Figure 13.12 – Adding an explicitly named ISAPI module

10. There's one more step to tell the web server to actually execute your DLL rather than try to download it. Click again on the new website under **Connections**, then double+click on **Handler Mappings**. Notice that **ISAPI-dll** files are listed in the **Disabled** section? Click on that line and select **Edit Feature Permissions**... under the **Actions** menu in the right side-bar. Check the **Execute** checkbox to enalbe ISAPI DLLs to be executed as web modules then click **OK**.

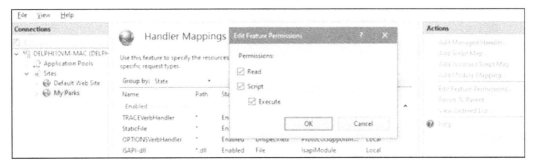

Figure 13.13 – Enabling execution of ISAPI DLLs

11. Copy the compiled DLL from your Delphi ISAPI project to the new website's folder (you'll probably need administrator rights).

Many tutorials stop right there and tell you to browse to your new web module. In our case, it's `localhost:8080/MyParks.dll`. This works, but typing that in or giving that URL to someone else is a bit unwieldy. It would be much better to just stop at `localhost:8080` (or a real URL when we finally deploy). However, web servers only serve pages they're told to or ones that have been configured to be default pages if a specific page is not listed. IIS also has a set of default pages that it looks for – all we have to do is add our custom module to that list and we can shorten the URL. Continue with the next step to learn how to do this.

12. Click on the new website in IIS Manager, and then double-click on the **Default Document** feature icon. You will see all the standard documents IIS was expecting to find. Click **Add...** to enter the filename of your ISAPI DLL and then click OK to add it to the list:

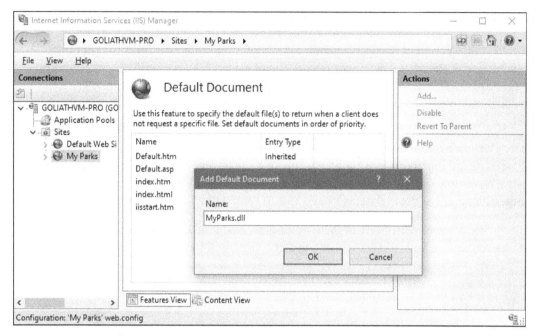

Figure 13.14 – Specifying your ISAPI module as a default document for this website

Restart the server. You can do this in IIS Manager by clicking on the server in the left column and then clicking **Restart** in the right column under **Manage Server**. Note that all websites defined on that server will be unavailable until the server comes back up.

Now, testing the shorter version of the address in a browser (just the hostname and port) automatically pulls up the default page in our ISAPI web module (which for me is the **About** page):

Figure 13.15 – Default page in the completed ISAPI web module

Great! Now, let's test out our links by clicking on one. Oops! Why did we get HTTP Error 404? Have you noticed that the address line of the browser shows localhost:8080 but not the filename of the ISAPI module? The links to other pages are subpages of the main URL that includes the filename. You can verify this by browsing to the full URL, localhost:8080/MyParks.dll/about; now, clicking on a link works. By adding that Default Document in *Step 11*, IIS "conveniently" hid that part of the URL for us.

There's a simple fix for this without reverting to always typing in the module's filename. We need to add one more web action and make it the new default, but instead of returning a page, it's simply going to redirect us to the full and explicit URL (complete with the module filename) of the *real* default page, just as if we had manually typed in the full URL:

1. In the Actions property of the web module, add another TWebActionItem, make it the new default, and leave both the PathInfo and Producer properties blank:

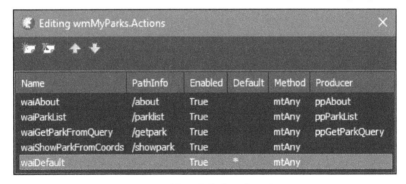

Name	PathInfo	Enabled	Default	Method	Producer
waiAbout	/about	True		mtAny	ppAbout
waiParkList	/parklist	True		mtAny	ppParkList
waiGetParkFromQuery	/getpark	True		mtAny	ppGetParkQuery
waiShowParkFromCoords	/showpark	True		mtAny	
waiDefault		True	*	mtAny	

Figure 13.16 – New default web action item

2. Create an OnAction event handler for the new default action item that redirects to the page we want to be our default page:

```
procedure TwmMyParks.wmMyParkswaiDefaultAction(Sender:
TObject; Request: TWebRequest; Response: TWebResponse;
var Handled: Boolean);
begin
  Response.
SendRedirect(ExtractFileName(WebApplicationFileName) + '/
about');
end;
```

3. Compile the project, shut down the server (or turn off the Application Pool), copy the DLL, and start the server back up (or restart the Application Pool).

That's it! With our ISAPI module set as a default document in IIS and with the default page in our web module redirecting to a fully specified file and page, our web server at `localhost:8080` now returns the full address and default page with working links.

> **Tip**
> If you work on ISAPI web modules often, you don't have to manually launch a web browser and type in the address. Instead, select the website in IIS Manager, then find the **Browse Website** heading in the right-hand column in the **Manage Website** section. There will be a clickable link for each port you've bound to the website. For ours, we only assigned one, port 8080, so there will only be one link, which for me says **Browse *:8080 (http)**. Clicking this brings up the browser with the address already entered.

Before we move on, we should mention logging.

Logging from an ISAPI web module

IIS web servers automatically log incoming requests. including the browser and IP address the request came from, the request URL (including any URL parameters), the return code, and a date/time stamp. You can configure the format and where log files are stored in IIS Manager by clicking on the server under **Connections**, and then double-clicking on the **Logging** feature. There's also a link to **View Log Files...** in the right column under **Actions**, but these logs are not controlled or generated by your ISAPI module – they're a feature of IIS.

The best way for us to include custom logging from our web module is to log to a file, as we've done with other server applications. The problem is that our ISAPI DLL is not running as a standard user but as the default web user, which is a member of the IIS_ IUSRS group and which, by default, has read-only access to the folder where our web module exists. We need permission to write to our web application folder to create log files; however, you should never modify a built-in Windows account such as the IIS_ IUSRS group as it could open up unintentional security holes. So, what should we do?

It turns out the custom application pool we created for our ISAPI DLL also plays a security role that we can use to grant ourselves (and only ourselves) the access we need. Follow these steps from IIS Manager to do this:

1. Right-click on the new website we created in the previous section for our web module and select **Edit Permissions...**.

2. Switch to the **Security** tab and click the **Edit...** button.

3. Do not edit the permissions for the **IIS_IUSRS** group; instead, click the **Add...** button.

4.   In the **Enter the object names to select** box, type `IIS AppPool\MyParks`
     (where `MyParks` is the name of the application pool you created for this app):

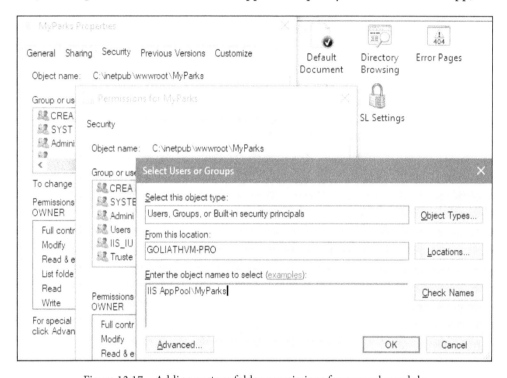

Figure 13.17 – Adding custom folder permissions for our web module

5.   Click the **Check Names** button to verify the validity of your app pool's name and
     click **OK**.

6.   Now that your app pool has been added to **Permissions for MyParks** (or whatever
     you named your website), you can check the **Allow** box for the **Modify** line in the
     Permissions list:

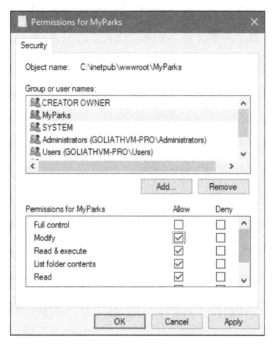

Figure 13.18 – Modifying the permissions that have been granted for our app pool

7.   Click **OK** a couple of times and restart the web server.

Our ISAPI web module now has the permissions it needs to write to files in its folder. The default logging settings assume the folder to write to is based on the currently running application, but we're logging from a module, so we have to change the initialization of the logging mechanism in the web module's OnCreate event handler by changing one line:

```
Log := BuildLogWriter([TLoggerProFileAppender.Create(5, 1000,
                TPath.Combine(WebApplicationDirectory,
                'logs'))]);
```

The WebApplicationDirectory function that's provided for us in the Web.HTTPApp unit conveniently gives us the directory where our module is called from; appending logs separates those files nicely for us. After updating and restarting the web server, you should see log files appear in the logs subfolder of our web application directory.

Now, let's turn our attention to the most ubiquitous web server in the world: Apache.

Getting started with the Apache HTTP server

The Apache Software Foundation has a plethora of open source software projects. Their HTTP server is probably the most well-known. Oftentimes, when people hear the Apache web server being mentioned, they automatically associate it with Linux, and while that may be the overwhelmingly largest platform on which it can be found, it also runs on Windows.

The Apache server is frequently paired with the MySQL database engine and the PHP scripting language. In fact, this is so common that these parts are often packaged together. When discussing these on the Linux platform, it's call **Linux, Apache, MySQL, and PHP (LAMP)**; on Windows, it's called **WAMP**. We won't be using MySQL or PHP, so when you're looking at download options, note that these packages will contain more than what we will work with in this book. However, if you have the time and curiosity, feel free to install this suite of tools and explore – there are many resources to get you started.

Let's get the Apache web server installed on both Windows and Linux, starting with Windows.

Installing and starting Apache on Windows

To install the Apache web server for Windows, go to the *Downloading Apache for Windows* link in the *Further reading* section at the end of this chapter and select one of the options listed. I found that the Apache Lounge (`https://www.apachelounge.com/download`) provides straightforward instructions with a download link. You'll likely need to install the Visual Studio C++ Redistributable library first – this process is quick and simple.

If you downloaded the Apache server by itself as a `.zip` file, there is no install procedure – just unzip the file to a folder and start the server! But I'm getting ahead of myself: read the included `ReadMe.txt` file, where you'll quickly see that starting and stopping the web server is really simple – just run `httpd.exe` from Apache's `bin` folder. Running it with the `-k install` parameters installs it as a service; if it's installed as a service, you can start it with the following command line:

```
net start apache2.4
```

You can also find Apache 2.4 in the Windows services and click **Start** to start it with a GUI.

> **Note**
>
> Both IIS and Apache web servers can be installed and running on the same machine at the same time, but they need to be configured for different ports. The default port for both servers is 80. When we created a new website in IIS, we assigned it port 8080 but left the default site listening on port 80. If your IIS is still running from the previous section, make sure to either disable its default website or configure it for a different port.

Once it's started, go to a web browser and browse to `http://localhost`. If all is well, you'll see a big **It works!** displayed on the screen.

Unlike IIS, there is no GUI to configure the settings; instead, a text file in the `conf` folder called `httpd.conf` is used to configure the web server. Fortunately, it has a lot of comments; unfortunately, it can be rather daunting if you're not familiar with it. We'll need to modify a few of these settings, turning the server off and back on when we do.

For example, let's change the default port that Apache listens on. Pull `httpd.conf` into a text editor and find the line that shows `Listen 80`. Change this to `8081` so that it's different than our IIS server, and then save the file.

> **Note**
>
> Just like IIS, we could create a new website and be listening to two different ports. We won't cover how to do that in Apache so instead, just change the default port.

To activate the new port setting, you must restart the server:

```
net stop apache2.4
net start apache2.4
```

Now, go back to your browser and browse to `localhost:8081` to see the default page again. Once we start writing web modules, you'll need to be familiar with editing the configuration file and restarting Apache.

Before we start writing code, let's get the Apache web server up and running on Linux – there are many similarities but also important differences.

Installing and starting Apache on Linux

Nearly every Linux system has a package manager, and we'll use that to get the Apache HTTP server installed under Linux.

> **Note**
>
> We'll continue working with the Ubuntu Linux subsystem on Windows that we set up in *Chapter 12, Console-Based Server Apps and Services*. If you have a different scenario, adjust the following instructions for your Linux distribution.

Using your system's package manager, getting Apache installed is as simple as typing two lines:

```
sudo apt-get update
sudo apt-get install apache2
```

The `sudo` command raises the permission level of the standard Linux user to a root level for the given command to install software or perform other aspects of system maintenance. This will require you to enter your user password.

After it downloads, unpacks, sets up, and enables several files and modules, it's ready to be started. The default port for all web servers is 80, so if you have either IIS or Apache for Windows still running on the default port, you should shut them down before running this command:

```
sudo service apache2 start
```

(Again, if you're using a different Linux distribution, starting Apache may work differently.)

As before, once it's started, go to a web browser and browse to `http://localhost`. The default web page for Apache on Ubuntu Linux is a full page of information with the Ubuntu logo and an overview of the configuration:

Apache2 Ubuntu Default Page

It works!

This is the default welcome page used to test the correct operation of the Apache2 server after installation on Ubuntu systems. It is based on the equivalent page on Debian, from which the Ubuntu Apache packaging is derived. If you can read this page, it means that the Apache HTTP server installed at this site is working properly. You should **replace this file** (located at /var/www/html/index.html) before continuing to operate your HTTP server.

If you are a normal user of this web site and don't know what this page is about, this probably means that the site is currently unavailable due to maintenance. If the problem persists, please contact the site's administrator.

Configuration Overview

Ubuntu's Apache2 default configuration is different from the upstream default configuration, and split into several files optimized for interaction with Ubuntu tools. The configuration system is **fully documented in /usr/share/doc/apache2/README.Debian.gz**. Refer to this for the full

Figure 13.19 – Apache2 default web page on Ubuntu Linux

Similar to the Windows version, Apache on Linux can be configured by modifying a configuration file. On Linux, this file is apach2.conf and can be found in /etc/apache2. Open the file in a text editor and change Listen 80 to Listen 8082. Then, execute these two commands to restart Apache:

```
sudo service apache2 stop
sudo service apache2 start
```

All three servers can now run simultaneously as we have configured them to listen on three different ports:

- Port 8080: IIS for Windows

- Port 8081: Apache for Windows

- Port 8082: Apache for Linux

This is really handy for testing and learning about the different peculiarities of each web server.

Now that we have the Apache web server installed on both Windows and Linux, let's start the web project.

Writing cross-platform Apache web modules

The first few steps in creating an Apache web module are very similar to creating an IIS module; just step through the **New Web Server Application** wizard:

1. Create a new **Web Server Application** project.
2. If you have Delphi Enterprise, check **Linux** in addition to **Windows** for the platform and click **Next**.
3. Select **Apache dynamic link module** for **WebBroker Project Type** and click **Next**.
4. You can leave everything as their defaults on the **Apache Module Options** page, both **Apache version** and the name of our new **Apache module**.

Figure 13.20 – Final step of creating an Apache web module project

After clicking **Finish**, the Apache web module will be created for you. If you're using Delphi Enterprise or higher, you will have both the Windows and Linux platforms enabled.

> **Note**
>
> Apache on Windows cannot support both 32-bit and 64-bit apps like IIS can
> – there's no configuration option to tell it what type of module to use; it is
> entirely dependent on which version of Apache you installed. If you installed
> the 32-bit version of Apache, you can only write 32-bit web modules for it; if
> you installed the 64-bit version, you can only write 64-bit web modules. You
> need to set the Windows **Target Platform** of your project accordingly; for
> Linux, there is only 64-bit.

We need to make a few changes to the project by performing a few more steps:

1. Save the project with a good name in a new folder. I called my project `MyParks`.
    Notice that the name is prefixed with `mod_` so that the fully deployed module name
    will be `mod_MyParks.dll`. If you switch the target platform to Linux, the name
    will change to `libmod_MyParks.so`. These names will be used when you deploy
    the modules.

2. From the previous project's folder, copy the web module, the MyParks data
    module, and the logging unit to the new web project folder and add them to the
    new Delphi project.

3. View the source of the main project and remove the line that added the creation of
    the data module.

4. Check the `uses` clause of the main project file. On some versions of Delphi, one of
    the system units needed for compiling on Windows is missing; that is, `System.`
    `Win.ComObj`. If it's not listed, add it.

5. Did you notice the `GModuleData` global variable just before the main project's
    `begin` statement? The name that gets exported should be modified to something
    unique that you'll use to reference the module within the Apache configuration. It
    should be all in lowercase with no spaces; I named mine `myparks_module`.

6. The line that calls `InitApplication` needs to be modified as well. That
    procedure has an optional parameter that I found was necessary to get things
    working right. That parameter names the module handler that's used in the Apache
    configuration. Here's what I added:

```
Web.ApacheApp.InitApplication(@GModuleData, 'myparks_
handler');
```

7. As one last change (completely optional), I added a compiler directive to modify the `APP_NAME` constant, depending on the platform, to confirm which one had been compiled. This is done in the web module, just after the `implementation` section:

```
{$IFDEF MSWINDOWS}
APP_NAME = 'My Parks - Apache Windows';
{$ELSE}
APP_NAME = 'My Parks - Apache Linux';
{$ENDIF}
```

You should now be able to compile the web module for both Windows and Linux. Let's discuss how to deploy each of these.

Deploying an Apache web module on Windows

Deploying a module for the Apache web server involves just a few steps. At a high level, these are as follows:

1. Copy the compiled module to the `modules` folder of the Apache installation.

2. Add the `LoadModule` and `SetHandler` directives to Apache's configuration.

3. Restart the server.

Let's go over these briefly.

The directory structure for the Apache 2 web server on Windows is pretty simple. I extracted the files to `C:\Apache24`. The `htdocs` subfolder contains the default web page; the `conf` folder, as mentioned previously, contains the configuration file. The `modules` folder is where we'll put our compiled module – it can be anywhere that the Apache server can find it as the location specifies the folder and filename, but by convention, this is where it should go. Copy our DLL to this folder.

With our new module in place, we need to tell Apache two things: where it is and how to handle it. Open the config file in a text editor – we've already looked in here once to change the default listening port. A little way further down from there are several lines starting with `LoadModule`, some of them commented out (indicated by a # at the beginning of the line). The ones that are not commented out are modules that are loaded by Apache when it starts. We need to add one for our new module – it doesn't matter where, but it's a good idea to keep it close to the other modules listed for consistency. Each line specifies the name of the module (the `GModuleData` name that was exported in the project) and the filename relative to the root Apache folder. Therefore, this new line will look like this:

```
LoadModule myparks_module modules/mod_MyParks.dll
```

Next, we need to tell Apache that if a specific path is on the URL, it should handle that path with this module. This set of lines starts with a `Location` directive, specifies the path, then lists the handler name we gave it in the call to `InitApplication` in our project:

```
<Location /myparks>
   SetHandler myparks_handler
</Location>
```

With these two additional pieces of information, Apache will be able to serve pages from our custom Apache web module.

Logging in Apache for Windows is done in the `logs` subfolder of the main Apache folder. This is the simplest location for our logging library so that we can revert the code in the web module back to the way the standalone server initialized it:

```
Log := BuildLogWriter([TLoggerProFileAppender.Create]);
```

Now, it's time to restart the Apache service in Windows:

```
net stop apache2.4
net start apache2.4
```

And with that, you can browse to `localhost:8081/myparks` and see the default page that was generated by our custom web module:

Figure 13.21 – Apache for Windows web module

Test the links to make sure everything works. Now, let's move on and look at the Linux version.

Deploying an Apache web module on Linux

Apache for Linux was installed with a package manager that set up the various files and directories in a much more elaborate manner, consistent with other Linux-based apps. This provides some power and flexibility but also requires some understanding of the Linux file structure.

Additionally, since many of these files are in protected areas of the Linux filesystem, you'll need to use the `sudo` command from the Linux command line to write to those folders. This means that most of the file copy procedures from your Windows development machine to the Linux subsystem will be a two-step process; that is, copying the files from Windows to a temporary folder on the Linux system, and then dropping to the Linux command line and running the `sudo` command from there to move them to their final destination. The following steps assume you have this understanding:

1. Copy your compiled module, `libmod_MyParks.so`, to Linux, and then move it to the Apache `modules` folder:

    ```
    sudo cp libmod_MyParks.so /usr/lib/apache2/modules
    ```

2. Under your `Public Documents` folder on Windows, drill down to the `Embarcadero\Interbase\redist\InterBase2020` folder and copy the redistributable InterBase support libraries from either the `linux64` or `linux64_togo` folders, depending on the license of InterBase in use, to Linux. Then, run the following commands from the Linux command line:

    ```
    sudo cp ib_util.so /usr/lib/apache2/modules/
    sudo cp libgds.so /usr/lib/apache2/modules/
    sudo cp interbase.msg /usr/lib/apache2/modules/
    ```

3. Create a `myparks.load` file in the `/etc/apache2/mods-available` folder with the `LoadModule` command:

    ```
    LoadModule myparks_module /usr/lib/apache2/modules/
    libmod_MyParks.so
    ```

4. Create a `myparks.conf` file in the `/etc/apache2/mods-available` folder with a `Location` directive:

    ```
    <Location /myparks>
        SetHandler myparks_handler
    </Location>
    ```

5. Create links to the two files we just created in the /etc/apache2/mods-enabled folder (execute these commands from the /etc/apache2/mods-enabled folder):

```
sudo ln -s ../mods-available/myparks.load
sudo ln -s ../mods-available/myparks.conf
```

6. Apache for Linux stores log files in the /var/log/apache2 folder, but granting access to custom modules there is strongly discouraged. To continue in the same manner we've taken with other servers, we'll create a logs folder as a subfolder where the module resides, in /usr/lib/apache2/modules, then grant access to the www-data Apache web user so that they can write to it:

```
sudo mkdir /usr/lib/apache2/modules/logs
sudo chown www-data /usr/lib/apache2/modules/logs
sudo chgrp www-data /usr/lib/apache2/modules/logs
```

7. The logging initialization in the Apache web module will need to differentiate between the Windows platform and Linux, so change those lines in the OnCreate event handler to this:

```
{$IFDEF MSWINDOWS}
    Log := BuildLogWriter([TLoggerProFileAppender.Create]);
{$ELSE}
    Log := BuildLogWriter([TLoggerProFileAppender.Create(5,
        1000,
                    TPath.Combine(WebApplicationDirectory,
                    'logs'))]);
{$ENDIF}
```

The big difference between Apache for Windows and Apache for Linux is that modules, configuration settings, and log files go in completely separate folder trees in Linux. Furthermore, all the load and handler commands are in separate files instead of lines that are added to one large configuration file, which occurs in Windows. This means that to disable a module, we must move the .load and .conf link files from /etc/apache2/mods-enabled to /etc/apache2/mods-disabled. As the main configuration file, /etc/apache2/apache2 includes all the files in the mods-enabled folder.

Once you've finished these steps, restart Apache for Linux:

```
sudo service apache2 stop
sudo service apache2 start
```

Now, browse to `localhost:8082/myparks` to view our **About** page on Linux:

Figure 13.22 – Apache for Linux web module

Going through these steps with all three web servers has given you experience that you can use to direct future development projects.

Summary

The ability to leverage existing Delphi skills to build modules for major web servers is a grand achievement. It opens the doors of opportunity to help you expand your product line, increase visibility, and support new mediums. It also adds to the complexity of your project.

But in these pages, we have lifted the veil shrouding some of the mystery behind web server technology and what it takes to move in that direction. You've seen the various available options, learned about the underlying framework, and configured the two most popular web servers on the planet to support your custom-written modules written in Delphi.

In the next chapter, our journey will take one more step toward fully embracing rapid application development by utilizing a powerful backend server. This will allow us to concentrate on the business logic and application architecture rather than managing security, developing code to support REST API endpoints, or dealing with multi-tenancy.

Join me as we explore RAD Server.

Questions

1. How long has WebBroker been part of Delphi?

2. What is a Page Producer's single event?

3. Do WebBroker action items have to return HTML in their Response object?

4. How do you get IIS onto a Windows machine?

5. How do you support a 32-bit ISAPI module in IIS?

6. Can IIS and Apache run on the same machine simultaneously?

7. How do you change the settings of Apache for Windows?

8. How do you change the settings of Apache for Linux?

9. What is the default prefix and extension for Apache modules for Windows? For Linux?

10. What global variable in an Apache web module project must be changed before it's deployed?

Further reading

To learn more about the topics that were covered in this chapter, take a look at the following references:

- Ultimate Web Frameworks For Ultra-Fast Web Application Development Using Delphi/C++ Builder: `https://blogs.embarcadero.com/ultimate-web-frameworks-for-ultra-fast-web-application-development-using-delphi-c-builder`

- Types of Web Server Applications: `http://docwiki.embarcadero.com/RADStudio/Sydney/en/Types_of_Web_Server_Applications`

- Internet Server Application Programming Interface: `https://en.wikipedia.org/wiki/Internet_Server_Application_Programming_Interface`

- Using WebBroker Index: `http://docwiki.embarcadero.com/RADStudio/Sydney/en/Using_Web_Broker_Index`

- Tutorial: DataSnap Application Using an ISAPI DLL Server: `http://docwiki.embarcadero.com/RADStudio/Sydney/en/Tutorial:_DataSnap_Application_Using_an_ISAPI_DLL_Server`

- The Apache Software Foundation: `https://www.apache.org/`

- Apache HTTP Server Project: `http://httpd.apache.org/`

- Downloading Apache for Windows: `https://httpd.apache.org/docs/current/platform/windows.html#down`

- Installing Apache2 with WSL on Windows 10: `https://tubemint.com/install-apache2-with-wsl-on-windows-10/`

- Embarcadero Developer Skill Sprint: Deploy to Apache: `https://youtu.be/p6SX9XWPeDs`

- How to Configure the Apache Web Server on an Ubuntu or Debian VPS: `https://www.digitalocean.com/community/tutorials/how-to-configure-the-apache-web-server-on-an-ubuntu-or-debian-vps`

14
Using the RAD Server

This book has been all about developing applications for various platforms. For the most part, the term *platform* means the combination of hardware and an operating system, such as **iOS**, **Android** phones, **macOS**, or **Windows** PCs. But we've also extended that idea to cover various types of applications on these platforms, such as **FireMonkey** or **VCL** apps on Windows or web servers running under **IIS** or **Apache**.

The last platform we'll look at in this book extends that idea one more time and builds on the server concepts presented in the last two chapters. **RAD Server** is the pinnacle of **Rapid Application Development (RAD)** for building multi-tiered, backend servers. It is a powerful application server platform that provides a foundation for writing modules to extend its functionality with **Delphi** packages, produces a REST interface for communicating JSON or XML data, handles multiple connections with ease either as a standalone server or running under IIS or Apache, includes robust user and group security, and allows us to concentrate on the business logic required in our application without spending time worrying about the infrastructure. RAD Server provides services any cross-platform client app can consume, opening doors of opportunity for development diversification and speeding time to market.

This chapter will teach you some of the benefits of this platform, how to build Delphi packages that work within RAD Server, and modifications to our *MyParks* application that will utilize the published data. We will go through these topics in the following sections:

- Establishing a use case for RAD Server
- Getting familiar with what's included
- Writing modules to extend your server
- Modifying MyParks for use with RAD Server

It's time to harness the full power of RAD development.

Technical requirements

The requirements for this chapter include **Delphi Enterprise** or higher and **InterBase 2020** running on your main Windows development machine. While RAD Server can be installed on either Windows or **Linux**, only Windows will be discussed in this chapter, with references to the *Further reading* section for information on deploying to Linux.

We will also make modifications to our cross-platform app, *MyParks*, and screenshots from both iOS and Android will be included—it is up to you which of those platforms you choose to use with RAD Server during your study of this topic.

This chapter makes heavy use of REST server concepts, specifically the four most popular HTTP request methods—**GET**, **POST**, **PUT**, and **DELETE**. If you're not familiar with these terms, they refer to the type of query made to a web server. When you browse a website, you're making a *GET* request to the server for a specific page. When you fill out a web form and submit it, the browser can make a *PUT* request to updated data, *POST* to insert new data, or *DELETE* to ask the server to delete data (there are other request methods defined in the HTTP specification, but these are the most common). This is what defines a typical REST server and replaces the older style of web-based data management, where all data was sent to the web server in *GET* requests using parameters on the URL separated by ampersands (&). *GET* requests are limited by the length of the URL, but the other methods use data packets that are unlimited in size. For an excellent discussion on this important topic, read Martin Fowler's classic article, *Richardson Maturity Model*, at https://martinfowler.com/articles/richardsonMaturityModel.html.

The code for the project we'll build in this chapter can be found on GitHub:

https://github.com/PacktPublishing/Fearless-Cross-Platform-Development-with-Delphi/tree/master/Chapter14

Establishing a use case for RAD Server

We've been building increasingly complex but more useful servers in this last section of the book, giving you first a working knowledge of the underlying technology with the low-level TCP communication protocols encapsulated in the **Indy** components, then building on that to provide web servers utilizing event-driven programming with **WebBroker**. Even with these building blocks, there's still a lot of scaffolding code that is needed to surround the specific functionality you're building in order to bring to life a full-featured, multi-client application.

Another layer in this client/server development tool stack is the **DataSnap** technology. Also requiring at least Delphi Enterprise, DataSnap builds on WebBroker and internet communication components adding remote invokable methods, secure JSON data transfers, encryption of data, and server change notification. However, you still need to build out quite a bit of functionality just to support your server, user authentication structure, and API endpoints.

With RAD Server, you can skip a large section of time-consuming database design and programming that would be necessary to build these pieces. Consider the following diagram:

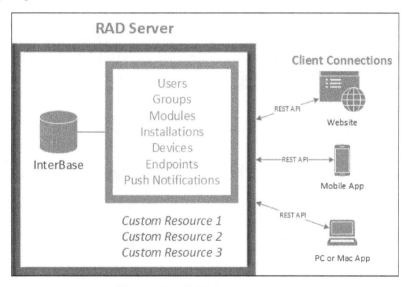

Figure 14.1 – RAD Server overview

This is an application framework in a box—all you do is provide the custom resources unique to your business; RAD Server configures and manages all the rest of the parts. We'll touch on just a couple of these parts, and one of the big ones is user management.

Considering an application's multi-user needs

Suppose the MyParks app we've been building will be made available to a wide variety of people and sharing park information, notes, and pictures will be a selling point of the app. Some people will only want to use it locally, but others may opt to subscribe and track their park visits across devices, download pictures from a friend's park visits, and rate their favorite parks. To provide various levels of functionality, you'll want a user hierarchy with various roles including paid and non-paid accounts, staff accounts for moderating comments and uploads, and various API authentication levels for accessing content from other sources. Building all that functionality into the backend, in addition to the basic user interface for adding, saving, and uploading park data, can be overwhelming.

User management and authentication is a common feature in these types of apps. It can also be a very large and complicated set of data structures and business rules to build and manage. RAD Server provides the technical details you need to get started:

- Database tables built to handle multiple groups, users, registered devices, and installations

- A FireMonkey application with a source to manage all these entities

- A web-based view of these resources and usage statistics

When you think about it, this is a huge piece of what would need to be designed and built: database tables, indexes, relationships, and queries. Not only is this set up for you, but there's also a full-featured management console for you with the source that you can extend:

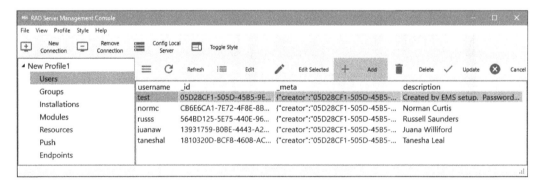

Figure 14.2 – RAD Server Management Console

As you add users and devices, and as people start using the application, you'll want to see how fast it's growing and how many devices are connected. You might want to be able to see this from several different devices or let support staff keep tabs on it as well. This would be best as a web application to enable you to get to it from anywhere without having to install a piece of software. This is provided for you as well with the web-based view of your application's resources:

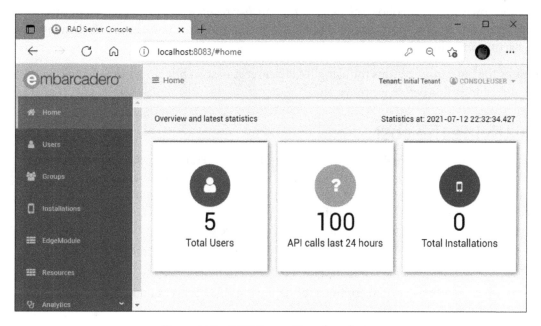

Figure 14.3 – RAD Server Console web view

RAD Server's support for applications and users doesn't end there. As people connect with various devices, they can register for additional services.

Enabling push notifications for registered devices

Another feature provided out of the box by RAD Server gives the ability to send notifications to devices that have been registered and have been granted permission to receive those notifications. Suppose you and your friend both have the MyParks app and you both want to be notified when the other rates a park or adds a picture. By creating an account in RAD Server, registering their device, then accepting push notifications, they can receive configured messages from the server. The installed devices, types of notifications, and other related resources can all be managed from the RAD Server Management Console without the need for extra coding—you can concentrate on the data your app deals with and set up the types of messages that should go out while RAD Server does the heavy lifting for you.

Not only does this save you a massive amount of time getting these vital aspects of your application ready for multiple users, but using the source to the management program shows you how to use the various user management API endpoints and allows you to extend it for your custom needs.

We've only touched on a couple of the major features of using RAD Server, but it's already obvious how much time can be saved using this extensible platform instead of rolling out your own solution. Follow the links in the *Further reading* section at the end of this chapter to watch some great videos about RAD Server and learn more about the benefits of starting with this solid foundation.

One of the big considerations for using RAD Server is cost. While prices can change, we should address this before we move on.

Justifying the cost

Deploying applications built with your own hand-crafted servers or using the DataSnap technology is royalty-free. This means the only software costs for each additional customer is the database license, if needed. There is merit in this if your user management needs are simple, or you have some of this infrastructure in place from another related application or service. If not, the cost of design and development to produce the out-of-the-box functionality provided by RAD Server can quickly eclipse the cost of deploying user licenses.

There are currently two pricing options for deploying RAD Server and they are based on user count:

- Unlimited users for USD 5,000
- Per-seat licensing at USD 99 each

If your organization is small and has a limited number of users, the per-seat licensing option would probably be best. If you're deploying multiple servers to various locations and each location will only have a few users, you would still likely be better opting for per-seat licensing, as each server requires its own license. Once you reach around 50 users on a server and see more growth coming, you can switch from per-seat licensing to unlimited. There are also multi-site licenses available—contact your **Embarcadero** sales representative to see what licensing model best fits your needs.

Let's turn our attention now to learning more about what you get with RAD Server and getting it up and running.

Getting familiar with what's included

RAD Server is all about saving you lots of time. It comes ready to run and ready to deploy! For developers, it's installed with Delphi, and for deploying to customers, you can download packages from **GetIt** to automate most of the steps.

Let's go over what comes pre-installed with Delphi.

Running RAD Server on a development environment

RAD Server is comprised of two parts:

- The **engine**, listening for requests and returning JSON data
- The **console**, a web server app

The engine provides application endpoints for accessing both built-in and extended functionality through packages that are installed. The console makes requests to the engine and presents a nice web view of your application resources and usage statistics.

There are both 32-bit and 64-bit versions of both of these applications residing in the `bin` and `bin64` folders of your Delphi installation, respectively. To start the RAD Server engine, simply run `EMSDevServer.exe`.

> **Note**
>
> You will see many names and references to **EMS** while working with RAD Server. EMS stems from an earlier name for this product, **Enterprise Mobility Services**.

If you have not started or configured RAD Server previously, you will be asked to run the configuration wizard when it is launched. This asks several questions to get you going:

1. The first question establishes the database:

Figure 14.4 – RAD Server configuration wizard – New Database

If you installed InterBase with its default settings, you don't need to make any changes on this screen—**Server Instance** should be left blank. If you have multiple instances of InterBase on your machine, enter the instance name you gave to InterBase 2020 when you installed it (which shows in parentheses in the title bar of **InterBase Manager**, the default being gds_db).

2. The next screen asks whether you want to generate **sample users** and **sample user groups** in the new database being created:

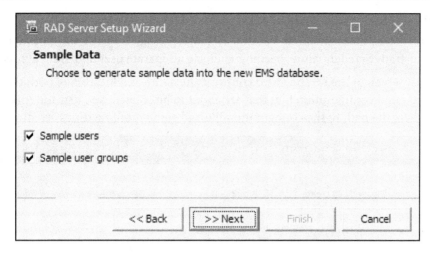

Figure 14.5 – RAD Server configuration wizard – Sample Data

I find this to be useful and leave both checkboxes checked. There is only one sample user in one sample group, but it allows you to see how data will look right away.

3. The next screen sets up the user that will access **RAD Server Console**:

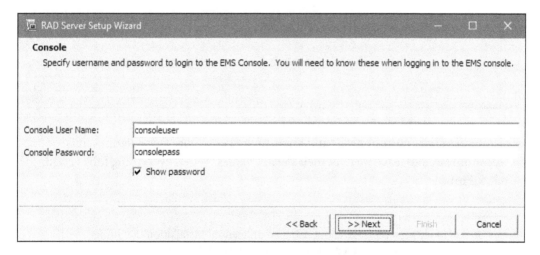

Figure 14.6 – RAD Server configuration – Console user

Establishing a separate username and password for console use will allow you to give read-only access to a web view for displaying usage statistics without exposing administrative credentials or needing to create a separate user and grant access.

4. The last screen of the setup wizard confirms the name and location of both the database and configuration files that are about to be created, and lists the registry key where the path to the configuration file and some profile settings are stored:

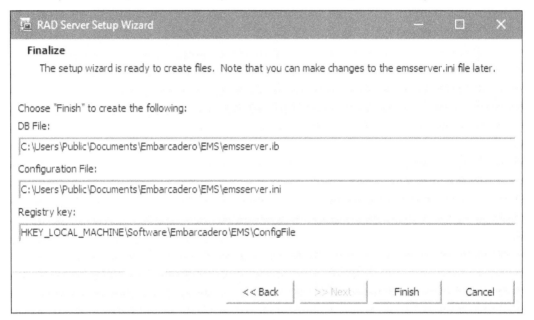

Figure 14.7 – RAD Server configuration – Finalize settings

Unless you have custom needs and know how to configure these settings, it is recommended to leave them at their default values. When everything looks good, click **Finish**.

> **Note**
>
> You will likely get a message about using an unlicensed installation that limits EMS features. This is expected for the five-user-limited development license.

5. Once the database is created and populated successfully, you'll see a confirmation message reviewing where your files are and the users that were created:

Figure 14.8 – RAD Server configuration complete summary

You're now able to see the log messages generated by the development version of RAD Server:

Figure 14.9 – Log view of the development version of RAD Server

These log messages, in JSON format, list a few pieces of information about the server then go through all the endpoints provided. Once all these are displayed, the server waits for incoming requests.

So, what can you do with this? Looking at the buttons across the top, it's obvious the first two **start** and **stop** the server listening for requests on the **port**, which cannot change unless the server is stopped. The next two buttons open your default web browser with URLs that make requests to the running server engine. Clicking **Open Browser** requests the version:

Figure 14.10 – Web browser requesting the version of the local RAD Server

You can change the URL to make different requests. If you look through the original list of APIs listed when the server first started (see *Figure 14.8*), you'll find one of the registered resources is **Users**. The first method in that array is `GetUsers`; in your browser's address line, replace `version` with that method's path value, `users`, to get the list of users:

Figure 14.11 – Web browser requesting users from RAD Server

There's only one user currently, but you can start to see how this works. The next method, named `GetUser`, has the same base path but adds a parameter, `{id}`. If that parameter is included in the URL request, the second method is called. We can test this out by copying the ID of a user from the result of the first request and modifying the URL request details for a single user:

Figure 14.12 – Web browser requesting details for a specific user from RAD Server

The `GetUsers` method returns a JSON array, whereas the `GetUser` method returns a single JSON object.

The last button on the development server app, **Open Console**, does the same thing as if you had manually launched `EMSDevConsole.exe`—it starts the console part of RAD Server, and opens a second tab in your browser pointing to the RAD Server console's configured port. However, instead of a list of JSON logs, you get a nice web app with a login:

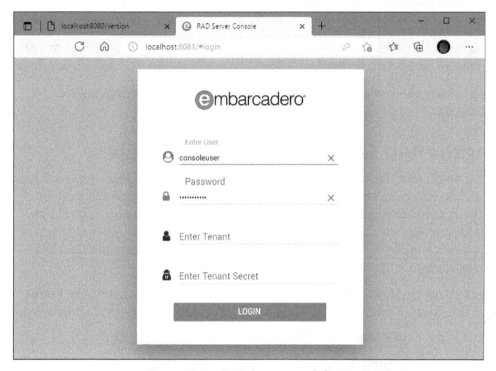

Figure 14.13 – RAD Server console login

The username and password are what was set up in the RAD Server configuration wizard (see *Figure 14.6*), with the defaults being `consoleuser` and `consolepass`, respectively. There is only one **tenant** when RAD Server is first installed, so that can be left blank.

> **Note**
>
> Additional tenants can be created to support completely separate applications with their own sets of users, groups, installations, and registered devices, all from one server and database. We will not cover multi-tenancy in this book.

Once you log in, you'll see a nice web view showing users, applications, installations, and much more. After adding a few users and making various requests, your view could look similar to what we showed in *Figure 14.3* earlier in the chapter.

At this point, you may be thinking two things:

- First, you have a greater appreciation of how much is built for you in the way of the database backend, API endpoints, and a web view of your user space.

- Second, you may be wondering how to actually add and modify the users and applications. Will you now have to write the user management application yourself?

Don't worry, RAD Server didn't leave you without a tool to manage the information you see in the web-based console. To quote a famous advertising line: "But wait, there's more!"

Read on to learn about one more part of this package provided for managing your infrastructure.

Using the RAD Server Management Console

Now that you've seen the sample data, you're probably eager to add your own. Look back in the Delphi folder, under either the `bin` or `bin64` folder, and launch `RSConsole.exe`. This is a FireMonkey application to manage the users, groups, and other aspects of your application suite that we've been talking about. The first thing you need to do is create a *connection profile*. You can use this tool to connect and manage multiple RAD Server installations.

Start by clicking the **New Connection** button (or select **New Profile...** from the **Profile** menu). This creates a connection profile and pulls up a dialog box to edit its properties:

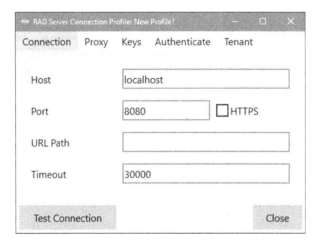

Figure 14.14 – RAD Server's Connection Profile editor

Enter the **host** and **port** configured when you started `EMSDevServer.exe`. You may specify a **proxy server** and **proxy port** on the **Proxy** tab—although this is probably not needed when working on and connecting to the development server running on your local computer. We will cover application keys in *Chapter 15, Deploying an Application Suite.*

The **Authenticate** tab lets you log in with a **username** and **password**, but when RAD Server is first installed, all API endpoints are all public (not restricted to certain users or groups); therefore, you don't have to log in to start managing groups and users. In fact, if you did not opt to create a sample user and group, you don't even have a user with which to log in.

That tab is also where you might enter a **master secret**, which, if defined, grants access to perform any action in RAD Server, regardless of other security in place. This is disabled by default and can only be set by modifying the configuration file. If used, remember that this acts as a "backdoor" admin-level password and should be kept highly secured.

Once a profile is ready (you can confirm this with the **Test Connection** button), close the profile editor. You can then expand the profile and browse users, groups, and other entities defined. You can rename the profile by selecting **Profile | Rename Profile...** from the menu. You can even switch to a dark theme by clicking the **Toggle Style** button. Here's a screenshot of the console using the dark theme showing two profiles, with the first one expanded and displaying some users added to the original test one:

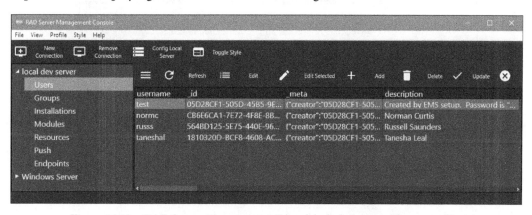

Figure 14.15 – RAD Server Management Console's dark theme with two profiles

The interface is pretty simple and it's quickly obvious how to add, update, and delete various entities. If you'd like to modify this application or peek at the code to see how it makes calls to perform its various functions, the source is included in your Delphi installation under its `source\data\ems\rsconsole` folder. This makes for a great learning tool for how to interact with the RAD Server API.

Speaking of learning how to interact with RAD Server, it's time to learn how to extend this platform with your own custom modules to provide interactions beyond just user management.

Writing modules to extend your server

Let's dive right into writing our first extension using a very simple example. All extensions we build will be Delphi packages that contain a **RAD Server resource**. A RAD Server resource is an extension of the functionality in RAD Server that groups API endpoints registered in RAD Server.

Using the wizard to create our first resource package

The steps to building a RAD Server resource are simple:

1. Select **File | New | Other...** from the Delphi menu.

2. Select **RAD Server Package** from the **RAD Server** section.

3. Select **Create package with resource** on the first page of the Package Wizard:

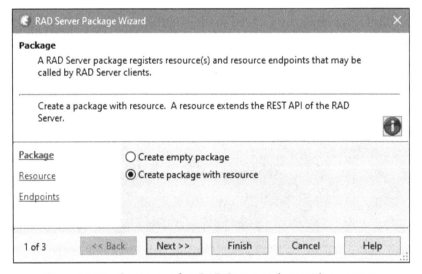

Figure 14.16 – Create your first RAD Server package with a resource

4. On the next page, set the **resource name** to something simple such as Test and select **Unit** for **File type**:

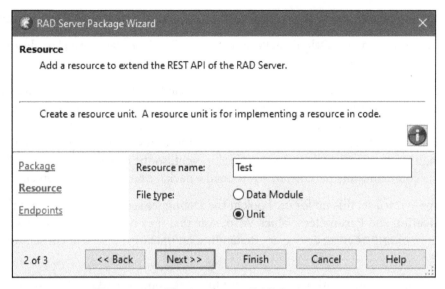

Figure 14.17 – Naming the new RAD Server resource

5.   From the list of **Sample EndPoints**, select only **Get** and **GetItem** (these differ only in the number of records returned; **Get** typically returns a list of records, whereas **GetItem** returns data for a single record), then click **Finish**:

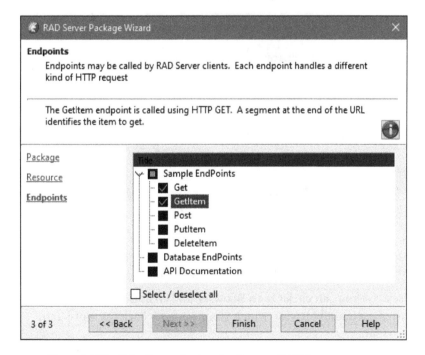

Figure 14.18 – Selecting endpoints for the new RAD Server resource

These endpoints correspond to the request methods provided by a typical REST server.

You should now have a new Delphi package with a small unit containing a class decorated with a couple of attributes.

Before adding any code, save the project and then look at the project's options. Since this Delphi project is a package and not an application, we can't run it directly. Instead, it must be launched by a host application—RAD Server, in this case. Normally, RAD Server packages must be installed into RAD Server and the server restarted before they can be used. Doing that would get tedious during development, so the developer edition of RAD Server has a command-line parameter, -l, to load a package temporarily for quick testing.

The Package Wizard set this up for us. Look at the **Debugger** section and check out the **Host application** and **Parameters** values. Also, note that they call either the 32-bit or 64-bit version of EMSDevServer.exe, depending on the target platform chosen for the project.

Cancel out of the **Project Options** window and hit *F9* to compile and run the sample RAD Server package. The **RAD Development Server** log window shows, and down at the bottom of the window you'll see the test package (Test.bpl) was successfully loaded and the *Test* resource registered with two endpoints, named Get and GetItem, with the test and test/{item} paths, respectively:

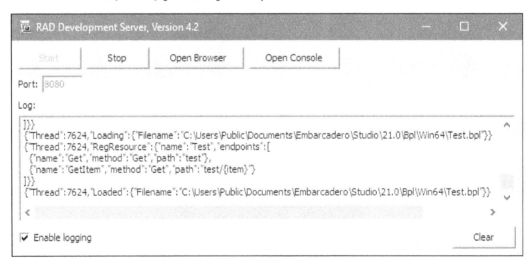

Figure 14.19 – First RAD Server package loaded

That was pretty easy! You can click **Open Browser** and change the URL from /version to /test to see the default behavior, which simply echoes back the word *Test*.

Now, let's add just a little bit of coding to see how simple it is to provide functionality from a RAD Server endpoint, and how to send JSON responses based on requests. Close the app and go back to the project in Delphi.

First, the Get procedure will define a list of elements. Our test one will simply return the letters of the alphabet:

```
procedure TTestResource.Get(const AContext: TEndpointContext;
const ARequest: TEndpointRequest; const AResponse:
TEndpointResponse);
var
  CharList: TJSONArray;
begin
  CharList := TJSONArray.Create;
  for var a := 1 to 26 do
    CharList.Add(string(Chr(64 + a)));

  AResponse.Body.SetValue(CharList, True);
end;
```

AResponse.Body must be a JSON object, in this case, an array of letters. Run it to see them:

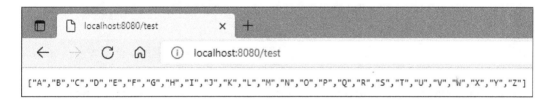

Figure 14.20 – RAD Server package returning an array of letters

If we add a / and something to the end of the URL, it calls the GetItem method, which is handled in our code by the method of the same name. For this test, it will just return the ASCII value of the letter submitted, so add this code:

```
procedure TTestResource.GetItem(const AContext:
TEndpointContext; const ARequest: TEndpointRequest; const
AResponse: TEndpointResponse);
var
  LItem: string;
  Ch: Char;
```

```
begin
  LItem := ARequest.Params.Values['item'];
  Ch := LItem[1];


  if CharInSet(Ch, ['A'..'Z']) then
    AResponse.Body.SetValue(TJSONString.Create(
        Format('ASCII(%s) = %d', [Ch, Ord(Ch)])), True)
  else
    AResponse.Body.SetValue(
        TJSONString.Create('Enter a letter [A..Z]'), True);
end;
```

Run this version and in the browser, when a valid uppercase letter is added to the URL, the output looks like this:

Figure 14.21 – RAD Server package returning the ASCII value of a letter

You can download this package from GitHub at https://github.com/ PacktPublishing/Fearless-Cross-Platform-Development-with- Delphi/tree/master/Chapter14/01_SimpleTest.

With an understanding of how to build and test RAD Server resource packages, we can build one to serve up park data from the database we've been working with in this book to see how this might work in a real-world scenario.

Implementing MyParks for RAD Server

To get started with setting up a MyParks resource, launch the RAD Server Package Wizard again and choose **Data Module** for **File type** in the wizard. Under **Sample EndPoints**, select all checkboxes; leave **Database EndPoints** and **API Documentation** unchecked.

> **Note**
>
> **Database EndPoints** uses a pre-defined **FireDAC** database configuration, which we will not use in this book. **API Documentation** decorates the RAD Server package classes with attributes that provide additional resource endpoints to generate industry-standard API documentation for your server's published endpoints. Follow the *RAD Server API Resource* and *OpenAPI Initiative* and *Swagger* links in the *Further reading* section at the end of this chapter for more information.

Once the package is created and saved, add the following components and code:

1. Add a `TFDConnection` component on the data module and connect it to your *MyParks* database. Add the appropriate driver link component, for example, `TFDPhysIBDriverLink` if you're using InterBase.

2. Add a `TFDQuery` component, hooked to the connection component, and add the following SQL to get the park ID and park name from all records:

   ```
   SELECT PARK_ID, PARK_NAME FROM Parks ORDER BY PARK_NAME
   ```

3. Replace the body of the `Get` procedure with the following to build a JSON array of the parks:

   ```
   var
     ParkList: TJSONArray;
   begin
     ParkList := TJSONArray.Create;
     qryParkList.Open;
     while not qryParkList.Eof do begin
       var ParkItem := TJSONPair.Create(
         TJSONNumber.Create(qryParkListPARK_ID.AsInteger),
         TJSONString.Create(qryParkListPARK_NAME.AsString));
       ParkList.Add(TJSONObject.Create(ParkItem));
       qryParkList.Next;
     end;
     AResponse.Body.SetValue(ParkList, True);
     qryParkList.Close;
   end;
   ```

Once the JSON array is built, the response is sent to the client. Viewing this in a browser might look like this (depending on the parks in your database):

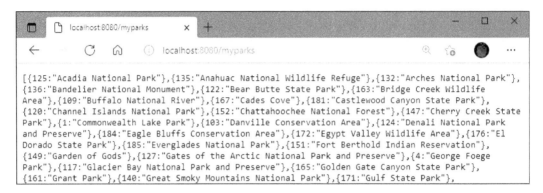

Figure 14.22 – MyParks RAD Server package returning a list of parks

Now, add another query component on the data module to get all the details for one park with this SQL:

```
SELECT * FROM Parks WHERE PARK_ID = :ID
```

Then, fill in the GetItem procedure to return the fields from one park:

```
var
  LItem: string;
  ParkID: Integer;
  ParkStream: TMemoryStream;
begin
  LItem := ARequest.Params.Values['item'];
  if TryStrToInt(LItem, ParkID) then begin
    qryParkByID.ParamByName('id').AsInteger := ParkID;
    qryParkByID.Open;
    if qryParkByID.RecordCount = 0 then
      AResponse.Body.SetValue(
        TJSONString.Create('Park not found for ID: ' + LItem),
  True)
    else begin
      ParkStream := TMemoryStream.Create;
      qryParkByID.SaveToStream(ParkStream, sfJSON);
      AResponse.Body.SetStream(ParkStream,
                CONTENTTYPE_APPLICATION_JSON, True);
```

```
      end;
   end else
      AResponse.Body.SetValue(TJSONString.Create('Invalid ''ID''
 parmeter: ' + LItem), True);
   end;
```

Before we look at the results, notice that in this method, I didn't manually create the JSON return value like I did with the `Get` method, but instead used FireDAC's `SaveToStream` method and specified `sfJSON` as the format. This builds the JSON result for us, which is convenient (especially if we decide to add more fields to the table), but it adds a lot more information for us to parse through:

Figure 14.23 – RAD Server MyParks package returning details for one park

Seeing this much information for one small record seems like overkill. There is, however, a structure to the JSON results that must adhere to the guiding principles of REST (see the link to *What is REST?* in the *Further reading* section at the end of this chapter). In this particular example, a lot of it is metadata that can be used by other components and helps define how it should be displayed. Download this short example project from GitHub at `https://github.com/PacktPublishing/Fearless-Cross-Platform-Development-with-Delphi/tree/master/Chapter14/02_MyParks`.

Now that you know how to build a server and concoct JSON data in a couple of different ways, let me show you a new component that does all the hard work for you.

Building a REST server without code

Let's review the four most common HTTP request methods used in a REST server and what their purpose is:

- GET: Queries data, returning either multiple records or just a single record, and supports parameters for filtering, sorting, and paging.

- POST: Sends data for a new record to the server. The data is typically in a JSON data packet embedded in the HTTP request.

- PUT: Sends data for updating a record to the server. The data is typically in a JSON data packet embedded in the HTTP request with the ID of the record to update as a parameter in the URL of the request.

- DELETE: Sends a request to the server to delete a record, with the record identifier as a parameter on the URL of the request.

The task of implementing these endpoints in a REST server is such a prevalent task that a new component has been added to the Delphi palette that encapsulates a lot of the work for you. This powerful component is TEMSDataSetResource.

Let's create a new package from scratch and show how all the work we just did previously can be done in a few steps by writing SQL with special syntax supported by FireDAC and setting some properties of this awesome component. Follow along with me:

1. Create a new **RAD Server package** with a resource.

2. Name it MyParksData and select **Data Module** for **File type**.

3. Don't select any checkboxes for **Sample EndPoints**—just click **Finish** in the wizard.

4. Drop a TFDConnection component and the driver link for your database and configure the connection parameters.

5. Add a TFDQuery to the data module, hooked to the connection component, and enter the following SQL into the **Query Editor**:

```
SELECT * FROM PARKS
{IF &SORT} ORDER BY &SORT {FI}
```

6. Click **Execute** to verify it returns data from your database.

7. Place a TEMSDataSetResource component on the data module and set its DataSet property to the query component you just added.

8. Expand the AllowedActions property and check all actions.

9. Set the KeyFields property to the primary key of your parks table, for example, PARK_ID.

10. Click the ellipses button for the ValueFields property and move all the fields to the **Included fields** list.

11. Add the following five ResourceSuffix attributes just above the declaration of TEMSDataSetResource:

```
type
  [ResourceName('MyParkData')]
  TMyParkDataResource1 = class(TDataModule)
    FDConnection: TFDConnection;
    FDPhysIBDriverLink: TFDPhysIBDriverLink;
    qryMyParksData: TFDQuery;
    [ResourceSuffix('list', '/')]
    [ResourceSuffix('get', '/{PARK_ID}')]
    [ResourceSuffix('post', '/{PARK_ID}')]
    [ResourceSuffix('put', '/{PARK_ID}')]
    [ResourceSuffix('delete', '/{PARK_ID}')]
    EMSDataSetResource: TEMSDataSetResource;
  published
  end;
```

These correspond to the four options under the AllowedActions property.

Now, simply select your target platform, compile and run, then click the **Open Browser** button and navigate to the myparks path to see your list of parks, then append one of the park IDs to see it list the details of just one of the parks.

In just 11 steps and a few minutes, you've created a complete RAD Server resource that implements the four major operations of a typical REST server! Download this simple RAD Server package from GitHub at https://github.com/PacktPublishing/ Fearless-Cross-Platform-Development-with-Delphi/tree/master/ Chapter14/03_MyParkData. We've only tested one type of endpoint so far, GET, but we'll be looking at the others soon.

Before we move on, there are a few other common practices most REST servers support with GET requests. If your dataset returns a large number of records, the response time for your client may exceed a timeout period or test the patience of an end user waiting for the hourglass to stop spinning. The concept of **paging** is frequently implemented, which limits the number of records returned and separates them at the server level into pages with a parameter on the URL to specify which page of data to access.

If the Options parameter of the TEMSDataSetResource component includes rsEnablePaging, two properties come into play:

- PageParamName: This names the parameter that can be included on the URL to request a specific page of data.

- PageSize: This lists the number of records to be returned on each page.

To see this in action with my small dataset, I set PageSize to 3 in the server then ran it and requested page 1 from the browser:

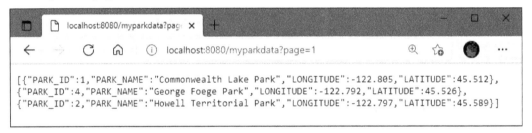

Figure 14.24 – Paged park list from RAD Server

Another common feature of REST servers is sorting. When the Options parameter of the TEMSDataSetResource component includes rsEnableSorting, the URL can include the value of the SortingParamPrefix property prepended to one of the field names listed in the ValueFields property to sort either ascending (using an A) or descending (using a D) with the named field. For example, here are my parks listed by PARK_NAME in reverse alphabetical order:

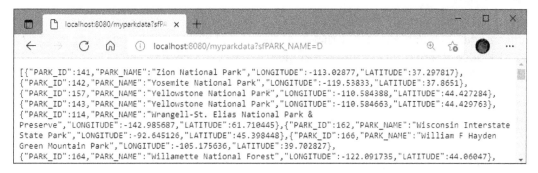

[{"PARK_ID":141,"PARK_NAME":"Zion National Park","LONGITUDE":-113.02877,"LATITUDE":37.297817},
{"PARK_ID":142,"PARK_NAME":"Yosemite National Park","LONGITUDE":-119.53833,"LATITUDE":37.8651},
{"PARK_ID":157,"PARK_NAME":"Yellowstone National Park","LONGITUDE":-110.584388,"LATITUDE":44.427284},
{"PARK_ID":143,"PARK_NAME":"Yellowstone National Park","LONGITUDE":-110.584663,"LATITUDE":44.429763},
{"PARK_ID":114,"PARK_NAME":"Wrangell-St. Elias National Park &
Preserve","LONGITUDE":-142.985687,"LATITUDE":61.710445},{"PARK_ID":162,"PARK_NAME":"Wisconsin Interstate
State Park","LONGITUDE":-92.645126,"LATITUDE":45.398448},{"PARK_ID":166,"PARK_NAME":"William F Hayden
Green Mountain Park","LONGITUDE":-105.175636,"LATITUDE":39.702827},
{"PARK_ID":164,"PARK_NAME":"Willamette National Forest","LONGITUDE":-122.091735,"LATITUDE":44.06047},

Figure 14.25 – Sorted park list from RAD Server

This is another example of the time-saving features RAD Server offers, and how well FireDAC works with it.

The next methods we want to implement start going beyond just visualizing data and start sending updates back to the server. We can't do this with just the browser alone—we need a client tool that allows us to control the request method and can send more data than just parameters on the URL.

Testing RAD Server with the REST Debugger

Delphi comes with a tool for testing and troubleshooting REST servers, aptly named the **REST Debugger**. I use this client tool so often when working with web services that I added it to Delphi's **Tools** menu (using **Tools | Configure Tools...**). You can find RESTDebugger.exe in Delphi's bin folder.

> **Note**
>
> There are other tools for building and testing REST server APIs. A couple of the popular ones are **Postman** (https://www.postman.com) and **Insomnia** (https://insomnia.rest). We will use the REST Debugger for our examples in this chapter.

Start RAD Server with our MyParks module and launch the REST Debugger, then enter the same address into the **URL** line that was used in the browser to list multiple parks, and click **Send Request**. The results are in the **Response** section at the bottom of the application's window in the **Body** tab:

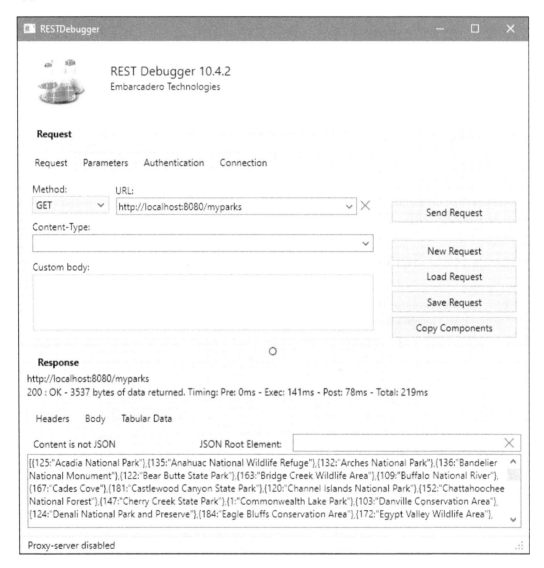

Figure 14.26 – Using the REST Debugger to get the list of parks

Notice that, directly under the URL in the **Response** section, it says **200 : OK**. Web requests return **status codes** you can check to see whether your request was accepted (as in this case), if the page was not there (which would return **404 : Not Found**), and so on.

You can modify the **URL** to append a park ID and add the sorting and paging parameters—these requests will return the same information in the **Body** tab at the bottom of the window we saw in the browser.

Instead of listing the parameters with question marks and ampersands (as you must when editing the URL in your browser), you can switch to the **Parameters** tab and enter them individually. You can also put the myparkdata resource name into the **Resource** box, leaving only the base URL (http://localhost:8080) listed in the URL on the **Request** tab. Here's an example of listing the first page of sorted parks:

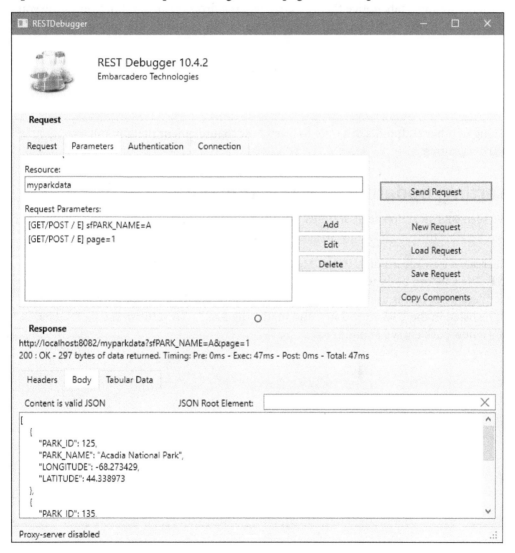

Figure 14.27 – Using REST Debugger with parameters

Sometimes, it's handy to enter the resources and parameters separately if you have a complicated query. The REST Debugger shows the full URL that was sent as the first line in the **Response** section.

The **Authentication** tab is used for communicating with various types of secure servers and the **Connection** tab is used to configure proxy servers.

Explore other features of the REST Debugger, such as switching the bottom view to **Tabular Data** or inspecting the **headers** of the response.

Back on the **Request** tab, notice that the **URL** box is a drop-down combo. Every time you send a request, it saves the URL and parameters and allows you to select them again. This saves a lot of time re-entering queries for testing.

Finally, notice that you can switch the **method** between GET, POST, PUT, or DELETE (we won't use PATCH) to test the other endpoints of our RAD Server module as we build them.

Speaking of other endpoints, it's time to learn how to implement them in our MyParks resource.

Inserting, updating, and deleting data

We've seen GET in action several times, which, for a REST server, only reads data from the server. It's now time to show how to add, change, or delete data in a REST server.

We'll create a new record first. In the **Method** dropdown of the REST Debugger, select **POST**. The **Custom body** textbox is enabled for you to type in. This is where we enter the embedded JSON that is used to insert a record into the database. In addition, the new ID must be listed in the **URL** dropdown and match the PARK_ID value in the JSON data. Here's a new park entry I submitted:

Figure 14.28 – Adding a park with the REST Debugger

Once you send that request, use the **URL** dropdown to select a previous query that lists all the parks (or that one specifically using the new PARK_ID value) to verify it was successfully added.

Next up is **PUT**—to modify an existing record. Let's say I added a park named *Tanner Springs Park*, with a PARK_ID value of 192, but didn't have the coordinates. When I run a query, it comes back with an entry: {"PARK_ID":192,"PARK_NAME":"Tanner Springs Park","LATITUDE":0,"LONGITUDE":0}. I want to update that entry with the correct coordinates. This is how I would do that with the REST Debugger:

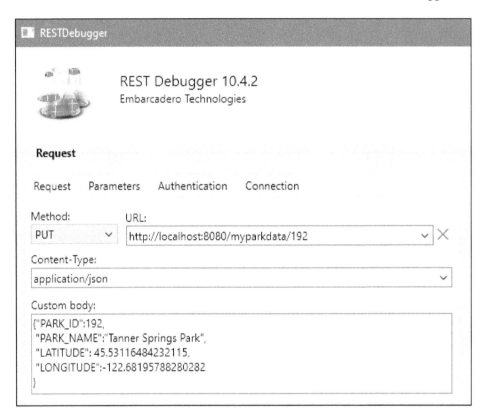

Figure 14.29 – Updating a record with REST Debugger

The URL and body are the same as a POST request for a new record, but RAD Server detects the method type and uses a different procedure to handle the request.

It's pretty simple to demonstrate **DELETE**. If you look back at *Figure 14.30*, you'll notice I have a duplicate. In the REST Debugger, simply switch the **Method** dropdown to **DELETE** and put the PARK_ID value on the **URL** dropdown:

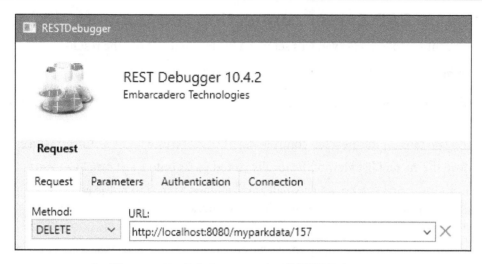

Figure 14.30 – Deleting a record with REST Debugger

Refresh the park list to see the duplicate is gone.

We didn't go through these so much to learn about the REST Debugger; the main goal was to test each of the endpoints of our MyParks RAD Server resource. By confirming each operation in the database after submitting the request with a working REST debugging tool, we are now confident our RAD Server module is working well enough to switch gears to build a client. Personally, I like to test each part of a system in isolation, rather than try to build both the client and server simultaneously and wonder where the problem lies when things don't work as expected.

So, now it's time to use our new server with a real application—our mobile MyParks app.

Modifying MyParks for use with RAD Server

I've copied the MyParks project, including the main form and the data module, from *Chapter 10, Cameras, the GPS, and More,* and will make the modifications starting with that finished, working version. (The modifications we made to this app in *Chapter 12, Console-Based Server Apps and Services,* and *Chapter 13, Web Modules for IIS and Apache,* were for specific server purposes, and we'd have to remove several parts to rework it for use with RAD Server, so we might as well start without them.)

Setting up RAD Server connection components

Before we connect to RAD Server, let's set up a new list view to hold the results.

Open the main form of the MyParks app and add the following:

1. Add a `TTabControl` behind `TListView` that holds the main list of parks on the `tabParkList` tab.

2. Add two tabs on this new tab control—`tabLocalParks` and `tabRemoteParks`.

3. Move the current list view of the locally stored parks onto `tabLocalParks`.

4. On `tabRemoteParks`, add a small panel aligned to the top of the tab with a refresh button and two checkboxes. The two checkboxes will be for sorting `A-Z` and `Z-A`.

5. Below the panel and filling up the remaining client area of `tabRemoteParks`, add a new `TListView` aligned to the client. This list view will contain the results of a call to RAD Server for its parks.

When you're done, the **Structure** pane for your main tab control should look something like this:

Figure 14.31 – Structure pane showing a new tab in the MyParks app for parks from the server

There are a few different ways to call a REST server and get results. The REST Debugger actually has a nifty button I like to use called **Copy Components**, which will set up the connection components, which you can simply paste to your application. Since we've been testing with the REST Debugger, select one of the GET options you've used to retrieve the list of parks from your RAD Server, click that button, and a message pops up telling you about the three components it's prepared for you.

Just paste them onto the form of your app and spread them out so you can see the different components and select them individually. There are only two changes we need to make.

First, when the mobile app is running, it won't have the same address as the REST Debugger which was using LOCALHOST because RAD Server was on the same Windows machine. This means we need to change the BaseURL property of the TRESTClient component to point to the actual IP Address of the machine instead of LOCALHOST. On my machine, running ipconfig from the command line revealed my machine's IPv4 address is 192.168.1.95. and so entered http://192.168.1.95:8080 for the BaseURL. Make the similar discovery and adjustment in your project.

Second, on the TRESTClient component, uncheck the AutoCreateParams checkbox, as we'll be creating those in code based on which way the user wants the list sorted.

> **Note**
> This is one advantage the REST Debugger has over other REST API tools—the REST Debugger knows about your Delphi code!

Create a TAction in your TActionList component to get the parks from RAD Server, and write the following code to call RAD Server:

```
procedure TfrmMyParksMain.
actGetRADServerParksRESTClientExecute(Sender: TObject);
begin
  RESTReqRADParks.Params.Clear;
  if radParksAZ.IsChecked then
    RESTReqRADParks.Params.AddItem('sfPARK_NAME', 'A',
      TRESTRequestParameterKind.pkQUERY)
  else
    RESTReqRADParks.Params.AddItem('sfPARK_NAME', 'D',
TRESTRequestParameterKind.pkQUERY);
  RESTReqRADParks.Execute;
end;
```

This action sets the parameter for the sorting option selected, then calls the `Execute` method of the `TRESTRequest` component, and fills the `TRESTResponse` component with the result JSON data. To see the data, we can either write some code to populate the new list view or add a couple more components to parse the JSON for us. I like the second idea:

1. Add a `TFDMemTable` component to the form and then right-click and select **Fields Editor...** to create the same fields that are coming back in the JSON data (check the REST Debugger for their exact names). If you've been following the example here, those field names are PARK_ID, PARK_NAME, LONGITUDE, and LATITUDE (case insensitive). Set the field types appropriately, then check the `Active` property.

2. Add a `TRESTResponseDataSetAdapter` component, set the `Response` property to point to your `TRestResponse` component, and set the `Dataset` property to the new memory table you just added.

3. Hook up the memory table's PARK_NAME field to the list view's `Item. Text` property with **LiveBindings**. After the `TBindSourceDB` component is automatically created and hooked up, also connect its asterisk to the `Synch` property of the list view:

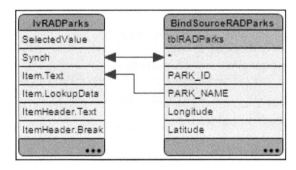

Figure 14.32 – The list view synchronized with the memory table in LiveBindings

4. Right-click the `TRESTRequest` component and select **Execute...**, and if your RAD Server's park module is active, you should see the list view fill with the server's parks.

Let's make a couple of adjustments to the list view:

*   In the **Object Inspector**, expand the `ItemAppearance` property to see a sub-property also called `ItemAppearance`, and set it to `ListItem`.

*   Expand `ItemAppearanceObjects`, `ItemObjects`, and `Accessory` to uncheck the `Visible` property.

Before you run it on your mobile device, you may have to make one adjustment. During development, your RAD Server is likely configured without SSL. In other words, connections are using HTTP rather than HTTPS. Both Android 8 and newer and iOS 9 and newer expect encrypted connections by default, but allow exceptions to be specified. Delphi adds the exception for iOS applications in a .info.plist file, which isn't generated until you run the app. For Android, you can optionally add an exception in a template manifest file in your Delphi project folder, namely AndroidManifest.template.xml. Delphi's build tools generate the real manifest file that accompanies your deployed Android app every time it is built. Open up this XML template file, locate the application section (where you'll see several lines starting with android:), then add the following line: android:usesCleartextTraffic="true". Follow the links in the *Further reading* section at the end of this chapter for more information.

You can now run this on a mobile device that has access to your local server on your network, click the refresh button, and see the list of parks. Also, try reversing the sort order:

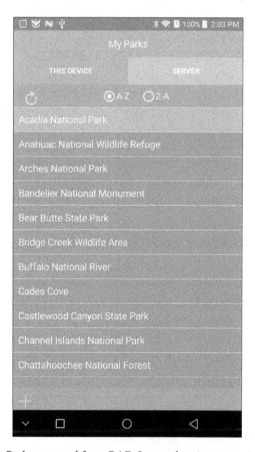

Figure 14.33 – Parks returned from RAD Server showing on an Android phone

You're now able to see both locally stored parks and parks stored on the server. The next logical thing is to add those items from the server to your own device. Let's build a checklist and do this for several at once. The list view provides a convenient way to do this:

1.　In the panel above the list view, add a couple of buttons next to the refresh button—one to toggle the park selection, and the other to download the selected parks.

2.　Create a new action, `actToggleRADParksEditMode`, to toggle the `EditMode` property of the list view. Assign this action to the first button created in the previous step.

3.　Create another new action, `actDownloadRADPark`, which will perform the actual insertion of new local parks, and assign it to the second button. Set this new action's `Visible` property to `False` because we only want to see this button when we're selecting items.

4.　Write the code for `actToggleRADParksEditMode` to set `EditMode`, and toggle visibility of the other action, which shows the second button:

```
procedure TfrmMyParksMain.
actToggleRADParksEditModeExecute(Sender: TObject);
begin
  lvRADParks.EditMode := not lvRADParks.EditMode;
  actDownloadRADPark.Visible := lvRADParks.EditMode;
end;
```

5.　When the list view's `EditMode` is enabled, checkboxes appear before each item in the list view. After the user has checked all the parks from RAD Server to import, they would click the download button to confirm and add the parks:

```
procedure TfrmMyParksMain.
actDownloadRADParkExecute(Sender: TObject);
begin
  TDialogServiceAsync.MessageDialog(
        'Would you like to save the selected parks to
          your device?',
      TMsgDlgType.mtConfirmation,
      [TMsgDlgBtn.mbYes, TMsgDlgBtn.mbNo],
      TMsgDlgBtn.mbYes, 0,
    procedure (const AResult: TModalResult)
    begin
      if AResult = mrYes then
```

```
        AddRADParksToLocal;
    end);
end;
```

6. Add the procedure, AddRADParkToLocal, which goes through all the checked
    items and inserts a record for each (this book's example will only save the name, but
    a real-world application would save as much information as is available):

```
procedure TfrmMyParksMain.AddRADParksToLocal;
begin
  for var i in lvRADParks.Items.CheckedIndexes(True) do
begin
    dmParkData.tblParks.Insert;
    dmParkData.tblParksParkName.AsString := lvRADParks.
      Items[i].Text;
    dmParkData.tblParks.Post;
  end;
  actToggleRADParksEditMode.Execute;
end;
```

Here's how selecting parks for download looks on an iPhone:

Figure 14.34 – Selecting parks from RAD Server for local saving on an iPhone

The last thing we'll show is updating a remote park with local changes.

Sending updates back to RAD Server

There are so many cool things that we could do to synchronize data between your device and a remote server, but we don't have time or space in this book to cover them all. The ideal way to update a remote park from a local list is to keep track of the remote server's ID as a separate field in the local database so return updates can simply use that field to reference the remote record.

Since our **SQLite** database doesn't have a field for this (and we're not going to take the time to go and modify it), we're going to approach this very simply by allowing only a change to the park name from the list showing in the **Server** tab of the mobile app. Even this, though, will require all four fields to be returned to RAD Server in its update method call. Instead of passing them in individually, we may add to the fields in the future, which would necessitate a change to the parameters. To simplify the code a little and to keep the parameter passing simple, create a record for these update fields in the form's private declaration section:

```
type
  TParkUpdateRec = record
    ParkID: Integer;
    ParkName: string;
    Longitude: Double;
    Latitude: Double;
  end;
```

For the user interface, instead of using the standard `OnItemClick`, we'll make sure the user really wants to do this by forcing a long-click to illicit the name change. To do this, first enable the `LongTap` option of the list view's `Touch.InteractiveGestures` property. When the `OnGesture` event fires, it will keep firing until the touch gesture is done, which causes the event handler to be executed multiple times. The trick to prevent this is to set a Boolean variable to `False`, and when we're in the name-change code, set it to `True`. After creating this variable, `FInNameChange`, in the form's private declaration section, write the event handler code:

```
procedure TfrmMyParksMain.lvRADParksGesture(Sender: TObject;
const EventInfo: TGestureEventInfo; var Handled: Boolean);
begin
  if FInNameChange then
    Handled := True
```

```
  else if EventInfo.GestureID = System.UITypes.igiLongTap then
begin
    Handled := True;
    FInNameChange := True;
    TDialogServiceAsync.InputQuery('Change a Park',
        ['Change the name of the park:'],
        [tblRADParksPARK_NAME.AsString],
        procedure (const AResult: TModalResult; const AValues:
          array of string)
        begin
          if AResult = mrOK then begin
            var ParkUpdateRec: TParkUpdateRec;
            ParkUpdateRec.ParkID := tblRADParksPARK_
              ID.AsInteger;
            ParkUpdateRec.ParkName := AValues[0];
            ParkUpdateRec.Longitude := tblRADParksLongitude.
              AsFloat;
            ParkUpdateRec.Latitude  := tblRADParksLatitude.
              AsFloat;
            UpdateRADParkName(ParkUpdateRec);
            FInNameChange := False;
          end;
        end);
  end;
end;
```

The prompt looks like this on an iPhone:

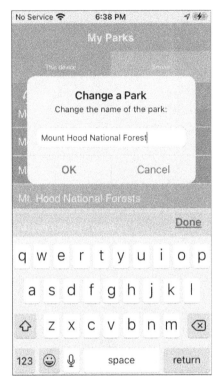

Figure 14.35 – Changing a park name to send to RAD Server from an iPhone

The real work of updating the RAD Server with the new name is done in the new
UpdateRADParkName procedure. As with getting the parks from RAD Server,
sending an update to RAD Server will also involve REST components. We can use
the existing TRESTClient unchanged but should copy the TRESTRequest and
TRESTResponse components—name the new ones RESTReqParkNameUpdate and
RESTRespParkNameUpdate. Set the Method property of the first one to rmPUT
and set its Response property to RESTRespParkNameUpdate, then write the
following code:

```
procedure TfrmMyParksMain.UpdateRADParkName(AParkRec:
TParkUpdateRec);
var
  UpdateJSON: TJSONObject;
begin
  UpdateJSON := TJSONObject.Create;
  try
```

```
    UpdateJSON.AddPair('PARK_ID', TJSONNumber.Create(AParkRec.
      ParkID));
    UpdateJSON.AddPair('PARK_NAME', AParkRec.ParkName);
    UpdateJSON.AddPair('LONGITUDE', TJSONNumber.
      Create(AParkRec.Longitude));
    UpdateJSON.AddPair('LATITUDE', TJSONNumber.Create(AParkRec.
      Latitude));

    RESTReqParkNameUpdate.Body.ClearBody;
    RESTReqParkNameUpdate.Body.Add(UpdateJSON);
    RESTReqParkNameUpdate.ResourceSuffix := AParkRec.ParkID.
      ToString;
    RESTReqParkNameUpdate.Execute;

    TDialogServiceAsync.ShowMessage('Park update result: ' +
      RESTRespParkNameUpdate.StatusText);
  finally
    UpdateJSON.Free;
  end;
end;
```

After creating a JSON object, this procedure fills it with the values from the TParkUpdateRec record, sets ResourceSuffix with the ParkID value to notify RAD Server which record it is updating, and calls the Execute method of the REST request component. Receiving a simple **OK** in response signifies success—refresh the server list to confirm it works.

It's a simple extension from this to send a park from your local device to the server as a new park by using a TRESTRequest component with Method set to rmPOST. By adding that, you're well on your way to building two-way data synchronization. If you go down that route, remember to keep in mind that the remote server will have different PARK_ID values than your local list unless you carefully manage them. I would suggest adding a field in the local database to store the remote PARK_ID value so that you can send updates whenever you need them without first getting the remote park. If you get really fancy, you could implement background updates to keep them constantly in sync. Of course, you also have to keep in mind other people sending updates, user authorization to do so, and data collision. Download this updated version of the MyParks app (called MyParksMobile) from GitHub at https://github.com/PacktPublishing/Fearless-Cross-Platform-Development-with-Delphi/tree/master/Chapter14/04_MyParksMobile.

Summary

This chapter has introduced you to the RAD Server technology stack by going through the benefits of using a pre-built infrastructure to manage users, security, and devices with the source code to a full-featured console application. This can jump-start your business application, saving you far more in development costs than the price of deployment licenses. We examined the various developer tools that come with Delphi and RAD Server, what they're used for, and showed how to use them as you build and test. Then we modified a mobile application to communicate with published endpoints and successfully shared data between the two systems.

The hands-on examples we walked through gave you a thorough working knowledge of the power provided by RAD Server, and how this can help you with your own product idea. As you refine the features and polish your user interfaces, you'll be faced with a final set of challenges: how to make sure your application is ready for widespread use.

The last chapter in this book will share ideas for testing your application, highlight areas to address for putting the final touches in your code, show how to deploy your RAD Server modules to a production server, and provide some tips for getting your mobile apps deployed successfully to a variety of devices.

Questions

1. What operating systems and web servers does RAD Server run under?
2. What database does RAD Server use?
3. What formats are used for communicating with RAD Server?
4. What is multi-tenancy?
5. What tool comes with RAD Server to manage users and groups?
6. What is the meaning of an HTTP 404 status code?
7. What is an API endpoint?
8. What are the four most common HTTP request methods?
9. What does the **Copy Components** button in the REST Debugger do?
10. How do you get around Android's restriction for accessing HTTP servers with plaintext?

Further reading

- Richardson Maturity Model: `https://martinfowler.com/articles/richardsonMaturityModel.html`

- DataSnap Overview and Architecture: `http://docwiki.embarcadero.com/RADStudio/Sydney/en/DataSnap_Overview_and_Architecture`

- RAD Server Overview: `http://docwiki.embarcadero.com/RADStudio/Sydney/en/RAD_Server_Overview`

- RAD Server Deep Dive Webinar: `https://youtube.com/playlist?list=PLwUPJvR9mZHgccq4EfTcsCngRqpTmm_wn`

- RAD Studio Enterprise Webinar Part 4 - RAD Server Development Lifecycle: `https://youtu.be/_pfxfiLsXJI`

- Learn How to Create a RAD Server with "David I" Intersimone in Delphi And C++: `https://blogs.embarcadero.com/learn-how-to-create-a-rad-server-with-david-i-intersimone-in-delphi-and-c/`

- Setting Up Your RAD Server Engine: `http://docwiki.embarcadero.com/RADStudio/Sydney/en/Setting_Up_Your_RAD_Server_Engine`

- RAD Server Management Console Application: `http://docwiki.embarcadero.com/RADStudio/Sydney/en/RAD_Server_Management_Console_Application`

- What is REST?: `https://restfulapi.net`

- API Endpoints – What Are They? Why Do They Matter?: `https://smartbear.com/learn/performance-monitoring/api-endpoints/`

- Best practices for REST API design: `https://stackoverflow.blog/2020/03/02/best-practices-for-rest-api-design/`

- RAD Server API Resource: `http://docwiki.embarcadero.com/RADStudio/Sydney/en/RAD_Server_API_Resource`

- OpenAPI Inititiative and Swagger: `https://swagger.io/docs/specification/about/`

- HTTP Status Codes: `https://www.restapitutorial.com/httpstatuscodes.html`

- Fix Android's "Cleartext HTTP Traffic not permitted": `https://www.tldevtech.com/fix-androids-cleartext-http-traffic-not-permitted/`

- App Transport Security Has Blocked My Request: `https://cocoacasts.com/app-transport-security-has-blocked-my-request`

- RAD Server Multi-Tenancy Support: `http://docwiki.embarcadero.com/RADStudio/Sydney/en/RAD_Server_Multi-Tenancy_Support`

15
Deploying an Application Suite

It is one thing to write applications that you use yourself in your safe development environment. It is quite another to send your carefully crafted programs out into the wild where people will install it on hardware you didn't plan for, use it in ways you weren't expecting, and encounter problems for which you did not test. This final chapter of the book will attempt to prepare you for several of these issues, to think from the perspective of the user without any development tools, and to take some of the guesswork out of the final processes of developing and deploying cross-platform apps.

We will look at the following subject areas as we complete our study of cross-platform development:

- Configuring for a wide audience
- Securing data
- Adding the graphical touch
- Establishing product identity
- Testing for deployment
- Distributing the final product

Let's get our server and mobile app ready to meet the real world!

Technical requirements

This chapter will utilize most of the tools and platforms discussed earlier in the book—except Linux. We'll work mostly on Windows for developing with Delphi 10.4, modify an ISAPI web module for IIS, add security to RAD Server, and update our mobile *MyParks* app to use the updated RAD Server services.

The code for the project we'll build in this chapter can be found on GitHub at `https://github.com/PacktPublishing/Fearless-Cross-Platform-Development-with-Delphi/tree/master/Chapter15`.

Configuring for a wide audience

As you've been working with your own local databases, you know right where everything is and it's been quick and easy to just set the database connection string directly. But your users will install your software in different folders or on older operating systems that have different default paths. Hardcoding a connection string with a path and port is always a setup for problems.

We have not addressed this throughout the book as we've been focusing on other concepts but as we look toward distributing our servers and apps to other systems, we have to take this into consideration and allow flexibility in where files and databases will be and which ports may or may not be available.

Enter configuration settings. There are many approaches to how and where to store these. Some apps create a small database file in a publicly accessible area of the local device. Others store settings in the Windows registry. Since the Windows registry is not cross-platform, we'll avoid that technique in this book.

Where you store application settings is somewhat dependent on the application type and how much data you need to store. For our purposes, all we need is a connection string that is set up once during installation and everything else is in the database. Here are my recommendations for various platforms, starting with desktop applications.

Getting settings in desktop applications

One of the simplest and most common techniques for loading configurable settings in desktop applications is to use a `.ini` file (a text file with sections of NAME=VALUE pairs) in the same path as the executable with this simple statement:

```
ConfigFileName := ChangeFileExt(Application.ExeName, '.ini');
```

Typical application installers run as an administrator so they can create folders and place applications, libraries, and other necessary files under C:\Program Files or C:\Program Files (x86), which is normally read-only for general users. The installer could prompt for the database location and write the configuration file in the application folder. This makes it simple to view the settings and if they need to be modified, an administrator-level user would know how to change them.

If it makes more sense to allow changes to the .ini file after installation, then a good place is under the C:\ProgramData folder with a new folder created for your application. Here's the code to do that using the application name as the sub-folder path:

```
var AppName := ExtractFileName(Application.ExeName);
var AppDataPath := TPath.Combine(TPath.GetPublicPath,
  AppName);
ForceDirectories(AppDataPath);
var ConfigFilename := TPath.Combine(AppDataPath, AppName+'.
  ini'));
```

Using this technique, an application called MyApp.exe would use settings stored in C:\ProgramData\MyApp\MyApp.ini on Windows.

If your app will need different settings depending on the user running the app, you'd want to use the user's Documents folder instead by calling GetDocumentsPath instead of GetPublicPath.

To see what is returned by these and other functions of the TPath class, you can run a simple FireMonkey application on various platforms that lists them upon startup in a TListView.

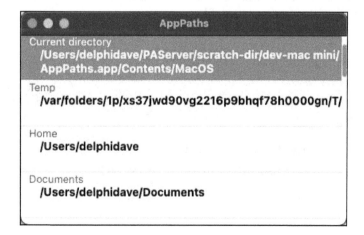

Figure 15.1 – Various application paths on a Mac

Get this app on GitHub at `https://github.com/PacktPublishing/`
`Fearless-Cross-Platform-Development-with-Delphi/tree/master/`
`Chapter15/01_AppPaths`.

A slightly more challenging scenario is dynamically loaded modules.

Updating a web module with dynamic settings

Loading a configuration file from the same path as the application isn't reliable from
within a library because a `.dll` (or `.dylib` on Mac or `.so` on Linux) file is not
necessarily in the same folder as the application that loaded it. For example, a web
module's `Application` path is the web server, not the path of the library itself.
Fortunately, there's a simple solution.

In *Chapter 13*, *Web Modules for IIS and Apache*, we used a data module to access a local
database of park data to serve up in a web browser. The database connection string was
simply hardcoded into the `TFDConnection` component's properties, which makes it easy
to test in a development environment but completely unusable in nearly any other situation.

To make the data module flexible, it should read the database connection properties
from an external source but the data module won't know what type of project it's linked
with—it needs to work from both applications and libraries. The simple solution is to
provide a property for the configuration filename and let whatever loads the data module
tell it where that file is.

Let's make these changes:

1. Copy the MyParks ISAPI library project from `Chapter 13` to a new folder.

2. Open up the data module unit, `udmParksDB.pas`, and add a public property to
    hold the name of the configuration file:

    ```
    private
      FConfigFileName: string;
    public
      property ConfigFileName: string read FConfigFileName
        write FConfigFileName;
    ```

3. Add a `BeforeConnect` event handler to the `TFDConnection` component that
    reads the database connection settings from the config file:

    ```
    procedure TdmParksDB.FDParkCnBeforeConnect(Sender:
    TObject);
    ```

```
const
  DBSection = 'Database';
var
  Cfg: TIniFile;
begin
  Cfg := TIniFile.Create(FConfigFileName);
  try
    FDParkCn.Params.AddPair('Server', Cfg.
      ReadString(DBSection, 'Server', EmptyStr));
    FDParkCn.Params.AddPair('Port', Cfg.
      ReadInteger(DBSection, 'Port', 3050).ToString);
    FDParkCn.Params.AddPair('User_Name', Cfg.
      ReadString(DBSection, 'Username', 'SYSDBA'));
    FDParkCn.Params.AddPair('Password', Cfg.
      ReadString(DBSection, 'Password', 'masterkey'));
    FDParkCn.Params.AddPair('Database', Cfg.
      ReadString(DBSection, 'DBFile', EmptyStr));
  finally
    Cfg.Free;
  end;
end;
```

This assumes the calling application, or web framework in this case, will pass in the name of the configuration file before it is needed—and if not, the data module should raise an error that is properly caught and gives a descriptive error message to help identify the source of the problem.

The calling module in this project is the data module inheriting from TWebModule, or uwmMyParks. We should set the name of the config file at the earliest possible opportunity and since the data module is created in the initialization section of its unit, we can set that property as soon as the web module is created. The OnCreate event handler for this web module already sets up the logging, so just add this one line:

```
  dmParksDB.ConfigFileName :=
ChangeFileExt(WebApplicationFileName, '.ini');
```

WebApplicationFileName is, conveniently, the name of the web module .DLL file. So, with this code in place, build the DLL file, replace the one in the IIS folder that we set up in *Chapter 13, Web Modules for IIS and Apache*, and restart your IIS web server. Now that the database connection is created at runtime with parameters loaded from the configuration file instead of hardcoded at design time, the installation is more flexible but requires the existence of the configuration file with the right settings to point to your database. Without it, you'll get an error on your web page when it tries to give you the list of parks. You'll also get an error if it's not configured correctly, such as in this example where I purposefully gave it the wrong folder name for the file:

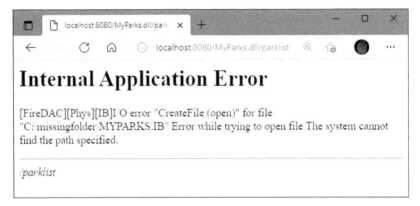

Figure 15.2 – Error message in the browser when the WebBroker server cannot find the database

Once you correct the error and restart the web server, you'll see the nice list of parks as we did previously. Here's an example of a good configuration file where the database is on the same machine as the web server and listening on the default InterBase port:

```
[Database]
Server=127.0.0.1
Port=3050
Username=SYSDBA
Password=masterkey
DBFile=C:\Users\Public\Documents\db\MYPARKS.IB
```

You can download the updated MyParks ISAPI web module and the sample configuration file from GitHub at https://github.com/PacktPublishing/Fearless-Cross-Platform-Development-with-Delphi/tree/master/Chapter15/02_MyParksISAPI.

While we're talking about loading settings from a library, we also need to update the RAD Server package.

Updating a RAD Server package with dynamic settings

In *Chapter 14, Using the RAD Server*, we wrote a small RAD Server package to access the same park data. Similar to the web module, the RAD Server package's database connection needs to be flexible. The changes are similar but we'll use a different technique to get the package filename.

Copy the MyParks RAD Server package project from Chapter 14 to a new folder. In the data module, add an `OnBeforeConnect` event for `TFDConnection`. Instead of being given the library name, we'll derive it ourselves. We can do that by using the `HInstance` global variable, which stores a handle to the current application, library, or package. We could've used this with the ISAPI DLL also but it uses low-level Windows API calls so isn't as friendly to work with; still, it gets us the information we need. We'll write these as separate functions for clarity:

```
function GetPkgName: string;
var
  PkgName: Cardinal;
  TempName: string;
begin
  PkgName := MAX_PATH;
  SetLength(TempName, PkgName);
  PkgName := GetModuleFileName(HInstance, PChar(TempName),
    PkgName);
  SetLength(TempName, PkgName);
  Result := TempName;
end;
```

We call the `GetModuleFilename` function to look up the RAD Server package file name, which returns a character array and its length; this, in turn, is used to derive the configuration filename:

```
function GetConfigFilename: string;
begin
  Result := ChangeFileExt(GetPkgName, '.ini');
end;
```

Finally, the `OnBeforeConnect` event handler is this simple one line:

```
procedure TdmMyParksList.FDParkCnBeforeConnect(Sender:
TObject);
begin
  LoadDBSettings(GetConfigFilename);
end;
```

Once we have the complete configuration filename, we pass it to a new private procedure that loads the settings exactly as we did in the web module (we don't need to list the code for that as it would be redundant). Copy the `.ini` file we created for the web module to the location of your compiled RAD Server package, make sure the base filename matches, and test it out.

> **Tip**
>
> If you want to test to make sure you're getting the right filenames, you can create a small GET resource that returns a JSON string containing the name. It can be added to this data module—just make sure to register it. The sample project for this section includes such a resource named `debug` that shows how to do this.

You can download this version of the RAD Server package from GitHub at `https://github.com/PacktPublishing/Fearless-Cross-Platform-Development-with-Delphi/tree/master/Chapter15/03_MyParkData`.

Finally, let's review configuring data storage paths on mobile apps.

Reviewing mobile data storage locations

In *Chapter 9, Mobile Data Storage*, we showed that you should use the `TPath.GetDocumentsPath` function from the `System.IOUtils` unit to get the default platform-friendly path for application data. If you ran the *AppPaths* project (from earlier in this chapter) on an Android or iOS device, you may have noticed that there's no "Public" path; using `GetPublicPath`, which might be viable in desktop applications, won't work on mobile devices, which expect to only be used by one person.

In addition to local device storage configuration, we also need to determine where the remote server is to which the app will be connecting. This is seldom a configurable item—a published resource should be accessible from a domain name so that users never have to bother with URL addresses and ports. If there are options, such as selecting the server closest to the user's physical location, a simple list should be available for the user to pick from. Always keep the options short and simple if you can.

With an understanding of where to store data on various platforms and how to provide configuration settings for different environments, let's turn our attention to security.

Securing data

If your app is completely self-contained, meaning no interaction with web services or saving data to a remote server, you might think you don't need to worry about securing it. If it's a simple utility measuring or reporting on some aspect of your device, that may be true, but if there's any personal information at all, your users will want to know that no other app can get at their data and that if their device is stolen or hacked, interpreting whatever is stored should be very difficult at the least.

This is one drawback of using an unsecured database such as **SQLite** or **IBLite**—there's no built-in encryption. You can encrypt data before it gets stored, which is a good start, but the table structures are still visible and it's far more troublesome to manually encrypt and decrypt yourself for every data operation. By using **IB ToGo** on mobile devices and a database such as **InterBase 2020** on a server, you'll have full table and column encryption—and peace of mind.

Securing database information is important but the effort is wasted if your multi-tier app communicates via plaintext between the endpoints. Securing data transmitted across the internet minimally involves three areas of security to consider:

- Securing your server's data transmission
- Securing access to resources
- Securing your hardware

Covering these in depth is far beyond the scope of this book but the following sections will give you a start for thinking through what's at stake.

Securing your server's data transmission

People seldom think about how data travels between them and resources on the internet—they just browse a website or use a mobile app and assume their data is safe. One of our responsibilities as developers is to think about all the possible vulnerabilities in our apps and in the flow of data going to and from our apps—and how to protect it.

Websites that provide encrypted connections use **Transport Layer Security** (**TLS**), more commonly known as **Secure Sockets Layer** (**SSL**), to encrypt data across the wires. Remember in the previous chapter when we modified the mobile app to connect to RAD Server by putting in an exception on Android devices to use cleartext traffic? Modern versions of mobile devices expect server communications to be encrypted. To do this, you'll need to install an SSL certificate on the server and then force all traffic to use only the secured ports.

Many tools exist on every platform to create self-signed certificates for free; there are also many companies that create and sell various levels of certificates. The difference is whether you need public verification that you really are who you say you are. Anyone can create their own certificate and if only you and a few people who know and trust you use it, then they simply install the certificate you create and everyone is happy. However, if you need to publish a server for new customers and don't want them getting prompted to accept a certificate from an unknown source, then you need a validating company that marks your certificates as "trusted" so they'll be automatically installed. These companies charge for the time it takes to validate you, which can take anywhere from a few minutes with simple domain name verification to several days for a full company, address, and phone validation. The costs vary accordingly.

We won't go into the details of getting and installing SSL certificates; follow a couple of links in the *Further reading* section at the end of this chapter to begin your search. There are instructions on the websites for the certificate you choose.

In addition to securing the flow of information over the wires, you need to also control who can access the data.

Controlling access to resources

In both *Chapter 12*, *Console-Based Server Apps and Services*, and *Chapter 13*, *Web Modules for IIS and Apache*, we wrote read-only servers and used only the GET HTTP method for accessing them. That is, they were used for querying existing data only—no support was made available for updating, deleting, or adding new data from the client.

In *Chapter 14, Using the RAD Server*, we expanded this with PUT, POST, and DELETE methods, which could be used just as well with our prior servers. Since we already have these additional methods supported in RAD Server, let's take off right where we left it at the end of *Chapter 14, Using the RAD Server*, and discuss how RAD Server makes securing your application quite simple. If you're not using RAD Server, read along anyway as implementing these techniques in your own code should be done to some level—and this might give you implementation ideas.

There are multiple ways to restrict access to RAD Server's published API endpoints:

- **Users and groups**: You can restrict resources to specific groups and if a user is not a member of that group, they will be denied access; you can also control access at the user level.

- **Application keys**: By setting an app secret or application ID in RAD Server's configuration file, all requests to the server must contain this parameter.

- **Master key**: This is an administrator-level override key that allows access to any endpoint of any resource on the server regardless of any other user security or application keys that may be defined. If it is blank in the configuration file (or commented out), this feature is disabled. If this is enabled, it should be carefully guarded—or only temporarily allowed.

Let's look at the application ID and show how to set that up:

1. First, start the server and run RSConsole to make sure you can connect and see the list of users as usual. Optionally, use the REST Debugger to view a list of parks.

2. Now stop the server and in RSConsole, select **File | Config Local Server**. A dialog pops up to confirm the filename of the configuration file; click the **Configure** button. (Optionally, you can navigate to the folder shown and edit the configuration file with your favorite text editor—there are many comments that explain the various options.)

3. On the **Server** tab, as shown in the following figure, enter some unique identifying text in the **Application ID** box, such as `MyParks1234`, then click **Save And Close** and then **Close** once more:

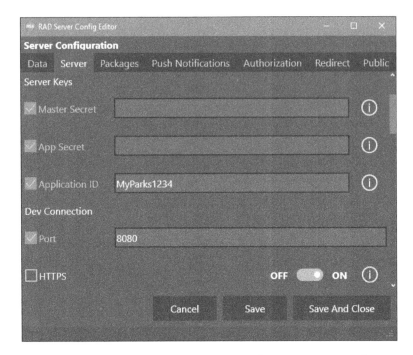

Figure 15.3 – Adding an application ID to the server in RSConsole

4. Close and restart the server so it rereads all the settings.

This immediately limits the scope of applications that use your service, providing one layer of security.

Close and restart RSConsole to force it to establish a new session with the server. Click on **Users** and notice the error message: **Application id missing from request**. In order to work with this server in RSConsole now, you must edit the connection profile. Click once on the root node of the local profile and select **Profile | Edit Profile** from the menu, switch to the **Keys** tab, as shown in the following figure, enter the application ID you gave in *Step 3*, and click **Close**:

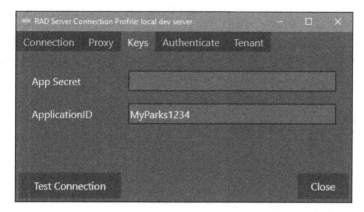

Figure 15.4 – Setting the application ID of a connection profile

Now you can continue working with RAD Server through the RSConsole app using the application ID; however, you can no longer get to the RAD Server resources with a web browser as the browser will not be sending the "header" parameter that is now required.

This is where the usefulness of the REST Debugger starts to shine. You can add a header parameter for the application ID or other server keys in the **Parameters** tab. When adding a parameter, switch **Kind** for the parameter to **HEADER** and in the **Name** drop-down, scroll down the list of header key names until you get to the **X-Embarcadero** keys.

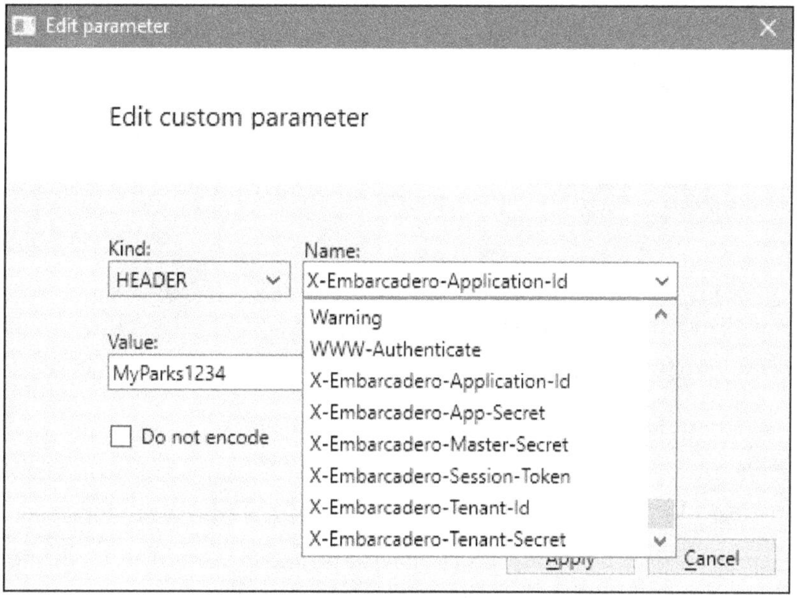

Figure 15.5 – Sending header keys from the REST Debugger

We'll use **X-Embarcadero-Application-Id** to set the application ID that will allow us to connect to the server; put the same application ID that you configured the server with into the **Value** field and click **Apply**. This allows you to successfully call RAD Server from the REST Debugger.

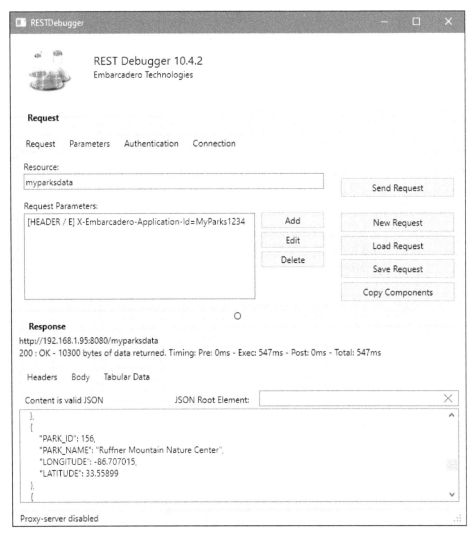

Figure 15.6 – Using the REST Debugger to test RAD Server with an application ID

Client testing can now only be done with applications that can send the application ID header parameter—and only with the *right* application ID value.

One of those applications affected is our MyParks mobile app. Let's fix that.

Adding application security for RAD Server clients

If you try running the MyParks mobile app from the previous chapter now, you'll notice no data shows up for the list of remote parks when you click the refresh button on the new **Server** tab we added. If you run it in debug mode, you'll see an error in Delphi as it tries to connect to RAD Server but since it isn't passing the application ID, it is not allowed access to any RAD Server resources. Here's how to add an application ID to the REST request.

Adding a header parameter is as simple in our mobile app as it is in the REST Debugger:

1. First, copy the project from the previous chapter and open up the main form.
2. Find the TRESTClient component and click the ellipses button for the Params property in the **Object Inspector**.
3. Add a parameter and set its **Kind** property to pkHTTPHEADER.
4. Set the **Name** property to the same value we selected in the REST Debugger, X-Embarcadero-Application-Id.
5. Set the **Value** property to the value of your application ID—we used MyParks1234 in our example.

Run the app again and click the refresh button and the list of parks should show for you as they did before.

You can download this version of the project from GitHub at https://github.com/PacktPublishing/Fearless-Cross-Platform-Development-with-Delphi/tree/master/Chapter15/04_MyParksMobile.

Even though we've limited the scope of applications that can connect, all the endpoints are still public. If you want to protect some aspects of your service, for example, providing free read-only access but limiting who can update your data, you should also add user or group security. RAD Server also provides resource-level security, which means you could add one resource that only provides the list of parks to free subscribers but provides additional resources that enable more advanced features. I would suggest studying the comments in the configuration file and reading the documentation further for the various types of security available with RAD Server and REST servers in general.

Protecting the data both inside and outside of your application is a big part of the overall security strategy and relies on the servers they connect to being constantly available. You can manage as much or as little of that piece as you want, as we'll see next.

Protecting your hardware and operating system

If you're hosting the server for your multi-tiered app at home or in your office, keep in mind the number of users that may be connecting and get a server powerful enough to support them all. Additionally, your internet connection will need to be stable and fast, for both upload and download. If you want to provide any sense of reliability, you should at the very least use a big enough **Uninterruptible Power Supply** (**UPS**) so that if your power goes out or even blinks, your server continues working. Also, make sure the machine cannot be stolen as best you can by either putting it in a locked but breathable cage or using cables securely fastened to the computer and attached to the wall or something unmovable. Your users would not be happy to find out the machine on which their data lies is now in unknown hands.

Decent server computers will take an initial outlay of cash but save money over the long run because you own the hardware. However, you have to maintain a good environment and make all repairs and updates yourself. If you rent remote server space from places such as Amazon Web Services, Rackspace, or Microsoft Azure, the fees you pay help to provide physical security and guaranteed up-time for the machines.

Because their business relies on keeping data safe and accessible for thousands of customers, they take physical security very seriously. They have live cameras everywhere, heavy doors and walls, locked cabinets, and an identification system that records the entry and exit of every person to the facility. In addition to preventing unauthorized access, they also have temperature-controlled rooms, fire retardant systems in place, and backup generators should they lose power. Using these **Infrastructure as a Service** (**IaaS**) providers can save you lots of time and worry about needing to secure your hardware but still give you the freedom to install and configure the operating system of your choice.

By managing your own servers, it's your responsibility to keep the system software up to date and provide protection against viruses, malware, and hackers. Taking it one step further, many of these same providers can also give you virtualized space where all you do is provide the custom middleware you support—they take care of keeping the operating system patched and web servers secure. As you might expect, these cost even more on an ongoing basis but are a great way to get started on a sure footing. This level is called **PaaS**, or **Platform as a Service**.

The following diagram helps visualize these layers and shows the increasing amount of personal involvement you will expend as you move down the pyramid, owning more pieces of the stack. You as the developer choose whether you utilize cloud-based providers or house your own hardware. All this is transparent to the user; their only visibility is the **SaaS**, or **Software as a Service**, you provide them:

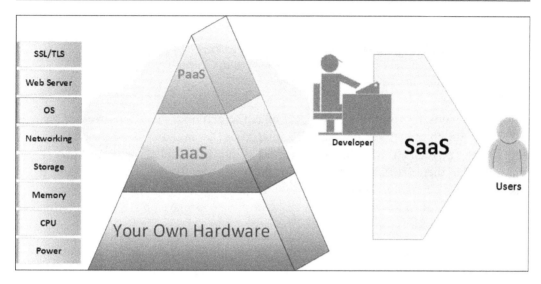

Figure 15.7 – Infrastructure management options powering SaaS for multi-tiered apps

These decisions are based on your budget, level of expertise, and availability but they don't have to be permanent. Internet providers can change, domain names can be rerouted, and servers can be moved or upgraded. Make sure your apps embed only names that can be updated remotely to point to new locations—never hardcode IP addresses or remote file paths in your distributed client applications.

The hard part is over; it's time to make your software look good before the final testing.

Adding a graphical touch

You're about ready to deploy your finished app but do you have a logo? Hopefully, you've at least thought of it. Your app needs some sort of graphical representation so when its icon is listed with dozens of other icons on a mobile device or in a menu, it can be quickly identified. If you don't have a logo, there are software and designers that can help. There are also many websites that offer royalty-free icons; some are free downloads, some require a subscription. The one I use is available for free from GetIt, **Icons8**, which provides many cross-platform icons; a paid option provides resolution sizes over 100 pixels and removes licensing restrictions. Whatever you use, be prepared to provide images that mobile apps expect to have in place—more than just one shortcut icon that you may be accustomed to providing with Windows apps.

> **Note**
>
> Console applications do not support icons. For Windows-based console apps, only the default console-mode (or "DOS prompt") icon is used when creating shortcuts.

In addition to icons, it's common to display a **splash screen** until the main screen of the app appears. Splash screens are popular for applications that take a while to start up, giving the user a visual cue that something is happening. This is most useful for slow devices or when there are databases to open or processes to initialize. If the user doesn't think anything is happening, they may try to start the app a second time. While splash screen support is added for you on mobile apps, desktop apps have no such option—you must add your own form that displays temporarily.

Let's explore the support for icons and splash screens built into the Delphi project options.

Iconifying desktop apps

Desktop operating systems such as Windows and macOS can launch applications from a menu by title but can also use icons; indeed, many people only know how to start an application by picking out its icon from a sea of other graphical images on their desktop.

Setting the icon for a Windows or Mac project is pretty simple. Pull up the project options page and go to **Application | Icons**. Both platforms list both icon file types but `.ico` files are used only on Windows and `.icns` files are used only on Macs. Simply click the **Load Icon...** or **Load Icns...** button to select the file to be associated and linked into your compiled application.

> **Note**
>
> For Windows, you'll also see logo image options for **Universal Windows Platform**, a special type of Windows application for building apps on Microsoft devices such as Windows 10, Xbox One, and HoloLens.

Mobile projects have a lot more options.

Iconifying mobile apps

As you've been building various mobile apps, you've undoubtedly seen the FireMonkey logo.

Figure 15.8 – The default logo on FireMonkey apps

The Android and iOS devices I use for testing the apps in this book are littered with many icons all using this same logo. On mobile devices, you need to have a custom, unique icon as that is the standard way to launch the app. But instead of providing one icon file as we did for each of the desktop platforms, you have to provide multiple .png files on each platform to support the variety of screen sizes. Additionally, both Android and iOS platforms provide options to set *notification* icons while iOS goes even further to define icon fields to identify your app in its *settings* and the *Spotlight* app. There are different screen sizes for Android, iPhone, and iPad, which all require different resolutions. In total, there are no less than 16 different image files you need to provide—just for the icons for one mobile app!

I selected an icon from Icons8 that I feel represents a park and saved several different sizes of .png files. Here's what the settings page on my iPhone looks like, showing the new MyParks icon at the bottom:

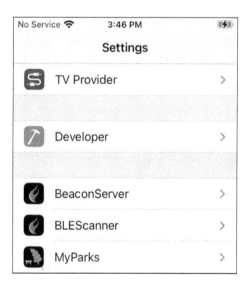

Figure 15.9 – The Settings page on an iPhone showing the new MyParks icon

The resolution of my largest image was 96 pixels so I used a paint program to resize the image to larger versions—losing transparency in the process—and painted the background of these larger images with a default green. The following screenshot shows a definite need for a professional graphic artist:

Figure 15.10 – App icons on an iPhone with a new one for the MyParks app

While we're still on the same **Icons** section of the project options, let's cover splash screens, starting with the iOS platform.

Setting splash screen options on iOS

iOS creates a splash screen for you based on your logo and a light or dark background color. On the **iPhone** tab, scroll down to the bottom of the list of images to find several **Launch image** fields.

Figure 15.11 – Splash screens on iOS are made up of launch images over a colored background

I simply selected the largest transparent icons I had for each of them. Notice there are two different scales of images depending on your device size. The background color is set with the bottom two options and the one that will be used is automatically selected based on your phone's current display mode when the application starts. The **iPad** options have only the **2x** scale for each of the color modes; the background color is taken from the **iPhone** tab's settings.

Creating splash screens for Android

Android options require you to create `.png` files for each of four resolution sizes.

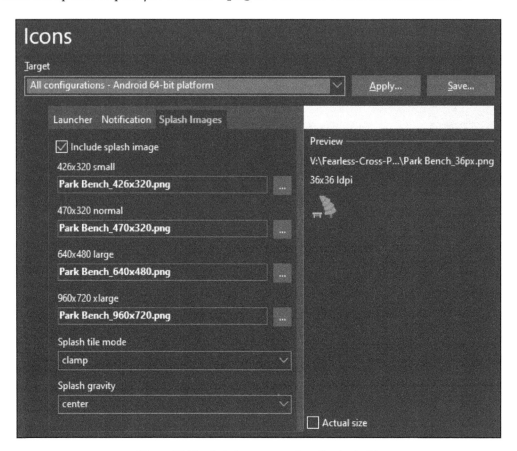

Figure 15.12 – Splash screen settings for Android

Four image sizes don't cover the variety of display sizes or whether the phone is in portrait or landscape mode when the application starts, so there are two additional options to handle these situations:

- **Splash tile mode**: This can be set to **disabled**, **clamp** (repeats the color at the edge of the image to fit the screen), **repeat** (repeats the image to fill the screen), or **mirror** (repeats the image to fill the screen, mirroring the image alternatively).

- **Splash gravity**: If tile mode is disabled, the options for this field determine the position of the image if the dimensions are less than the size of the screen.

If you're creating the images yourself, invest in an image program that allows quick scaling and saving of your logo to different sizes—it will save you time assigning them all here.

The MyParks mobile app we looked at earlier in this chapter has been updated with icons and images; again, the project can be downloaded from GitHub at `https://github.com/PacktPublishing/Fearless-Cross-Platform-Development-with-Delphi/tree/master/Chapter15/04_MyParksMobile`.

Looking good is great but app stores need several pieces of application identification as well.

Establishing product identity

It's always a good idea to set the version information of a project that will be deployed to multiple users. The **Application | Version Info** section of a project's options changes depending on the platform selected. For Windows, Mac, and iOS platforms, you can choose whether or not to include version information in the distributed application with a checkbox—for Android platforms, there's always at least default information provided. Let's go over each of these as they serve somewhat different purposes on each platform.

Including Windows version information

When you use the Windows API functions `GetFileVersionInfo` or `VerQueryValue`, all the version information you have entered on the **Version Info** project options page is returned for the file. This is also the same information that is shown when you view the properties of a file in Windows Explorer. This includes the module version number (major, minor, release, and build) and any values listed on the **Key-Value** table on the bottom half of that page. While they don't affect how the program is run or give any information to Windows, some installers use this information to determine whether to overwrite an installed application with a newer version or not.

Version information for Mac and iOS devices gets a little more involved.

Identifying your Apple product

Including version information for Mac and iOS projects is necessary when deploying apps to devices not configured for Developer mode. In addition to the module version number, the values in the **Key-Value** table are added to the `info.plist` file, providing critical application identification to Apple.

The two most important of these key-value pairs are the CFBundleName and CFBundleIdentifier keys. CFBundleName is the short name for your project and by default is the project name; we showed how to create CFBundleIdentifier in *Chapter 4, Multiple Platforms, One Code Base*.

On iOS platforms, the UIDeviceFamily field is important if you need to restrict your app to either iPads or iPhones; by default, it's set to iPhone & iPad. The macOS platform does not use this key.

The bottom part of the list has several optional NSxxx keys that allow you to define descriptive reasons for requesting various types of permissions should your app need them. We looked at some of these in *Chapter 10, Cameras, the GPS, and More*.

Finally, let's look at what version information fields are needed for Android.

Identifying your Android product

Similar to iOS projects, version information is required for deploying Android apps, and values in the **Key-Value** table are used for this purpose. For Android projects, the information is stored in the AndroidManifest.xml file and the primary key is package, which must uniquely identify your application. There are keys for label, versionCode, and versionName; some of the others deserve special mention:

- installLocation: You can set this to internalOnly, auto, or preferExternal (default) to direct where the app should be installed.
- theme: This can be set to TitleBar (default) or No TitleBar; if the latter, your app runs in fullscreen mode (similar to setting BorderStyle = None in iOS apps).
- apiKey: You can specify your Google Maps API key here (refer to *Chapter 10, Cameras, the GPS, and More*).

Now that you think you have everything ready on your machine, it's time to make sure it'll work for others.

Testing for deployment

It's quite embarrassing to send a compiled and nicely working program to a customer only to get a response back that it won't even start because a required library was not found or it can't access a database file it was expecting. In your development environment, you have everything set up to work with files, packages, and databases in expected paths, firewall ports you opened and forgot about, and configurations primed for optimal use. But a customer is unlikely to have any of this in place. Even when you're just giving a customer an update, at some point that customer may need to get a new computer or upgrade their phone and need to reinstall your software—and possibly restore a database backup.

One of the best ways I've found to test for these cases on Windows is to use **virtual machines**.

Using virtual machines

With a virtual machine, you get a completely separate and customizable platform with an operating system to run server and desktop applications. With some versions of virtual machine software, you can take **snapshots** of the current state of your system, installed applications, path settings, and so forth; then, after testing, revert to that saved snapshot so you can run the same tests again. It's also a way to cut down on the number of computers on your desk as you can run Windows in a virtual machine on a Mac, thus combining uses into one piece of physical hardware; however, you must have both a fast hard drive and lots of memory. The following figure shows Delphi running on Windows 10 in VMware Fusion on a Mac mini:

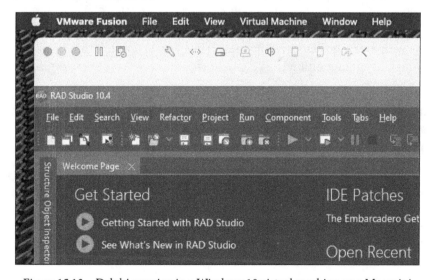

Figure 15.13 – Delphi running in a Windows 10 virtual machine on a Mac mini

I usually keep several snapshots, one just after Windows is installed where I've got a brand-new system with nothing else on it, then another one after a web server is installed but not yet with my web server modules, or just after the database is installed. Another good snapshot point is after version 1.0 of your software is installed and working but before an upgrade.

Microsoft offers free, limited-time virtual disks you can download, which makes a great testing facility for Windows apps and services without getting a separate license for Windows. They do require you to have virtual machine software in place but Hyper-V is one of the choices and free on Windows 10. Check out the *Microsoft Edge Developer* link in the *Further reading* section at the end of this chapter.

If you don't have enough space on your computer or enough memory to run both your main environment and a virtual machine, get an old computer and reinstall the operating system. Do whatever you can to provide a clean testing environment that, initially, has none of your support files to make sure that after proper installation, your software works.

Finally, there are cloud-based solutions that allow you to rent virtual machines for temporary use, such as **Microsoft Azure** and **MacinCloud**. You'll find these links in the *Further reading* section as well.

For mobile devices, it's not so simple.

Testing the wide range of mobile devices

There is such great variability in mobile devices that it can be daunting to think of testing for every scenario. You might keep old phones or go buy a couple of used ones for development and testing but that will only give you a false sense of victory. There *will* be someone else with a strange device or that uses it in a way you didn't think will break your app—and complain about it online!

Unless you're a very large corporation with deep pockets, it's unlikely you'll have access to even a fraction of the variety of mobile devices that exist. You'll need to study the various capabilities of different editions of the major brands, pick a few, and thoroughly test on the ones you are able to.

You may need to enlist friends or give discounts to select customers for beta testing in exchange for feedback on how well your app works. You need to be aware of screen sizes, whether users will likely be using a physical keyboard or a pen or just finger swipes, and what happens when they're trying to connect to your server and they're out of internet range. Making a checklist of testing points to give to those helping you will increase the likelihood you'll get useful results.

For even greater testing coverage and peace of mind, but also at greater cost, there are companies that provide both testing software and testing services. If you're serious about deploying the next great app and expect it to go viral, using a service like this is imperative.

While testing is taking place, you can put into place the steps for the final distribution of your application suite.

Distributing the final product

Here we are! The final piece of the journey to developing cross-platform applications is to actually get them deployed to your end users for which they were designed.

If you're familiar with distributing Windows applications, you've probably used an installation building program such as **InstallAware** or **InstallShield** to create a self-contained program that extracts itself, asks the user a few questions (such as where to install), creates the destination folder with your app inside, and adds a shortcut icon to either the desktop or the Start menu—or both. My favorite install builder for these types of apps is **Inno Setup**. This is a widely used free tool and I have not come across any Windows installation task it cannot handle.

Deploying application servers is different than end user apps. Servers are often installed by skilled IT people that set up firewalls, redirect ports, and manage user permissions—they usually prefer to extract applications and library files to specific locations and edit configuration files manually to explicitly control every facet of the critical machines for which they're responsible.

We've covered a few different server types in this last section of the book, from standalone console apps to web server modules to RAD Server packages, and have explained several steps to take along the way for each. Sometimes you can simply package the files necessary into a `.zip` file and give it to the right person with a set of instructions and they'll make it happen. Other times, you'll be setting it up yourself. For small in-house database server installs on Windows, using your favorite install builder will suffice.

RAD Server deserves a little more explanation.

Installing RAD Server modules to production

Installing your custom-written packages to RAD Server is really pretty simple but obviously requires RAD Server itself to already be installed on the production server—which is quite a bit more involved than just running it from your development machine. For RAD Server to be able to load your custom Delphi packages means all its libraries must be separate package files as well (instead of being linked into the `.exe`). Therefore, you need to ensure it comes with all the support libraries needed for TCP/IP communications, encryption, database drivers, and so forth, plus automatically starts with the server and has any database or file permissions it may need.

There are detailed instructions for installing RAD Server on both Windows and Linux in the *Further reading* section at the end of this chapter but I would highly recommend downloading the installer you need from the GetIt portal as this automates many of the steps for you. For Windows, go to `https://getitnow.embarcadero.com/ RADServerInstallerforWindows-104-1.0` and for Linux, go to `https:// getitnow.embarcadero.com/RADServerInstallerforLinux-104-1.0`. If your Linux installation is the server edition of Linux and does not include a graphical environment, pay special attention to extra command-line parameters you may need to include that use console-based prompts rather than windowed prompts.

> **Note**
> In addition to RAD Server itself, you may also need to install a database server. We have used InterBase with our sample RAD Server package but when deploying to a production installation of RAD Server, your custom package cannot use the same instance of InterBase that RAD Server is using. The RAD Server license adds a secret **system encryption password** for the InterBase database instance it uses, which prohibits that database from general use by any other application. You can install a second instance of InterBase and configure it to listen on a different port but you would also need to add a server license of InterBase for that instance in order to use InterBase from your package. Your RAD Server module could, of course, use a different database product as long as it is accompanied by the appropriate Delphi packages.

Once RAD Server is installed, follow these three steps to deploy:

1. Copy your compiled RAD Server packages, configuration files, and any support libraries to a directory accessible by RAD Server (such as a sub-folder of `C:\ inetpub\RADServer` if running as an IIS module under Windows).

2. Edit the `emsserver.ini` file and add the names of your packages to the `[Server.Packages]` section.

3. Restart RAD Server.

> **Tip**
>
> After compiling your RAD Server project, select **Project | Information for**
> *ProjectName* (where *ProjectName* is the name of your RAD Server project)
> from the Delphi menu and it will list all the Delphi library packages that are
> required to deploy with your RAD Server package. Many of these files are
> already in the RAD Server folder, which makes adding your custom modules to
> that folder quite simple. Any files your package depends on need to be in your
> deployment folder.

Once your server is ready, test it with your favorite REST server testing tool (such as
the REST Debugger that we've used in this book). Then, the focus shifts to getting your
client apps deployed. The rest of this section highlights things you'll need to do to deliver
finished apps to your users.

Selecting deployment configuration

As we've been developing and testing mobile apps, the project configuration has been,
by default, set to **Development** for Android and iOS and **Developer ID** for the macOS
platform. By switching the configuration to **Ad hoc** on iOS and **Normal** on macOS,
it allows us to distribute our release-mode applications to a wider audience. And
switching the configuration to **Application Store** enables submission of our apps to the
platform's corresponding app store. The app stores must approve your app before it will be
available publicly.

> **Note**
>
> The following sections are high-level overviews of the steps needed to deploy
> your mobile app. There are many other details that may need to be performed
> when preparing and submitting apps for final deployment to the various app
> stores. Additionally, the app stores (especially Apple) have constantly changing
> rules through which the application must pass; this could take quite a bit of
> time and possibly some negotiation with the company before being approved.
> Please follow the links in the *Further reading* section at the end of this chapter
> to read—in greater detail—all you need to know about the complete process
> and how it applies to your application suite.

Here are some steps to take for each platform to prepare your apps for final delivery.

Deploying macOS and iOS applications

In *Chapter 4, Multiple Platforms, One Code Base*, we walked through the steps for using your Apple Developer account to create a signing certificate, an application identifier, and finally a profile that is associated with your apps. We went through that exercise for iOS apps, creating a wildcard profile that allowed us to build multiple apps under one profile, but we need explicitly named profiles when submitting to the App Store; this also applies to Mac apps when you're ready to deploy. Review those steps to create a profile in your Apple account that includes deploying to the macOS platform, then apply the bundle ID created there to the CFBundleIdentifier key on the **Version Info** screen of the project options, as described earlier in this chapter in the *Establishing product identity* section.

For **Ad hoc** (iOS) or **Normal** (macOS) deployment, select **Project | Deployment** from the Delphi menu, and click the **Deploy** button. This deploys the finished application to your Mac through the Platform Assistant server to a sub-folder of the scratch-dir folder corresponding to your Delphi profile.

Mac application distribution is complete at this point and you can launch the application or copy it to another location. iPhone and iPad apps require one more step. The deployment process placed a .ipa file that needs to be deployed to your iPhone or iPad. Bring up Xcode, select **Devices and Simulators** from the **Window** menu, and select your device. Now drag and drop the .ipa file onto the **INSTALLED APPS** section of the window—and watch the app get installed onto your connected device.

Figure 15.14 – Apps installed on an iPhone via Xcode

Setting your Delphi project's **Configuration** to **Application Store** prepares the app to be submitted to the Apple App Store and once compiled, cannot be run as is from your devices—you have to switch **Configuration** back to either **Development** or **Ad hoc**. Once generated, there are a few more steps for Apple store submission:

1. Browse to `https://appstoreconnect.apple.com` and select **My Apps**.

2. Click the plus sign (+) on the upper left of the screen and fill in the form, then click **Create**.

3. Fill in the name, description, keywords, and other information that help identify and classify your app; choose whether the app will be limited to certain users or allowed full access; upload screenshots and perform other application preparatory steps as found on the app submission pages.

4. Once everything looks good, click **Submit for Review**. This creates an app record with a status of **Prepare for Submission**.

5. From your Mac, use the Transporter app (download it from the Apple App Store if you don't already have it) to prepare and initiate the app transfer of your app archive that Delphi compiled.

After an app is submitted, you can view its status in your Apple Developer account.

Deploying an Android app

When we first introduced the steps for building and testing Android apps back in *Chapter 4, Multiple Platforms, One Code Base*, we didn't need to sign our apps because it was just for our development purposes. Now that they'll be distributed to devices other than our own, we need to code-sign them so that the devices (and their owners) will trust our software.

There are multiple ways to sign your apps. The Android documentation shows how to do so with Android Studio; the Embarcadero documentation shows how to do it with Delphi's built-in tool. There's also an open source tool called KeyStore Explorer (`http://keystore-explorer.org`) that some use. Whichever way you choose, you'll need to provide some identifying information about you as the developer, your organization, and your location. You'll provide both a *keystore* and an *alias* password and when you're done, you'll have a `.keystore` file that you should keep safe.

> **Note**
>
> Do not use spaces in your password! If you do, when Delphi tries to use the keystore tool, you'll get back a strange error message as it tries to use the characters after the space as part of the command line. Also, don't use punctuation in any of the location fields as they may cause problems as well.

You'll use this file in the next steps to sign your Android app:

1. In your Delphi project options window, expand **Deployment** and select **Provisioning**.

2. Switch **Build type** to **Application Store** and the rest of the screen will present you with places to enter the information about your new keystore file.

3. If you have not yet created your keystore file, you can click the **New Keystore...** button to create one; otherwise, select the one you have previously prepared.

4. Select **Project | Deployment** from the Delphi menu and click the **Deploy** button and a `.aab` file (Android Application Bundle) should be created in the `bin` folder of your project.

5. Now, log in to your Google Play Developer Console (`https://play.google.com/apps/publish`) and click **Create app**.

6. Fill in the information about your app and upload the `.aab` file.

You may have noticed that there are only two Android build configuration types, **Development** and **Application Store**, as opposed to iOS, which also includes **Ad hoc**. Apple apps that are specifically designated as "store" apps cannot be manually delivered and installed; Android apps that are marked as "store apps" are not restricted to Google Play submission but can be distributed in the same manner as "ad hoc" iOS apps.

With your store apps successfully submitted, you can take a vacation—you have deserved it!

Summary

This final chapter has been a conglomeration of many topics, pulling together several aspects of application development, marketing, testing, and deployment. We added configuration to both server and client applications, providing flexibility for our installations. We talked about the importance of security, not only locally but also when communicating over the internet, and how to restrict access to key resources. Finally, making sure our apps have the proper icons and version information and having performed final testing, we then showed how to submit our finished products to both the Apple and Google app stores.

This has not been an exhaustive study but hopefully has given you the insight and tools to put you well on the road toward building a wide range of cross-platform applications. With these increased capabilities, you can take on new clients, upgrade legacy applications, break into new markets, and use your enriched skills with fresh ideas for developing the next great app!

Thank you for reading this book and trying out the examples. I hope you enjoyed the journey, have a renewed and greater appreciation for the Delphi programming environment in the cross-platform arena, and learned a great deal in the process! Always keep learning and growing—fearlessly!

Questions

1. What simple function call returns the filename of the current web module?

2. What Windows API function call returns the filename of the current package or DLL?

3. How do you dynamically set the parameters for a FireDAC database connection?

4. When should you use `GetDocumentsPath` and when should you use `GetPublicPath`?

5. Which embedded version of InterBase supports encryption?

6. In what ways can the REST Debugger help test REST services that a web browser cannot?

7. What's the difference between PaaS and IaaS?

8. What type of icon file does Windows require? And Mac?

9. How are splash screens created for iOS apps?

10. How do you restrict your apps for iPad use only?

11. Where do you store your Google Maps API key in an Android project?

12. What are the purposes of the three types of iOS configurations?

Further reading

- *RAD Server Resource Overview*: http://docwiki.embarcadero. com/RADStudio/Sydney/en/RAD_Server_Resource_ Overview#Declaring_an_Endpoint_Method

- *What is SSL?*: https://www.ssl.com/faqs/faq-what-is-ssl/

- *Let's Encrypt*: https://letsencrypt.org

- *Web Server SSL Certificates*: https://ksoftware.net/ssl_certs.html

- *RAD Server Engine Authorization*: http://docwiki.embarcadero.com/ RADStudio/Sydney/en/RAD_Server_Engine_Authorization

- *Installing the RAD Server or the RAD Server Console on a Production Environment on Windows*: https://docwiki.embarcadero.com/RADStudio/Sydney/ en/Installing_the_RAD_Server_or_the_RAD_Server_Console_ on_a_Production_Environment_on_Windows

- *How to Deploy Your RAD Server Project on Windows with IIS*: https://blogs. embarcadero.com/how-to-deploy-your-rad-server-project-on- windows-with-iis/

- *How to Deploy the Production Version of RAD Server To Linux*: https://blogs. embarcadero.com/how-to-deploy-the-production-version-of- rad-server-to-linux/

- *Infrastructure as a Service (IaaS)*: https://searchcloudcomputing. techtarget.com/definition/Infrastructure-as-a-Service-IaaS

- *Application Options*: https://docwiki.embarcadero.com/RADStudio/ Sydney/en/Application_Options

- *What is an ICNS file?*: https://fileinfo.com/extension/icns

- *Icons8 and Delphi GUIs*: https://blogs.embarcadero.com/icons8-and- delphi-guis/

- *Version Info*: https://docwiki.embarcadero.com/RADStudio/Sydney/ en/Version_Info

- *Preparing a macOS Application for Deployment*: https://docwiki. embarcadero.com/RADStudio/Sydney/en/Preparing_a_macOS_ Application_for_Deployment

- *Customizing Your info.plist*: https://docwiki.embarcadero.com/ RADStudio/Sydney/en/Customizing_Your_info.plist_File

- *About Info.plist Keys and Values*: `https://developer.apple.com/ library/archive/documentation/General/Reference/ InfoPlistKeyReference/`

- *What is Hyper-V and How Do You Use It?*: `https://www.cloudwards.net/ hyper-v/`

- *Microsoft Edge Developer*: `https://developer.microsoft.com/en-us/ microsoft-edge/tools/vms/`

- *Best virtual machine software for Mac 2021*: `https://www.macworld.co.uk/ feature/best-virtual-machine-software-3671133/`

- *Microsoft Azure*: `https://azure.microsoft.com/en-us/`

- *MacinCloud*: `https://www.macincloud.com`

- *Software Testing Help*: `https://www.softwaretestinghelp.com`

- *InnoSetup*: `https://jrsoftware.org/isinfo.php`

- *Deploying Your Multi-Device Apps*: `https://docwiki.embarcadero.com/ RADStudio/Sydney/en/Distributing_Your_Multi-Device_Apps`

- *App Store Connect Help*: `https://help.apple.com/app-store-connect`

- *Android Studio - Sign Your App*: `https://developer.android.com/ studio/publish/app-signing#generate-key`

- *How to create an Android Keystore file*: `https://headjack.io/tutorial/ create-android-keystore-file/`

Assessments

This section contains answers to questions from all the chapters.

Chapter 1 – Recent IDE Enhancements

1. The **Structure** pane is a hierarchical list of components but also shows procedures, variables, used units, and so on and is mostly used for quickly jumping to the declaration in code, whereas the **Class Explorer** is a TreeView of classes that when clicked shows the fields, methods, and properties in a lower pane.

2. Right-click and select **Show in Explorer** from the context menu.

3. In the __recovery folder of the project.

4. Right-click on the form and select **Hide Non-Visual Components** or simply hit *Ctrl + H*.

5. When your project requires a style or library installed only via the GetIt Package Manager.

6. Yes, with the drop-down **Desktop** menu from the title bar.

7. *Language Server Protocol*, the new way that Code Insight is handled, which allows the editor to remain responsive while the code is being parsed and indexed, and also allows Code Insight features while debugging.

Chapter 2 – Delphi Project Management

1. Console, Package, Dynamic Library, RAD Server Package, Web Server Module.

2. The *Tabbed* template has one TitleBar across the top with four tabs; the *Tabbed with Navigation* template also has four tabs but each tab has a unique, embedded TitleBar, plus the first tab has two sub-tabs with a button that activates tab navigation animation.

3. **Dynamic Link Library (DLL)**.

4. **File | New | Customize**.

5. In **Project | Options | Building | Build Events**, set **Target** to the release configuration for the selected platform, then modify the **Post-build events Commands** field.

6. The **History** pane.

7. `dccaarm.exe`.

Chapter 3 – A Modern-Day Language

1. When the `Halt` procedure terminates an application abruptly.

2. `(* and *)`.

3. Yes.

4. Since Delphi 2009 (late 2008).

5. A negative integer, typically -1.

6. Because the main form is not thread-safe and some other process may be updating that property.

7. When you want to inspect an object or type at runtime.

8. Delphi 10.3 Rio.

Chapter 4 – Multiple Platforms, One Code Base

1. Blank Application.

2. `Checked` was changed to `IsChecked`; the `OnClick` action is now handled by `OnChange`.

3. LiveBindings.

4. FmxLinux.

5. In the `PAServer` folder in your Delphi installation path.

6. It is a combination of an app ID, device ID, and signing certificate by Apple.

7. A device driver specific to the device type.

8. There are form-specific sizes and device type properties that define it.

9. `CPU64BITS`.

Chapter 5 – Libraries, Packages, and Components

1. Windows: `.dll`, Mac: `.dylib`, Linux: `.so`.

2. Any unit—but not in the project file itself.

3. On the **Description** page of **Project Options**.

4. *Runtime packages* can be compiled for any supported platform and contain distributable code; *design-time packages* work only with Delphi to register components or augment the functionality of Delphi.

5. By adding `ComponentPlatformsAttribute` above the component class.

Chapter 6 – All about LiveBindings

1. No.

2. Layers and element hiding.

3. `TDataGeneratorAdapter`.

4. Select it from the **Binding Components** list (by double-clicking on the **BindingsList** component).

5. No, if the display is one way or if the changed value will be interpreted the same, there's no need to "undo" the display format.

6. The unit in which it was coded and registered.

Chapter 7 – FireMonkey Styles

1. With a `StyleBook` component and by using the `TStyleManager` class.

2. No.

3. A default style customizes the style for all controls of the selected type whereas a custom style customizes the style only for the selected control.

4. With the **Structure** window.

5. RCDATA.

6. By using `TStyleBook`.

Chapter 8 – Exploring the World of 3D

1. DirectX on Windows, Metal on Mac, OpenGL on Mac, iOS, and Android.

2. `Position.Z`; positive to push it further away, negative to bring it closer.

3. `TLayer3D`.

4. Set `RotationAngle.X` to `180`.

5. `Emissive`.

6. Set its `Opacity` property to `0`.

7. By setting the `MaterialSource` property of all the mesh components in `MeshCollection` (only in code at runtime).

8. Make the camera a child of the object.

9. Enable **Custom orientation** in its project options, then check `Portrait`.

10. Set the `Enabled` property to `False`.

11. `TDummy`.

12. Temporarily set the `Position.Z` value to a negative value to pull it forward, then set it back to a positive value to push it back out of the way.

13. iOS.

Chapter 9 – Mobile Data Storage

1. The Developer Edition and IBLite.

2. The character string gets stored as if it were a text field.

3. Through triggers and generators.

4. Only when the table is created.

5. `TPath.GetDocumentsPath`.

6. Yes.

7. The upper half of the screen.

Chapter 10 – Cameras, the GPS, and More

1. Check off the appropriate permissions in **Project | Options**, and ask for permission from the user at runtime.

2. Accessing the camera, writing to external storage, and reading from external storage (which is implied if writing is granted).

3. Use the image passed in from the `OnDidFinishTaking` event.

4. `TLocationSensor`.

5. Apple's MapKit framework and the Google Maps API.

6. Check **Maps Service** under **Entitlement List**, check **Access network state**, **Access course location**, and **Access fine location** in **User Permissions**, and set the **apiKey** value in **Version Info**.

Chapter 11 – Extending Delphi with Bluetooth, IoT, and Raspberry Pi

1. Call the `StartDiscovery` method of the current `TBlueToothManager` (which can be accessed with `TBlueTooth.CurrentManager.Current`).

2. Classic requires pairing and more energy and doesn't work on iOS.

3. Clearly defined published services for Bluetooth low-energy.

4. A BLE that sends out data on a periodic basis using the **Generic Access Profile** (**GAP**) in advertising mode without the need for connecting.

5. No, a beacon server is `TBeaconDevice`, while a beacon "client" that scans for beacon servers is `TBeacon`.

6. A term that refers to a set of Delphi components that adhere to specific BLE profiles for accessing IoT devices.

7. Specialized versions of Android.

Chapter 12 – Console-Based Server Apps and Services

1. Enterprise or higher.

2. In Windows Explorer, navigate to `\\wsl$`.

3. It's so much easier to step through code and debug.

4. Indy (Internet Direct).

5. The Windows event log.

6. Syslog.

7. In the Windows registry.

8. 127.0.0.1.

9. Redirecting an internet request through a router to a computer that can handle the request.

Chapter 13 – Web Modules for IIS and Apache

1. Since Delphi 3.

2. `OnHTMLTag`.

3. No, they can return a wide array of data types.

4. Enable the feature in **Turn Windows Features on and off**.

5. In the application pool's advanced settings, enable 32-bit support.

6. Yes, if they're configured to listen on different ports.

7. Modify the `httpd.conf` file in the `conf` sub-folder where Apache for Windows is installed.

8. Modify the `/etc/apache2/apache2.conf` file, add, remove, or change the `.load` and `.conf` files in the `/etc/apache2/mods-enabled` folder, modify the `ports.conf` file, or modify other `.conf` files under `/etc/apache2`.

9. `mod_XXX.dll` for Windows, `libmod_XXX.so` for Linux.

10. `GModuleData`.

Chapter 14 – Using RAD Server

1. Windows or Linux; standalone or under IIS (Windows only) or Apache (Windows or Linux).

2. InterBase.

3. JSON (default) or XML.

4. The concept of multiple applications with completely separate sets of users, security, resources, and devices.

5. `RSConsole`.

6. Not found.

7. An application programming interface with a specific URL method call to RAD Server that results in either data returned or an action taken.

8. GET, POST, PUT, and DELETE.

9. Takes the current URL and parameters in REST Debugger and composes a set of components for making that request that you can then paste into your application.

10. Add `android:usesCleartextTraffic="true"` to the manifest template.

Chapter 15 – Deploying an Application Suite

1. `WebApplicationDirectory`

2. `GetModuleFileName`

3. By setting the `Params.Values[]` properties

4. `GetPublicPath` on Windows to support any user that might use the application, `GetDocumentsPath` in most other cases.

5. IB ToGo

6. By passing header parameters and also with PUT, POST, and DELETE HTTP methods.

7. **Platform as a Service (PaaS)** provides more updates and management for you than **Infrastructure as a Service (IaaS)**, which is pretty much just the raw hardware.

8. Windows: `.ico`; Mac: `.icns`.

9. Your image icon is displayed on a fullscreen solid-color background.

10. Set the `UIDeviceFamily` field in the project's **Version Info** to `iPad`.

11. In the **apiKey** key of **Version Info**.

12. **Development** is for testing and debugging; **Ad hoc** is for deploying a release-mode project without going through the Apple Store; **Application Store** is for submitting to the Apple App Store.

Packt.com

Subscribe to our online digital library for full access to over 7,000 books and videos, as well as industry leading tools to help you plan your personal development and advance your career. For more information, please visit our website.

Why subscribe?

- Spend less time learning and more time coding with practical eBooks and Videos from over 4,000 industry professionals

- Improve your learning with Skill Plans built especially for you

- Get a free eBook or video every month

- Fully searchable for easy access to vital information

- Copy and paste, print, and bookmark content

Did you know that Packt offers eBook versions of every book published, with PDF and ePub files available? You can upgrade to the eBook version at packt.com and as a print book customer, you are entitled to a discount on the eBook copy. Get in touch with us at customercare@packtpub.com for more details.

At www.packt.com, you can also read a collection of free technical articles, sign up for a range of free newsletters, and receive exclusive discounts and offers on Packt books and eBooks.

Other Books You May Enjoy

If you enjoyed this book, you may be interested in these other books by Packt:

Delphi GUI Programming with FireMonkey

Andrea Magni

ISBN: 9781788624176

- Explore FMX's fundamental components with a brief comparison to VCL
- Achieve visual responsiveness through alignment capabilities and layout components
- Enrich the user experience with the help of transitions and visual animations
- Get to grips with data access and visual data binding
- Build exciting and responsive UIs for desktop and mobile platforms
- Understand the importance of responsive applications using parallel programming
- Create visual continuity through your applications with TFrameStand and TFormStand
- Explore the 3D functionalities offered by FMX

Delphi High Performance

Primož Gabrijelčič

ISBN: 9781788625456

- Find performance bottlenecks and easily mitigate them
- Discover different approaches to fix algorithms
- Understand parallel programming and work with various tools included with Delphi
- Master the RTL for code optimization
- Explore memory managers and their implementation
- Leverage external libraries to write better performing programs

Hands-On Design Patterns with Delphi

Primož Gabrijelčič

ISBN: 9781789343243

- Gain insights into the concept of design patterns
- Study modern programming techniques with Delphi
- Keep up to date with the latest additions and program design techniques in Delphi
- Get to grips with various modern multithreading approaches
- Discover creational, structural, behavioral, and concurrent patterns
- Determine how to break a design problem down into its component parts

Packt is searching for authors like you

If you're interested in becoming an author for Packt, please visit `authors.packtpub.com` and apply today. We have worked with thousands of developers and tech professionals, just like you, to help them share their insight with the global tech community. You can make a general application, apply for a specific hot topic that we are recruiting an author for, or submit your own idea.

Share Your Thoughts

Now you've finished *Fearless Cross-Platform Development with Delphi*, we'd love to hear your thoughts! Scan the QR code below to go straight to the Amazon review page for this book and share your feedback or leave a review on the site that you purchased it from.

`https://packt.link/r/1-800-20382-9`

Your review is important to us and the tech community and will help us make sure we're delivering excellent quality content.

Index